S0-CAS-612

A Practical Approach to Quantitative Metal Analysis of Organic Matrices

A Practical Approach to Quantitative Metal Analysis of Organic Matrices

MARTIN C. BRENNAN

A John Wiley and Sons, Ltd, Publication

This edition first published 2008

Registered office
John Wiley & Sons Ltd, The Atrium, Southern Gate, Chichester, West Sussex, PO19 8SQ, United Kingdom

For details of our global editorial offices, for customer services and for information about how to apply for permission to reuse the copyright material in this book please see our website at www.wiley.com.

Library of Congress Cataloging-in-Publication Data

Brennan, Martin, 1943-
A practical approach to quantitative metal analysis of organic matrices / Martin Brennan.
p. cm.
Includes bibliographical references and index.
ISBN 978-0-470-03197-1 (cloth : alk. paper)
1. Atomic emission spectroscopy. 2. Metals–Analysis. 3. Chemistry, Organic. I. Title.
QD96.A8B73 2008
543'.52–dc22 2008010999

British Library Cataloguing in Publication Data

A catalogue record for this book is available from the British Library

ISBN 978-0-470-03197-1 (H/B)

Typeset in 10/12 pt Times by Thomson Digital, Noida, India
Printed and bound in Great Britain by CPI Antony Rowe, Chippenham, Wiltshire

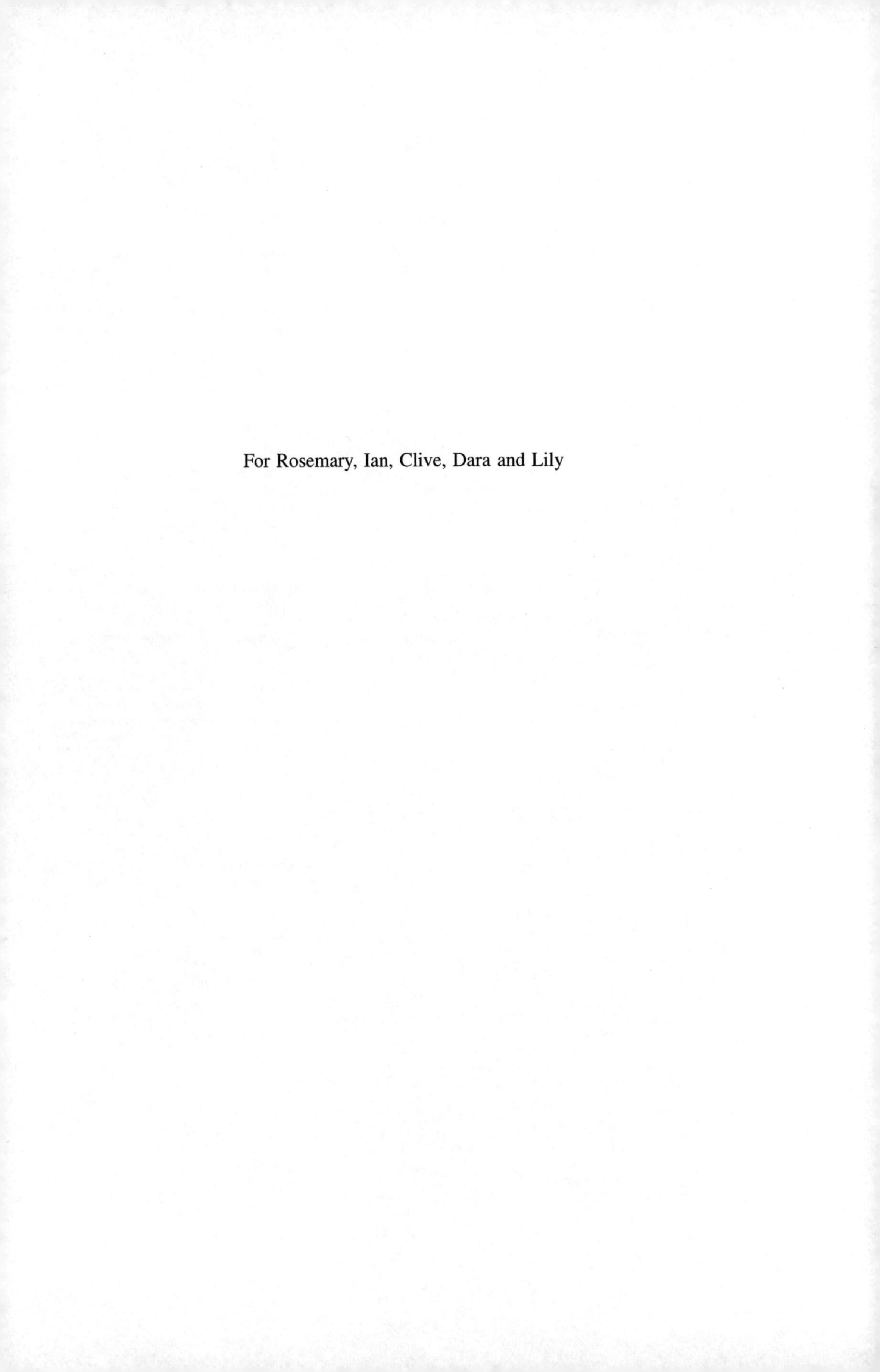

For Rosemary, Ian, Clive, Dara and Lily

Contents

Preface

'Theory Guides, Experiment Decides'

Izaak Maurits Kolthoff (1894–1993)

In the last 20 years atomic spectroscopy has made great strides, particularly with the introduction of new improved optic designs and detection methods. These improvements have led to superior resolution of the wavelengths of the excited atoms and detection techniques measuring lower levels of metals with ease. After a slow and problematic start, inductively coupled plasma optical emission spectrometry (ICP-OES) has become an established technique in most laboratories analysing a wide range of sample matrices reporting accurate and precise results.

Most chemists are familiar with atomic emission spectroscopic techniques for metal analysis of aqueous solutions and are equally aware that most of these methods cannot be readily applied to non-aqueous samples. In recent years atomic spectroscopy instrumentation has increased in sophistication allowing the analysis of a wide range of samples on a routine basis for metals content using manual or automated methods. This book aims to cover the importance of metal analysis for a range of organic samples.

ICP-OES continues to dominate the market because of its ease of use and relatively low maintenance cost. Inductively coupled plasma mass spectrometry (ICP-MS) is a very powerful state-of-the-art technique used for metal analysis of all kinds of samples but requires highly skilled operators. A vast amount of information is received that is not necessarily required as part of problem-solving or routine support. The cost difference and relative freedom from maintenance problems would favour ICP-OES. This book is aimed at practitioners requiring multi-elemental analysis in industrial, environmental, pharmaceutical and research laboratories, where information on identification and quantification is required on a regular basis. The main focus of this book will be on sample preparation, a topic overlooked in most books on atomic spectroscopy. It is aimed at most ICP-OES and ICP-MS users to show that the instrument is useless unless the sample is prepared in a suitable state that can be used to accurately and precisely quantify the metals present.

Despite the array of sophisticated instrumentation associated with atomic spectroscopy, non-invasive measurement is not possible in the majority of cases. Some samples may need multiple steps in their preparation. This presents an enormous challenge for accurate and precise analysis as part of quality assurance, and in environmental and health knowledge. The sample preparation stage can be referred to as an enrichment, clean-up, and signal enhancement process. This important step usually requires that the analyst appreciate the chemistry associated with sample preparation in order to achieve accurate and precise results. This step is often considered the slowing down of analysis, as it tends to be labour-intensive and requires extreme care.

Fortunately, in the last 20 years, improvements in sample preparation techniques have become easier and faster, leading to accurate analysis. The microwave acid digester, high pressure oxygen bomb combustion, preconcentration columns, rapid dry ashing, and solvent extraction with complex reagents are the common sample preparation techniques used in modern laboratories. Extraction methods and microextraction techniques are used in dedicated laboratories as part of sample preparation. Specialised miniaturised methods, such as automated workstations for rapid sample preparation-analysis of a large number of similar samples, can use a flow through microwave acid digester and pump the prepared sample directly to the plasma for analysis.

This book is devoted to the analysis of organic materials with emphasis on the compatibility of the ICP-OES torch with a range of solvents. Selected suitable solvents can be used as part of non-destructive sample handling for the metal analysis of a range of organic liquids and semi-solids and contributes to eliminating tedious sample-preparation procedures.

Unfortunately, this is not the case for a number of other difficult organic matrices which must be prepared using either low or high pressure oxygen combustion bombs, microwave acid digester, dry ashing or UV digester methods. The range of important organic-based metal salts used as catalysts in organic syntheses, petroleum products, plastics, polymers, pharmaceuticals and adhesives require metal analysis as part of quality assurance. Solvent extractions can be used as a method for removing matrix interference and preconcentration of trace metals in some samples and eliminate the need for concentrated corrosive acids.

Analysis of organic compounds using ICP-OES requires higher radio frequency power, a suitable nebuliser spray chamber and solvent resistant pump tubing for transporting the sample solution using a peristaltic pump.

This book is not intended to be an exhaustive study, but is aimed to provide readers with an insight and in some cases an alternative approach to the analysis of organic matrices using ICP-OES.

Biography

Dr Martin Brennan is an analytical scientist of more than 30 years standing with considerable experience in atomic spectroscopy. He has a MSc in analytical science from Queens University, Belfast and a PhD in atomic spectroscopy and electroanalytical techniques from University College Cork, Ireland. He is the author and co-author of several published articles in atomic spectroscopy and electrochemical sciences. His research interests include trace analysis of difficult matrices and improvements in sample preparation techniques. He has considerable experience in the analysis of a wide range of samples particularly organic type samples and is currently employed in the Research and Development Department of Henkel (Ireland) Ltd, manufacturers of adhesives and other organic compounds. He holds the position of honorary secretary of the Republic of Ireland sub-region Analytical Division of the Royal Society of Chemistry.

Acknowledgements

I would like to acknowledge permission to reproduce figures and diagrams from Perkin Elmer LAS, Beaconsfield, Bucks, HP92FX; E-Pond, SA, CP, 389, CH-1800, Vevey, Switzerland; Cetac Technologies Ltd, South shields, Tyne and Wear, UK; PS Analytical, Orpington, Kent, BR5 3HP, UK; CEM Technologies, Buckingham, MK18 1WA, UK; Scientific And Medical Products (Parr Instruments), Cheshire, SK8 1PY, UK; BDH, London, supplier of Conostan standard, metal free low and high viscosity oils, Henkel (Ireland), Tallaght, Dublin 24, Ireland; Polymer Laboratories now a part of Varian Inc., Amherst, MA 01002, USA; Dionex Corporation, Plato Bus. Park, Damaston, Dublin 15, Ireland.

I am grateful for the input of my colleagues at work who demanded precise analysis of a range of difficult organic matrices over the last 20 years. Special thanks to Professor Raymond G. Leonard, RD&E Associate Director of Material Testing and Analytical Services, Henkel Ireland, for reading Chapter 6 to check for any breach of the company's confidentiality.

I would like to thank the editors and staff of John Wiley & Sons, particularly Gemma Valler, Richard Davies and Jenny Cossham for their valuable suggestions. Thanks are also due to Jo Hathaway for copy editing and corrections. I would also like to acknowledge the expert assistance of typesetters Mr Tarun Mitra and Mr Poirei Sanasam of Thomson Digital, India.

Last, but by no means least, I wish to thank my wife Rosemary for her encouragement and support and for grammatical corrections.

1

A Practical Approach to Quantitative Metal Analysis of Organic Matrices Using ICP-OES

1.1 Introduction and Basic Overview

When salts of certain metals in solid or in solution are subjected to thermal energy associated with flames, characteristic mono- and multicolours are produced. This colour characteristic used for metal identification gave rise to the birth of a science commonly known as spectroscopy and was discovered by Isaac Newton (1643–1727) during his study of the solar spectrum and made possible by his invention of a triangular dispersing prism. The Isaac Newton prism was used to disperse the emission light from a flame into bands, which could be used to characterise two elements in the same solution (Figure 1.1). A common event seen on a wet day is the separation of white light from the sun passing through raindrops (a prism) high in the sky, being diffracted and separated into colours and forming attractive rainbows.

This procedure has long been used to qualitatively detect the presence of alkali and alkaline elements such as sodium, potassium and calcium. Henry Fox Talbot (1800–1877) noted that when the wick of a candle was dampened in a solution containing table salt (NaCl) an intense yellow colour formed in the flame which he correctly associated with sodium metal. The identification of colour(s) is attributed to the thermal energy caused by a heat source, which raises electrons in atoms to a higher energy state. These electrons cannot remain in this excited state for too long and will emit energy in the form of light to return to the more stable, ground state. The disadvantage of this procedure is that it is limited to single alkali or alkaline earth elements in the flame.

The flame will visually impart colours when selected elements such as sodium (yellow), potassium (blue) and calcium (predominantly red with a little green and blue)

A Practical Approach to Quantitative Metal Analysis of Organic Matrices Martin Brennan

are placed at the base of the flame with the aid of a previously acid cleaned platinum wire. A mixture of near similar levels of metals in the flame would emit colours that would confuse detection with the naked eye unless one element is predominant. Prussian chemist, Robert Wilhelm Eberhard Bunsen (1811–1899) and German physicist, Gustav Kirchhoff (1824–1887) discovered this metal/colour phenomenon when they studied the behaviour of metal salts in a flame made from a mixture of air and coal-gas and attributed these colours to line spectra from the elements rather than compounds using a simple apparatus that consisted of a prism, slits and magnifying glass. They soon discovered that elements when heated would emit light at the same wavelength they absorbed, producing bright lines in the spectrum. Thus by heating an unknown compound and examining the spectrum, scientists could identify the elements that made the compound. This formed the basis of the modern science of spectroscopy. They carried out tests of different salts containing the same metallic element to give the same colour. This simple analogy concluded that no matter what the element is compounded to, the same colour would result. Sodium chloride gave the same yellow colour as sodium sulphate, sodium nitrate, sodium phosphate, etc. Similarly consistent colours were obtained for other alkali and alkaline elements and their compounds.

Later, German physicist Joseph von Fraunhofer (1787–1826) discovered 'dark lines'* [1] and with the advent of more sophisticated optics he was able to attribute these lines (bands) to characteristic wavelengths. An early practical example of this is the separation and identification of lithium (Li) and strontium (Sr) in solution. In the Li and Sr example, one line or band is the wavelength for Li and the other for Sr. However, it is worth noting here that Li^+ emits only one waveband while Sr emits several and despite this they are still identifiable for each element. Elements giving multiple colours (as calcium, above) illustrated the presence of several lines associated with a particular element. Kirchhoff and Fraunhofer were the first to observe these lines that are now assigned to wavelengths in modern spectroscopy.

It was not until the early 20th century, with the advent of astronomy and atomic physics, that the science of quantitatively measuring metals in solutions became possible [2,3]. Atomic emission was the first to emerge, quickly followed by atomic absorption spectroscopy (AAS) and, later (~mid 1960s), by atomic fluorescence. These methods were attributed to the effects that occur when most metals achieve a sufficiently high temperature and most compounds decompose into atoms in the gas phase. In atomic spectroscopy, samples are vaporised sufficiently at temperatures as low as 1800°C and as high as 9000°C, and atomic concentrations are determined by measuring absorption or emission at characteristic wavelengths against calibration curves prepared from standards of the elements under testing. The high selectivity and sensitivity of wavelengths caused by the absorption or emission of the atoms, and the ability to distinguish between elements made possible by modern sophisticated optics in a complex matrix, was the beginning of the study of developing methods for the quantifying of metals in sample solutions by atomic spectroscopy.

---

*'Fraunhofer lines are when white light containing all wavelengths is passed through a cool version of a gas of an element and the photons from the light interact with atoms. Assuming some of the wavelengths from the light have correct frequency to promote an electron of that element to a higher energy level, photons at this frequency are absorbed by the gas causing "gaps" in the spectra giving rise to "dark lines".'

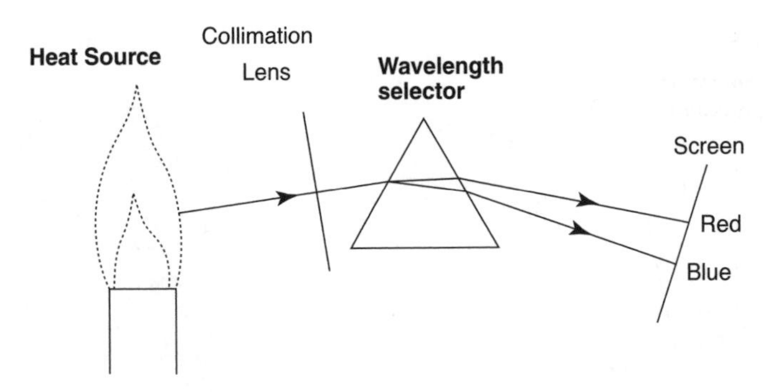

Figure 1.1 *Simple diagram showing separation of combined multiple wavelengths using a light dispersion prism*

The fact that light emitted from thermally treated elements caused interference patterns and could be diffracted, illustrated that they must behave with wave characteristics. In 1841, James Clerk Maxwell discovered the electromagnetic theory of radiation, which showed that light consists of photons with oscillating electric field (E) in a magnetic field (B) travelling rapidly through space. Maxwell also showed that the combined electric field and magnetic field vibrate at right angles to each other, to which the photon is propagating in a straight line (Figure 1.2). However, for the sake of simplicity it is easier to consider waves as sine waves and the photons can be explained in terms of properties of sine waves. The successive distances between the peaks or troughs are equal, and measured as lambda, λ.

The atomic spectra of most elements originate from the transition of electrons from the ground state to the excited state, giving rise to what are commonly called resonance lines [4]. The diagrams in Figure 1.3 are transitions – selected lines for sodium and potassium and the wave-numbers associated with each transition. Some elements in the periodic table contain very complicated electronic structures and display several resonance lines close together. The widths of most atomic lines are extremely small (10^{-6} nm), and when broadened in various ways the width never exceeds 10^{-2} nm [5]. Fortunately, the modern optics available on the latest instruments can isolate lower bandwidths.

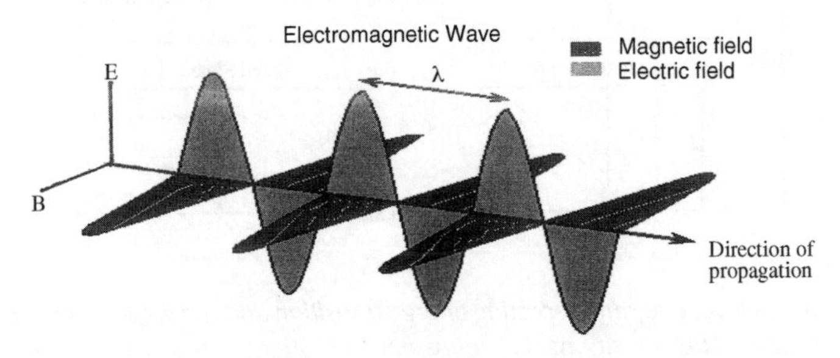

Figure 1.2 *Maxwell's electromagnetic radiation diagram*

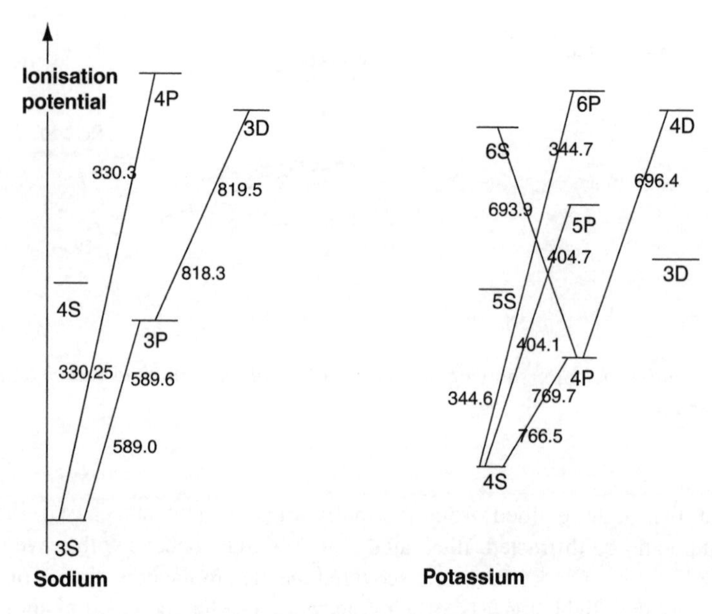

Figure 1.3 *Energy levels and wavenumber (nm) diagram for sodium and potassium*

1.2 Schematic Representation of the Energies Generated by Atomic Spectroscopic Methods

The three main types of energies applied for the excitation, ionisation and emission steps used for elemental analysis can be shown schematically, as in Figure 1.4.

The horizontal lines represent energy levels of an atom. The vertical arrows represent energy transitions. These energy transitions can be either radiations (i.e. absorption or emission of electromagnetic radiation) or thermal (energy transfer through collisions

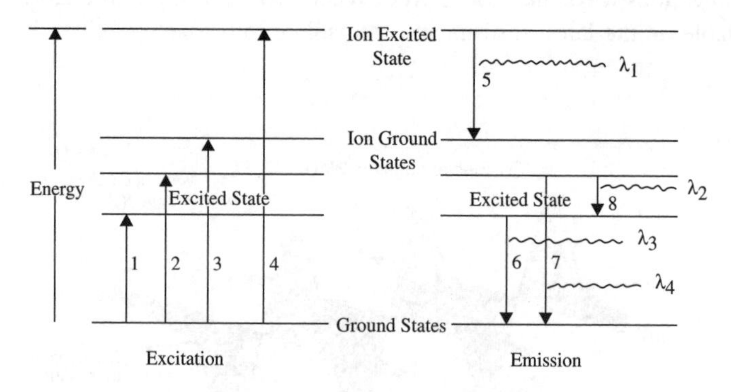

Figure 1.4 *Energy level diagram depicting energy transition and the wavelength (λ) associated with each transition. The transitions 1,2 represent excitation; 3 is ionisation; 4 is ionisation/excitation; 5 is ion emission; and 6, 7 and 8 are atomic emission*

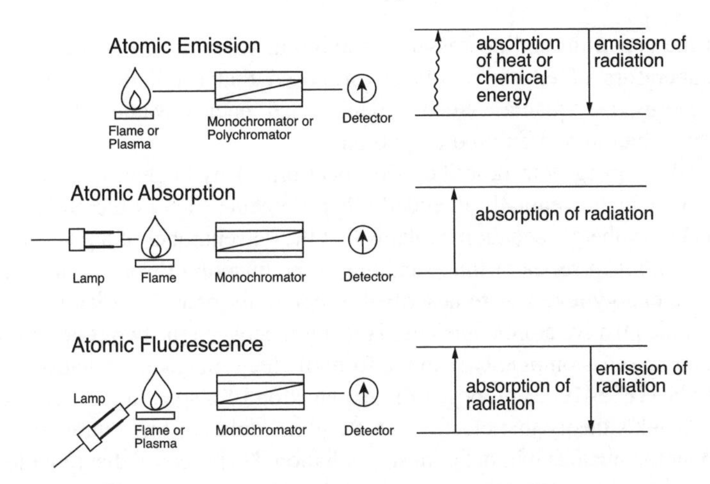

Figure 1.5 *Three types of atomic spectroscopy techniques shown diagrammatically. (Reproduced by kind permission: copyright © 1999–2008, all rights reserved, PerkinElmer, Inc.)*

with other atoms or particles). The difference in energy between the upper and lower levels of a radiation transition defines the wavelength of the radiation involved in that transition.

Atomic emission spectroscopy is applied to the measurement of light emitted by thermal energy caused by the thermal source from the chemical species present. Examples of emission, absorption and fluorescence spectroscopy can be shown schematically, as in Figure 1.5.

The colour of light obtained from excited atoms stems from the chromophore of the valence shell electrons emitting light as electromagnetic radiation. Photons are absorbed during the promotion of an electron between waves' mechanically allowed (i.e. quantised) energy levels. The ultraviolet (165–400 nm) and visible regions (400–800 nm) of the electromagnetic spectrum are the regions most commonly used for analytical atomic spectroscopy. Wavelengths from 700 nm upwards are in the infrared region and are inapplicable to atomic spectroscopy. The 165–700 nm region in the electromagnetic spectrum is generally referred to as 'light' although, technically, all electromagnetic radiation can be considered as light. Known wavelengths for elements can vary from 1 for Li to 5700 lines for Fe. Some lines are more sensitive than others and this fact is in itself useful because low levels would need very sensitive lines while less sensitive lines can be useful for higher concentrations of elements in samples.

1.3 Excitation Energy (Quantum Theory and Atomic Spectra) [7]

In 1900 Planck derived an empirical relationship for data from a 'black body radiation'; by introducing a concept of 'quantisation of energy' he was able to prove the relationship theoretically. It had been shown that at a specific temperature the spectrum of radiation from a 'black body' was unparalleled in its characteristics and the energy varied throughout the spectrum and possessed a maximum value at one particular wavelength for one temperature of emission. Planck proposed that oscillations were emitting or absorbing energy as 'quanta'. Such 'quanta' had energy values dependent on the frequency of radiation and emitted as photons.

Balmer and Rydberg both described the spectrum of hydrogen through mathematical interpretations, which eventually included other elements such as the alkali and alkaline earth elements. Rydberg's equation explains that the wavenumber of a given spectrum line is constant for a given series of lines, leading to the quantum theory of atomic spectra in which discrete energy levels were described as coulomb forces between the valence state electrons and the positive atomic nucleus. Transition between the two states corresponds to the absorption or emission of energy in the form of electromagnetic radiation of frequency 'v'. Niels Bohr (1885–1962) proposed the explanation of a spectrum of atomic hydrogen using Max Planck's theory that electrons in an atom could exist in a number of orbits and circulate about the nucleus without emitting radiation. He proposed that radiation was only emitted when an electron went from a higher orbital to a lower one. The magnitude of this energy is given by Planck's equation as follows:

$$\Delta E = E_o - E^* = hv = hc/\lambda \quad (1)$$

where h is Planck's constant (6.6×10^{-34} J s), v is the frequency of the radiation and $E_o - E^*$ is the energy difference between the two energy levels in the atom. The frequency is related to wavelength by λ (m) $= c$ (speed of light $= 3 \times 10^8$ m s^{-1})/v. To obtain energy in kJ mol^{-1}, we multiply the value of E by the Avogadro constant, $L = 6.02 \times 10^{23}$ mol^{-1}. Bohr also postulated that the further the electron is from the nucleus (i.e. greater orbital) the higher the energy level. When electrons of an atom are in orbit and close to the nucleus they are at the lowest energy levels sustained by the atom and in its preferred ground stable state. When thermal energy (other energies could be used) is added to the atom as the result of absorption of electromagnetic radiation or collision with other electrons, several events take place within the atom. One or all of these events cause excitation by absorbing energy causing electrons to move from the ground state orbital to an orbital further from the nucleus and to a higher energy level. This atom is said to be in an excited state and such an atom is less stable and will decay back to a less excited state losing energy in the process emitting this energy as photons (particles) of electromagnetic radiation. As a result the electron returns to an orbital closer to the nucleus.

In some cases if the energy absorbed by an atom is high enough, an electron may be completely lost from the atom leaving an ion with a net positive charge (see Figure 1.6). This energy is called ionisation, which is characteristic for each element in the periodic table. This form of energy has been assigned the *ionisation potential* for each element. Similar to atoms, ions can also have ground and excited states by being able to absorb and emit energy by the same process as a ground state atom.

Planck showed that a photon has particle properties and proved that the energy of a photon is proportional to its frequency. This shows that energy and wavelength are inversely related, i.e. as the energy increases, the wavelength decreases and vice versa.

In the ground state the electrons are at their lowest energy levels and by contacting the atoms with thermal or electrical excitation, the energy is transferred to the atoms causing the atoms to collide. These collisions cause the electrons in the atoms to change to higher

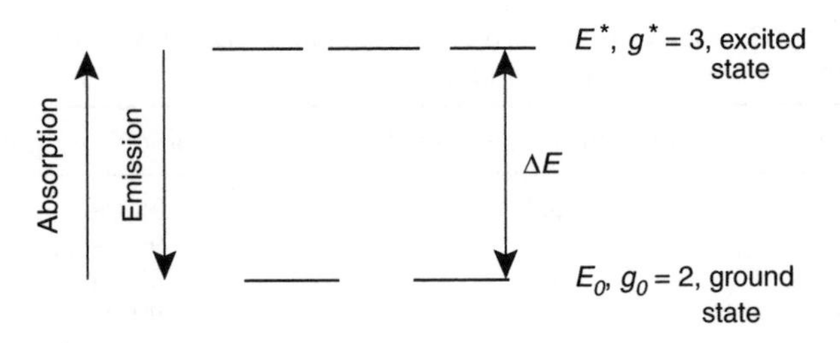

Figure 1.6 Energy levels at ground state and excited state of an atom showing degenerates

orbits. The quantity of energy transfer varies from atom to atom, sample matrices to sample matrices, resulting in a range of energy states. The result is radiation of a number of different species displaying a spectrum, which is highly complex. The proportion of excited and ground state atoms at a given temperature may be given by the Boltzmann distribution[1], which describes the temperature effect in atomic spectroscopy. It is well known that temperature determines the degree to which a sample breaks down to atoms and the extent to which a given atom is found in its ground, ionised or excited state. Consider a molecule with two possible energy levels, as shown in Figure 1.6.

The ΔE (positive number) is divided by the lower energy E_o and higher energy E^*. An atom may have several states at a given energy level. In Figure 1.6, there are three states at E^* and two at E_o. The number of states available at each energy level is called the degeneracy: denoted g^* and g_o. These are the statistical weights of. These the excited and ground state levels, respectively. The Boltzmann distribution describes the relative population of different states at thermal equilibrium.

1.4 Ionisation Energy and Number of Excited Atoms [7]

The intensity of atomic emission is dependent on temperature and is emphasised in the Boltzmann relationship. The relationship between the line intensity and temperature is derived from the energy transferred by collision of the argon ion with another atom or atoms. The ease with which an atom will form an ion depends on the magnitudes of its ionisation energy and its electron affinity. In the Bohr Theory of the hydrogen atom a certain amount of energy was required to completely remove the electron from its orbit to infinity. This energy is called the ionisation energy and the magnitude of ionisation potential depends on (i) the distance of the electron(s) from the nucleus, (ii) nuclear charge, less a correction for screening effect by inner shells and (iii) type of electron being removed i.e., s, p, d. In general, the further the electron is from the nucleus the less firmly it is held and the lower the ionisation potential. The approximate ionisation energy (in electron volts) of the argon ion is 15.9 eV and the ionisation energy of most elements detected by inductively coupled plasma atomic emission spectroscopy (ICP-AES) is in the order of

[1] Boltzmann is a distribution constant of a large number of particles among different energy states which could be discrete and quantified for particles under study. The Boltzmann constant is a constant per molecule and is given by $k = 1.38 \times 10^{-23}\ \mathrm{JK^{-1}}$.

Table 1.1 Relationship between ionisation energy of selected elements and the degree of ionisation

Element	Ionisation energy (eV)	Degree of ionisation (%)
Li	5.4	99.0+
Na	5.1	99.0+
K	4.3	99.0+
Rb	4.2	99.0+
Be	9.3	74.0
Mg	7.7	96.0
Ca	6.1	97.0
Ba	5.2	91.0
Cr	6.8	96.0
Mn	7.4	95.0
Fe	7.9	96.0
Co	7.9	93.0
Cu	7.7	90.0
B	8.3	59.0
P	10.5	36.0
S	10.4	16.0
As	9.8	49.0
F	12.9	0.1
Cl	12.7	0.6

5 to 14 eV of which most are closer to 6 to 8 eV, hence sufficient energy is still available for several excitations of an ion. Therefore many transitions from ions are possible. The alkali elements usually have the lowest ionisation energies, metalloids are slightly more difficult to ionise and the non-metals are even more difficult. The ionisation energies would range from ~4.5 for alkali, ~8.5 for metalloid to ~12 eV for non-metals. Table 1.1 illustrates the relationship between ionisation energy and the degree of ionisation for Group 1 – alkali, Group 2 – alkaline, common transition elements and metalloids.

At equilibrium, the relative population of excited and ground state atoms at a given temperature can be considered using the Boltzmann relationship. N^* and N_o of any two states is given by:

$$\frac{N^*}{N_o} = \frac{g^*}{g_o}\ exp[-E_1 - E_0/kT] \qquad (2)$$

where N^* and N_o are the number of atoms in the excited and ground states, respectively, T is the temperature (K) and k is the Boltzmann constant (1.38×10^{-23} J K^{-1}). The terms E_1 and E_0 are energies at higher and lower states, respectively. This illustrates that ground state atoms can absorb light to be promoted to the excited state. Excited atoms, conversely, can emit light returning to the ground state. This equation holds for ionisation, and excitation of atoms and molecules provided that the energy levels considered are non-degenerate.

Table 1.2 shows the ratio of atoms in excited state for temperatures ranging from 2000 K to 8000 K as calculated using the Boltzmann equation above. The number of excited Na atoms at 2000 K is approximately 10^{-5} while at 8000 K it is in the order of 10^{-1}. The fraction of atoms excited increases with increasing temperature which is almost negligible at lower temperature but significant at higher temperatures. AAS measurements are less dependent on temperature because they are based on the number of unexcited atoms, however, in the case of inductively coupled plasma optical emission spectrometry (ICP-OES) the number of excited atoms at higher temperatures gives greater sensitivity. Table 1.2 shows the effect of temperature on the number of excited atoms.

1.5 Width of Atomic Lines [8]

Spectral lines have a very small but finite width (broadening), over a wide range of wavelengths giving a variety of breadth and shape, with a maximum at a certain wavelength. Some atomic lines are very thin and precise while others are not and it is an accepted analogy that their width is taken as the width at half the signal height ($\Delta v/2$ cm^{-1}). Spectral interferences common are coincident line overlap, wing overlap from intense nearby lines of the same or another element and background shift. These signals lend themselves to several types of broadening, described below.

1.5.1 Natural Broadening

This type of broadening is the mean lifetime of an atom in an excited state when photons are absorbed in the atom. The absorption process is rapid and is in the order of 10^{-12} s while the excitation is longer and in the order of 10^{-8} s. This is short enough to support the Heisenberg Uncertainty Principle, which states that if we know the state of the atom, we must have uncertainty in the energy level, i.e the shorter the lifetime of the excited state, the more uncertain is its energy relative to the ground state. This uncertainty is expected at the level of the elementary particles, and yields a line broadening which is particularly noticeable at the base of the peak. This broadening is insignificant at 9000°C.

1.5.2 Doppler Broadening

The narrow natural line is broadened by motion of the atoms and ions in the plasma. This effect is due to the rapid motion in which atoms move and is based on the theory that if an excited atom in the process of emitting photons is moving towards a detector the resulting wavelength will appear to be shorter. If, on the other hand, the photon is moving away from the detector it will appear longer. This velocity observed in the line of sight will vary according to Maxwell distribution, for atoms moving in all directions relative to the observer.

1.5.3 Lorentzian Broadening or Pressure Broadening

This results from collision of atoms with atoms of other species. The energy level of both the ground and excited states of an atom will be influenced by interaction with

Table 1.2 *Effect of temperature on the number of excited atoms*

Element	Line (nm)	g_i/g_j	Excitation energy (eV)	2000 K	4000 K	6000 K	8000 K
Cs	852.10	2	2.340	4.21×10^{-4}	2.93×10^{-2}	1.21×10^{-1}	2.44×10^{-1}
Na	589.12	2	4.632	9.50×10^{-6}	4.37×10^{-3}	3.36×10^{-2}	9.41×10^{-2}
Ca	422.71	3	3.332	1.41×10^{-7}	6.47×10^{-4}	1.11×10^{-2}	4.33×10^{-2}
Mg	285.21	3	4.346	3.35×10^{-11}	7.65×10^{-6}	1.65×10^{-4}	5.55×10^{-2}
Co	338.29	1	3.664	5.85×10^{-10}	2.23×10^{-6}	6.73×10^{-5}	4.44×10^{-4}
Au	267.59	1	4.634	2.12×10^{-12}	7.88×10^{-7}	3.34×10^{-5}	6.61×10^{-4}
Fe	371.99	–	3.332	2.29×10^{-9}	5.43×10^{-5}	7.43×10^{-4}	3.12×10^{-3}
V	437.92	–	3.131	6.87×10^{-9}	7.34×10^{-5}	2.22×10^{-4}	5.89×10^{-3}
Zn	213.86	3	5.795	7.45×10^{-15}	1.94×10^{-7}	4.86×10^{-5}	7.75×10^{-4}

g_i/g_j = statistical weights of the corresponding levels, i.e. the atomic and molecular ground state degeneracies respectively.

surrounding particles. It is known that these collisions can shift, broaden and cause asymmetry in the line. Lorentzian broadening increases with pressure and temperature and is generally regarded as proportional to pressure and the square root of temperature. Therefore, $\Delta\nu$ increases with increasing temperature and pressure. It is accepted that Lorentzian broadening affects the wings of the signal profile. The Lorentzian half-width is of the same order of magnitude as the Doppler half-width (Table 1.3). The classical Lorentzian distribution predicts a symmetric line profile, but in practice there is an asymmetric profile and a red-shift of the maximum.

Table 1.3 *Comparative list of approximate half-widths of selected elements for Doppler and Lorentzian broadening*

Element	Wavelength	Doppler (4000 K)	Lorentz (4000 K)
Sodium	589.00	4.8	4.0
Calcium	422.70	2.9	1.3
Magnesium	285.21	2.8	—
Iron	371.99	1.9	1.7
Gold	267.59	0.8	—
Silver	328.07	1.5	1.4
Vanadium	437.92	2.6	—
Zinc	213.86	1.3	—
Cobalt	338.29	1.6	1.9

1.5.4 Holtsmark Broadening or Resonance Broadening

This is collision between atoms of the same element in the ground state and results in an intensity distribution similar to Lorentz broadening but without line asymmetry or shift. The effect depends on concentration and half-widths which are very small and negligible when compared with other collisions.

1.5.5 Field Broadening or Stark Broadening

This takes place in an electric or a magnetic field, where the emission line is split into several less intense lines. At electron densities above 10^{12} the field is relatively inhomogeneous, splitting varies for different atoms and the result is a single broadening line.

1.5.6 Self-Absorption and Self-Reversal Broadening

This is the sum of all factors considered so far because whilst the maximum absorption occurs at the centre of the line, proportionally more intensity is lost at the wings. As the concentration of atoms increases the intensity and profile change. High levels of self-absorption can cause self-reversal, i.e. a trough at the centre of the line. This is significant for emission lines in flames but is almost absent in ICP-AES which is a major advantage of this source.

1.6 Brief Summary of Atomic Spectroscopic Techniques Used for Elemental Analysis

With the development of electric spark or electric arc excitation, the spectra formed after a high voltage pulse are recorded using photographic plates, which form spectral lines characteristic of the element in the source. With the aid of a spectrograph, the spectra of the elements occur mostly in the ultraviolet region and the optical system used to disperse the radiation is generally made of quartz. The lines formed from an unknown sample are compared with calibration lines of known standard(s). Luckily these techniques have been replaced by atomic absorption, atomic emission and plasma emission spectroscopy, which have the ability to identify, measure and quantify up to 60–75 elements depending on the technique. There are three kinds of emission spectra: (i) continuous, (ii) band, and (iii) line. Incandescent solids for which sharply defined lines are absent and have little or no use in atomic spectroscopy emit continuous spectra. The band spectra consist of a group of lines that come nearer to each other until a limit is reached, and are caused by excited molecules. Line spectra, which have most use in emission spectroscopy, are definite lines and are characteristic of atoms or atomic ions, which have been excited to emit energy as light of constant wavelength.

Quantitative analysis by atomic spectroscopy is based on the measurement of radiant energy by free atoms in the gaseous state. The technique owes its selectivity to the fact that spectra of gaseous atomic species consist of defined narrow lines at wavelengths characteristic of the element of interest. The energy input into any of the techniques is converted to light energy by various atomic and electronic processes before being measured. The light energy is in the form of a spectrum, which consists of discrete wavelengths.

Regardless of the forms of energy the atom is subjected to, be it absorption, emission or fluorescence, they can all be used for analytical purposes. The following is a brief description of each of these thermal methods, commonly used to excite, isolate, identify and quantify metal concentrations in sample solutions.

1.6.1 The Atomic Absorption Spectrophotometer

This instrument is a very popular and versatile technique and finds use in many laboratories worldwide. However, limitation to certain elements in terms of poor sensitivity, detection and reproducibility means that some elements cannot be quantitatively determined by this technique. The type of samples that are difficult to measure at trace levels using AAS are refractory types, such as rocks, slags, lava, ceramic, cements, and ashes containing elements such as W, Mo, Si, P, B, Al, and Pb. The reason for this is that the use of the gas mixture acetylene/air to achieve temperatures of ~2100–2500°C or nitrous oxide/acetylene to achieve ~2500–3100°C are insufficient to fully excite these elements. However, this technique is very applicable to alkali, alkaline and first-row transition elements and can detect these elements with considerable ease. A further disadvantage of AAS is the limited linear range with which deviations from linearity occur for various reasons, such as unabsorbed radiation, stray light or disproportionate decomposition of molecules at high concentrations. It is desirable to work at the mid point of the generated linear straight line to achieve accurate results. A simple rule of thumb is to prepare five to eight standards and note the linear range and obey the Beer-Lambert law of the relationship between concentration and absorbance. Modern AAS will also measure emission energy of selected elements.

AAS is relatively free from elemental interferences because it determines elements using absorption by specific lamps for the metal under test. However, it is prone to background interferences caused by:

(a) absorbing molecular species of the sample, e.g. some absorptions by other species in the sample may occur at the same wavelength as the element under test;
(b) absorbing atomic species from other elements in the sample, e.g. phosphorus in the presence of calcium;
(c) particles from high salt concentration in the atom cell can cause light scattering.

Fortunately, methods for background correction are now a part of most modern AAS and the most commonly used background correctors are a deuterium source, the Zeeman effect and the Smith-Hiefte effect.

In atomic absorption most of the atoms in vapour phase are in the ground and unexcited state and therefore it might be expected that atomic absorption would be more sensitive than atomic emission. This is not true because the higher temperature achieved with plasma sources excites most of these atoms and it is easier to measure emission of small signals with good precision than a small difference between two large signals from the sample and reference beam in AAS. This ground state contains atoms that are capable of absorbing radiant energy of their own specific resonance wavelength, which is the wavelength of radiation that the atoms would emit if excited from the ground state. However, if light of resonance wavelength is passed through a flame containing atoms of similar wavelength the light will be absorbed. The absorption is proportional to the number of ground state atoms present in the flame. This is the principle of AAS.

1.6.2 Atomic Fluorescence Spectroscopy

These instruments analyse elements by observing the re-emission of absorbed energy by free atoms, and quantitative measurements are similarly monitored through fluorescence detection. This technique has not found favour in many laboratories but is used as a detector for liquid and gas chromatography for a limited number of elements such as Se, As, Cd, Sr, Ni, Ti, Sn, Pb, and Hg in their organo-metallic form, which are separated on columns for environmental samples. AFS detects resonance lines of the lowest frequency and its advantage is that it is easier to excite and is less affected by scatter when compared with resonance lines of higher frequency. The lack of instrument development because of the wide acceptance of AAS, which dominated the market since its inception, is unfortunate. If the market had responded with the same enthusiasm as that given to AAS, the commercial suppliers would certainly have responded with research, financial and development support.

1.6.3 Direct Current Plasma Optical Emission Spectrometry (DCP-OES)

This technique was first described in the 1920s having been thoroughly investigated since then as a source as an analytical tool. It was not until the late 1960s with the development of suitable optics that it has improved to such an extent that it could compete with ASS and later to some extent with ICP-OES. The DCP jet source (which excites the atoms) consists of three electrodes arranged in an inverted 'Y' configuration consisting of two tungsten electrodes at the inverted base and a graphite electrode at the top (Figure 1.7). The argon plasma formed is caused by bringing the cathode momentarily into contact with the anodes where a high electrical charge ~15 A, initiates the argon to form the plasma.

The viewing region of the plasma can achieve a temperature of 5000–6000°C and is reasonably stable. The sample solution is aspirated into the core area between the two arms of the 'Y' where it is atomised, excited and viewed. This technique keeps with the atomic spectroscopy theory in that the measurements are obtained by emission from the valence electrons of the atoms that are excited, and the emitted radiation consists of short well-defined lines. All these lines fall in the UV or VIS region of the spectrum and identification of these lines permits qualitative/quantitative detection of elements.

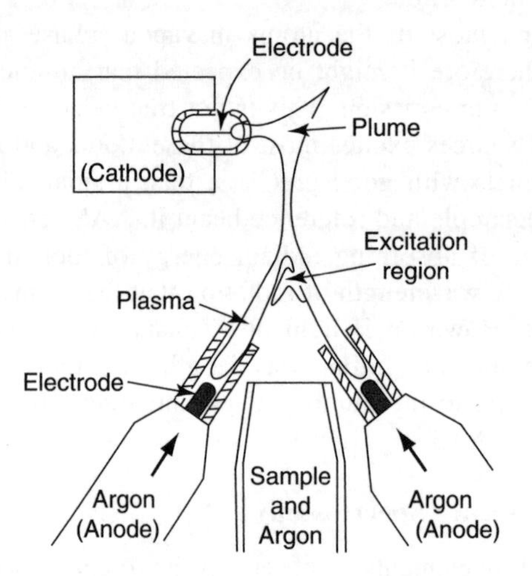

Figure 1.7 *Diagram of DCP-OES showing electrodes and plasma configuration*

The design of the DCP-OES allows the use of both aqueous and most non-aqueous solvents, providing standards and samples are prepared under similar conditions. It is more expensive to operate than AAS but cheaper than ICP-OES. The limitation of DCP-OES is the susceptibility to excitation interferences and increased signals from easily ionisable elements (EIEs). It has lower limits of detection and wider linear range for most elements but not as good as ICP-OES.

1.6.4 Microwave Induced Plasma (MIP)

This technique uses helium as the plasma gas which enables a higher temperature so that non-metals are excited. The MIP is hampered by matrix interferences, even water. Therefore it is used mainly for the analysis of gases, particularly in conjunction with gas chromatography.

1.6.5 Glow Discharge Optical Emission Spectrometry (GD-OES)

This technique is used mainly for surface analysis of electrically conductive materials provided that correction factors are applied if all components are known. The theory is based on the light emitted from a glowing discharge between a hollow cathode lamp and the sample (cathode) in an atmosphere of argon. Argon cations are formed which are accelerated in the direction of the negatively charged sample from which atoms are released, exited and quantified.

1.6.6 Inductively Coupled Plasma Optical Emission Spectrometry (ICP-OES)

This instrument was developed by Stanley Greenfield 1964 [9] and is an analytical tool used for the determination of 75 elements and their states (oxidation, isotopic, etc.) in a wide range of sample matrices. Elements that are not determined are those already in the plasma from sources not in the sample, argon, carbon dioxide, hydrogen, oxygen, and nitrogen that are in the air surrounding conventional plasma, and elements that require very high energy, such as the halogens. Some elements not detected using ICP-OES may be detected with ICP-MS, which measures their mass to electron ratio (*m/e*). The inductively coupled plasmas are designed to reach temperatures higher than ordinary combustion flames. The argon gas used to generate this high temperature has many advantages in that it forms a stable and chemically inert environment, which eliminates many of the interferences encountered with combustion flames. Plasmas are used for emission because the temperature is high enough to excite most of these elements. Detection of elements at its wavelengths is usually by photomultiplier tube(s) (PMT) or by charge coupled device (CCD). More expensive ICPs are designed to direct individual emission lines from different elements in the plasma to individual detectors. Such instruments allow simultaneous multi-elemental analyses that are rapid and carry out considerably more analysis in a shorter analysis time. They are applicable where hundreds of samples need to be analysed per day. The disadvantage of this instrument is that they are confined to built in elements. Research carried out by Wendt and Fassel [10] showed that higher temperatures can be achieved by argon plasma created by a magnetic field using a radio frequency generator. The neutral atoms, ions and electrons collide under the influence of the magnetic field causing excitation and emission of wavelength energy in the UV-VIS region of the atomic spectrum. Therefore, this instrument offers a procedure to enable analysis of most elements in the periodic table including refractory type elements in the region of 160–700 nm with ease. The latest sophisticated optics developed for ICP-OES allows excellent resolution to cater for multiple elemental analyses and shows reduced interferences by other elements, for example, the determination of aluminium and phosphorus in the presence of a high concentration of calcium.

The basis of ICP-OES is a sample solution introduced into the core of a hot (~9000°C) argon gas, i.e. highly energetic ionised gas. At this temperature, all elements in the sample become thermally excited and emit light at their characteristic wavelengths. The light is collected by the spectrometer and passes through a diffraction grating that resolves the light into a spectrum of its constituent wavelengths. These wavelengths are amplified to a signal that can be measured and used for quantitative purposes by comparing with calibrated standards prepared under similar conditions. The interaction is based on thermal dissociation of the elements causing emission of the free atoms. This technique is now prevalent in almost every field of chemical analysis, more so than any other technique in spectroscopy. Detection limits as low as 10^{-12} g can be achieved with reasonable accuracy and selectivity. Initially, these techniques were developed for aqueous and metallic samples but in latter years, with improvements in instrument design, elemental analysis of organic samples has also benefited. This instrument is now the most frequently used and most sensitive elemental analyser available and is adaptable to ancillary attachments to cater for difficult samples, to improved limits of detection and improved speed of analysis. The multi-element analysis feature offers considerable savings in analysis time and at the same time maintains excellent accuracy and reproducibility. Unfortunately, the technique is not without problems as it is very sensitive to trace analysis and extreme care must be applied in preparation of standards and samples in ensuring that the results obtained are true. The instrument parameters, such as gas flow, wavelengths, clean sample, tubing for transporting sample to the nebuliser must be correctly fitted and tested prior to use.

1.7 Summary: Applications of Atomic Spectroscopy

Flame atomic absorption and flame emission techniques were developed before inductively coupled plasma emission spectroscopy and are still used extensively for analysis of a wide variety of samples on a routine and non-routine basis. They are very useful techniques for elemental analysis of selected sample matrices; however, the lower temperature (~2800°C) of these techniques limits their sensitivity to a range of important samples. The development of plasma sources (1970s) capable of achieving temperatures of 9000°C has revived the use of emission instrumentations enabling improved sensitivity and multi-elemental analysis at major, minor and trace levels. The use of plasma as an excitation source has an added analytical advantage in its ability to use a wider linear dynamic range allowing little or no dilution and reducing chemical and physical interferences. The higher excitation temperature of plasmas compared with flames results in a more efficient atom excitation which leads to increased sensitivity especially for refractory elements such as B, P, W, Nb, Zr and U. The plasma source geometry and dynamics mean fewer sample atoms in the plasma and temperature profiles result in minimal line reversal and matrix interferences.

Both flame and plasma sources are sensitive and selective techniques measuring as little as 10^{-16} g of analyte solution in complex mixtures. However, understanding the techniques and the type of sample being analysed helps in deciding which technique is suited for a particular application.

Elements occur in natural and synthetic compounds at various levels and since the beginning of the development of atomic spectroscopy analytical instruments, more information about toxicity, benefits, etc., became known. Modern atomic spectroscopy instrumentation can determine from % levels to trace levels (ppm) and sub-trace low levels (sub-ppb) with a high degree of accuracy and precision. At whatever concentration, knowing the concentration of these elements plays a very important role in understanding more about products in terms of health issues, benefits, shelf life, stabilities, etc. Table 1.4 is a summary of elements of great importance that need to be monitored in order that any changes can be interpreted as part of behaviour that could be good or bad.

Table 1.4 *List of some elements of importance requiring quantification using atomic spectroscopy techniques*

Element	Area of importance
Mg, P, S, K, Si, V, Cr, Fe, Co, Ni Co, Zn, As, Sc, Mo, Sn and I	Biochemistry and medicine
Almost all elements in the periodic table	Environmental science
Contamination and wear metals e.g. Fe, Ni, Co, Mn, Cr, Mo, W, Na, K	Crude and virgin oil, and petroleum industries
Radioactive elements used for industrial and medicinal purposes	e.g. U^{238}, U^{235} Cs^{132}, Pu^{239}, B, Si, Cd, etc.
Si, Al, As, Fe, Cu, Mo, Hg, Cr, Se, Ba, Sr, Sb, Au Bi, Nb, Zr and Pb	Electronics and semiconductor industries
Almost all earth and clay containing elements	Geological research
Cu, Pb, Al, P, S, Ca, Mg, Si, Hg, Cr, Co, etc.	Works of art, paint mixtures, etc.
Si, B, Cd, Pb, Fr, Cs, U Hg, As, Sn, Sr, Ni, Ge, Ga, etc.	Ceramic industries
All elements	Forensic support

The importance of major and trace elemental analysis is paramount in biological, biochemical, medicine, environmental, forensic, pharmaceutical compounds, geological and gas samples. In most cases quantifying the level of metal can play a vital role in obtaining information on whether samples contain toxic metals or metals that are beneficial, acting as a catalyst or retarding effect on compounds or formulations. In the early part of the 20th century trace metal determination was not possible, hence the effects of a wide range of elements were unknown. A colour method for the determination of iron in aqueous samples was developed in the 1940s and is an example of a non-spectroscopic method used for determining this metal at sub-trace levels. The procedure involved complexing the iron with thioglycolic acid using a series of operations and reactions. This colour test is extremely sensitive and prone to errors caused by impurities in reagents and some samples. It is also time-consuming, taking several hours for confirmation of presence, and can only be used for one element at a time. The same measurement can be carried out using an atomic spectrophotometer against standard iron prepared the same way as the sample eliminating impurities or correcting for them in the method of analysis. Measurement by atomic spectroscopy of the same sample and many more could be analysed in minutes. Nowadays, modern inductively coupled atomic spectroscopy methods can scan a sample for 75 elements in 60 s using a multi-elemental scanning ICP-OES.

Presence or absence of trace elements in living organisms is essential for health information. Analysis of such samples by atomic spectroscopic techniques can make a distinction between vital and non-vital elements. Deficiencies in element(s) lead to deficient syndromes and if supplemented, the element(s) may prevent or cure the syndrome. Therefore the correct dosages are extremely important. In some cases deficiency of certain elements can lead to a decrease in specific biochemical functions that could be fatal. However, a high intake of some elements through foods, drinks, and air particulates can also seriously damage health. Modern day scientists work closely with health workers in sharing information and the detection of more elements at lower or higher concentrations is aiding in the better understanding of the behaviour of metals and determining their beneficial or non-beneficial effects.

To illustrate an example, monitoring the level of iron haemoglobin in blood, which acts as a binding agent for the oxygen molecule is an important test in terms of health control. Therefore, it would be assumed that the higher concentrations of iron in human blood would be beneficial. The pumping mechanism of the hearth transports blood containing the Fe-haemoglobin around the essential parts of the body. Monitoring the iron level can aid medical workers in diagnosing whether women have serious deficiencies of this metal due to monthly menstruation. Supplementing such deficiencies with the correct dosages established from analysis can assist the medical workers to correct this. Iron deficiencies can lead to anaemia, fatigue, headache and sometimes anorexia. However, on the other hand, consuming high dosages of iron can injure the alimentary canal, cause hepatitis, haemochromatosis and lead to cirrhosis that could be dangerous.

Elements such as oxygen, nitrogen and carbon which are available through normal chemical processes of the atmosphere and biochemical functions which are determined by these atomic spectrometric techniques are also as vital, as are Mg, S, P, Cl, Na, K, Ca, I, F, Si, Co, Ni, Cu, Zn, Se, Mo, Sn Cr, Mn, etc.

This book will be devoted to the practical approach of quantitative metal analysis using ICP-OES. There is a lack of analytical information and methodologies available on the analysis of simple and difficult organic matrices. In this book, I hope to present a few ideas in terms of sample preparation, quantification, and comparison of techniques of actual work carried out by the author over several years.

The plasma source is, at present, the most important method of atomic excitation and is compatible with organic solvents. As previously stated a definite wavelength can be assigned to each radiation, corresponding to a fixed position in the spectrum. However, as the colours for calcium, strontium and lithium are similar and with the advanced optics design it is now possible to differentiate between them with certainty by observing their spectra in the presence of each other. Similar elemental differentiation/identifications are carried out for 75 elements in the periodic table. By extending and amplifying the principles inherent in the flame qualitative test, an analytical application of emission spectroscopy has been developed using ICP-OES. The developments in atomic spectroscopy using plasma sources, instrument control and data processing by computers for multi-elemental determinations in a wide variety of inorganic and organic matrices are truly indicative of the trends in analytical analysis.

References

[1] Walsh, A. (1955) The application of atomic absorption spectra to chemical analysis, *Spectrochimica Acta*, **7**, pp108–117.

[2] L'vov, B.V. (1961) The analytical use of atomic absorption spectra spectrochimica, *Spectrochimica Acta*, **17**, p761.

[3] Greenfield, S., Jones, I.L. and Berry, C.T. (1964) High pressure plasmas as spectroscopic emission sources, *Analyst*, **89**, pp713–720.

[4] Bell, X. and Lott, X. (1966) *Modern Approach to Inorganic Chemistry*, 2nd edition, London: Butterworth & Co.
[5] Parsons, M.L., McCarthy, W.J. and Winfordner, J.D. (1966) Design stage of atomic spectroscopy, *Applied Spectroscopy*, **20**, p223; (1968), **22**, p385.
[6] Hassan, S.S.M. (1994) *Organic Analysis Using Atomic Absorption Spectrometry*, London: Ellis Horwood, p16.
[7] Rubeska, I. and Moldan, B. (1967) *Atomic Absorption Spectrophotometry*, London: Iliffe; Prague: SNTL.
[8] King, R.B. and Stockberger, D.C. (1940) *Astrophysics Journal*, **91**, pp488–492.
[9] Greenfield, S. (1980) Plasma spectroscopy comes of age, *Analyst*, **105**, p1032–1–44.
[10] Wendt, R.H. and Fassel, V.A. (1965) Inductively-coupled plasma spectrometric excitation source, *Analytical Chemistry*, **37**, pp920–922.

2

Instrumentations Associated with Atomic Spectroscopy

2.1 Instrumentation

Major and trace metal content of most inorganic, organic and biological samples can be achieved using optical emission spectrometric techniques and these techniques are the oldest and most developed available. All elements can be made to emit radiation characteristic of their state under thermally controlled conditions. Unfortunately, no single source will excite all elements in an optimal way, e.g. halogens, carbon, oxygen, nitrogen and some of the lanthanides and actinides, because they either need very high excitation energies, are outside the UV-VIS range, are present in the atmosphere surrounding the torch or are present in the argon gas. Some elements, particularly the halogens, require very high energies to excite them and the argon plasma is unable to perform this task. They can be excited using a MIP with helium gas because the ionisation energy of helium and the energies of helium plasma species are greater than that of argon. The flames/plasmas are mainly used for quantitative analysis and the arc and spark are used widely in certain limited applications for semi-quantitative and qualitative analysis. In the latter years, ICP-OES came into existence through the efforts of several early workers, e.g. Stanley Greenfield and his assistants [1] at Albright and Wilson, and Velmer Fassel and Richard Wendth [2] at Iowa State University. This radiation source was developed for emission spectrometric analysis in the early 1960s. However, it was not until the late 1970s that ICP-OES, DCP and MIP have become commercially available and are finding uses in many laboratories. DCP and MIP will only be briefly discussed in this book because of their specificity and limited applications.

The plasma sources commonly available for trace elemental analysis are DCP-OES and ICP-OES of which the latter is more popular due to being the most studied and user-friendly and its compatibility with hyphenated accessories. The MIP mainly uses helium

A Practical Approach to Quantitative Metal Analysis of Organic Matrices Martin Brennan

gas and achieves higher excitation energies so that non-metals can be excited. The MIP is severely restricted by matrix influences, even water and other common solvents, and because of this finds excellent use as a detector of gas chromatography (GC) which isolates the solvents from the species of interest for easy detection. When connected to GC, the detection limits for halogens are very low, as are those for carbon, sulphur, oxygen and certain volatile organometallic compounds.

Both the DCP-OES and ICP-OES are operated as stand alone instruments and controlled as atomisers and excitations sources for more than 70 elements in the periodic table. They are used successfully in many industrial, public analyst, medical and educational laboratories worldwide. These techniques are required so that optimisation of source conditions can be used for the determination of different elements present in a wide array of samples requiring variable excitation energies. Some elements such as the alkali and alkaline can be analysed routinely using flame emission spectroscopy (FES) in aqueous or non-aqueous solutions. The ease of excitation of such atoms are not as readily lost at lower temperatures as they are at higher plasma temperatures. Most elements of the alkali and alkaline groups tend towards the red end of the visible spectrum and are best detected with a cooler flame using an atomic absorption or flame photometer instrument equipped with specific filters. However, these elements can be detected by ICP-OES but are not as good as flame emission AAS. All other elements not readily detected by flame emission are detected by ICP-OES including the elements detected by flame only. The ideal environment for atomisation and excitation of elements is in an inert atmosphere that minimises background emission and other interferences, particularly where trace analysis is required. The source should be stable, reproducible and continuous in order to obtain precise and accurate elemental analysis.

The plasma sources can achieve temperatures in the range of 5000 to $\sim$10 000 K which are advantageous over flame emission that can only achieve temperature ranges of 1500 to $\sim$2500 K. The flame in AAS lends itself to self absorption, spectral, chemical and ionization interferences which gives rise to noisy background. These interferences including ionisation are not very severe in plasmas because the extra electrons released by EIEs have little effect on the ionisation equilibrium of other elements and the extra electrons form a small portion of the total electron concentration in the plasmas.

A major advantage of DCP-OES and ICP-OES is that chemical interferences are reduced as the temperature increases and are such that they are of minor importance. In the older ICP-OES spectral interferences were a problem due to poor isolation of signals caused by inferior optics design. In the latter years, the introduction of modern sophisticated optics, better detectors and improvements in computer control of instrument conditions have given rise to better resolution and sharper signals. The combination of the latest optics and detectors can resolve lines to form signal shapes that are neat and characteristic of the element. The routine use of echelle, holographic gratings and a range of electronic detectors e.g. charge coupled detectors (CCDs), charge injection detectors (CIDs) and photodiode arrays (PDAs) is so successful that chemical interferences are considerably reduced or virtually ignored.

Physical interferences are caused by matrix effects, which can change the physical properties of the solution being nebulised. An example of the suppression of calcium by proteins and fats in serum by high concentration of barium, chromium, cobalt and zinc is evident when a concentration of 10 $g L^{-1}$ of each element is added to the same

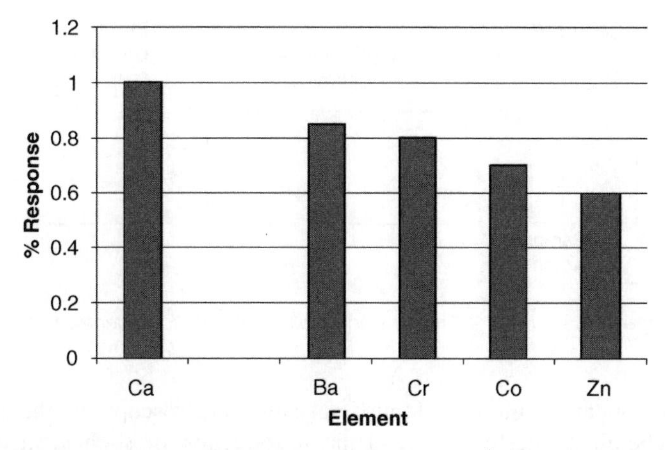

Figure 2.1 *Suppression of Ca atoms by the addition of 10 g L^{-1} of Ba, Cr, Co and Zn separately in solution of each individual element. The Ca at 100% response is in water only and none of the listed elements*

concentration of Ca (Figure 2.1) using standard plasma conditions [3]. Variable liquid densities, viscosities and different surface tension of solvents can affect the drop-size formation (Figure 2.2). The drop-size produced by the nebuliser plays a major part in the sensitivity of measurements and is governed by each or combinations of the above. There are many other examples of the behaviour of chemical and physical effects with other elements and such information is available from dedicated flame emission handbooks and literature. In conclusion, the higher temperature offers greater freedom from chemical interferences and greater sensitivity, particularly in the UV region. However, with physical effects they are not temperature related.

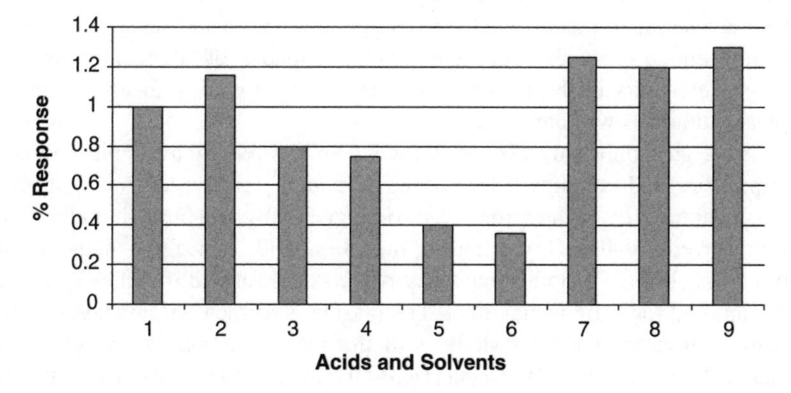

Figure 2.2 *Effect of 50 μg/ml Fe in (1) H_2O, (2) 100% CH_3COOH, (3) 50% HCl, (4) 50% HNO_3, (5) H_2SO_4, (6) 50% H_3PO_4, (7) 100% C_2H_5OH, (8) 100% C_3H_7OH, (9) 100% Kerosene*

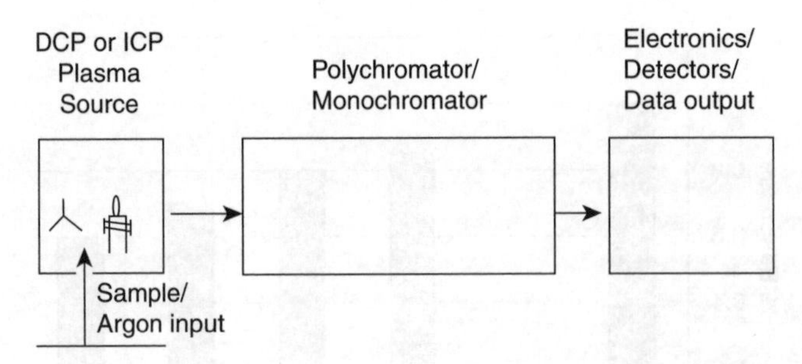

Figure 2.3 *Basic configuration of plasma emission spectrometers*

The improvement in plasma atomic emission spectroscopy in the reduction or removal of chemical interferences and the introduction of sophisticated optics and computer controlled instrumentation allow this technique to generate calibration curves several orders of magnitude greater than atomic absorption techniques. It is also possible to perform simultaneous multi-element analysis on multiple samples, gather a large pool of information and assist in rapid reporting of important results of a wide range of samples. Figure 2.3 shows the basic configuration of plasma emission spectrometers.

2.2 Types of Plasma Sources

The definition of plasma in the *New Oxford Dictionary of English* is:

> 'An ionised gas consisting of positive charged argon and negative electrons in proportions resulting in more or less no overall electric charge, typically at low pressures (as in the upper atmosphere and in fluorescent lamps) or at very high temperatures.'

To translate this into atomic spectroscopy terminology, plasma may be defined as: 'an ionised gas consisting of positively charged argon ions ($Ar+$) and negative electrons ($e-$) moving independently at the very temperature capable of atomising, ionising and exciting ~75 elements in the periodic table that can be readily measured analytically in samples' (author's own interpretation).

Argon is the gas commonly used in ICP-OES as it is readily available, can achieve a high temperature and contains a low concentration of reactive chemical species. The plasma is sustained by transferring electro-magnetically via a high radio frequency alternating current (~40 MHz) using an induction coil, hence the name inductively coupled plasma (ICP). The radio frequency plasmas are formed in a flow of a gas by an externally applied radio frequency field. The plasma is formed within and/or above a set of refractory tubes arranged coaxially with the induction coil, the whole forming a plasma torch. The direct current plasma (DCP-OES) uses a high current with two or three graphite electrodes and a tungsten electrode to sustain the plasma and the MIP uses an ultra high frequency Tesla coil producing alternating current (~2500 MHz) to form a plasma using argon or helium gas.

2.2.1 Direct Current Plasma Atomic Emission Spectrograph

The DCP plasma source consists of a high-voltage discharge where two or three pyrolytic graphite anode electrodes sustain the plasma and a toriated tungsten cathode electrode is arranged so that the stable plasma formed is an inverted Y-shape. The electrodes are disposable items, as are the ceramic sleeves that surround the electrodes. When the jet is in operation, the argon flows through the sleeves around the electrodes and the sample as an aerosol and enters the plasma through the sample introduction tube. Argon is also the carrier gas for the sample. The main excitation area is located below the plasma continuum where a temperature of 5500 to 6500 K is achieved and the noise ratio is considerably lower than in the continuum. The sample is nebulised at a lower flow rate than that for ICP using argon as the carrier gas. The argon ionised by the high-voltage discharge is able to sustain a current of ~20 A indefinitely. The DCP has a low limit of detection and is cheaper to purchase and operate. The graphite electrodes and sleeves need to be replaced frequently. A flow of argon is directed over each electrode at a relatively high velocity to cool the ceramic sleeves containing the electrodes to prevent melting. The sample is nebulised with the argon gas. At the maximum excitation temperature the sample volatilisation is not complete due to the relative short residence times in the plasma. However, the extent of volatilisation that does occur is sufficient to cover most elements in the periodic table. This may cause problems with samples that contain elements with high excitation energy. A further disadvantage is the small region where optimal line to background ratios occur and detection limits are inferior to ICP-OES. The rugged design of the DCP plasma source offers major advantages over ICP-OES: it is more tolerant of a wider range of solvents and samples with high salt content because of the wider bore at the injector. The requirements to form DCP discharge are simpler and cheaper to operate than ICP discharge. See Figure 1.7 for a diagram of the main components of the DCP-OES.

2.2.2 Microwave Induced Plasma

The MIP is designed to use either argon or helium as the plasma gas (Figure 2.4). It can achieve higher excitation energy with helium so that non-metals can also be excited and analysed. The MIP is subject to matrix interferences from common solvents–even water–and this is a major advantage when analysing elements in gaseous form and finds many applications when used as a detector on a gas chromatograph. The main reason for using helium in MIP is that the metastable state of the argon atom has less energy than the metastable state of the helium atom and argon plasmas give molecular emissions rather than atomic emissions. The high excitation temperature (7000 to 9000 K) is achieved by collisions with the metastable helium atoms and this highly excited state of helium is formed from electron-ion recombination. This high excitation energy and temperature will favour the formation of emitting states of non-metal atoms as they need more energy to excite than metal atoms. It is known that an ionised atom in the plasma source allows the recombination process to form highly excited atomic states. Many emission lines from excited states to less energetic excited states will be in the UV-VIS region and readily detectable. These plasmas have limited uses and are generally used for analysing elements capable of forming gases.

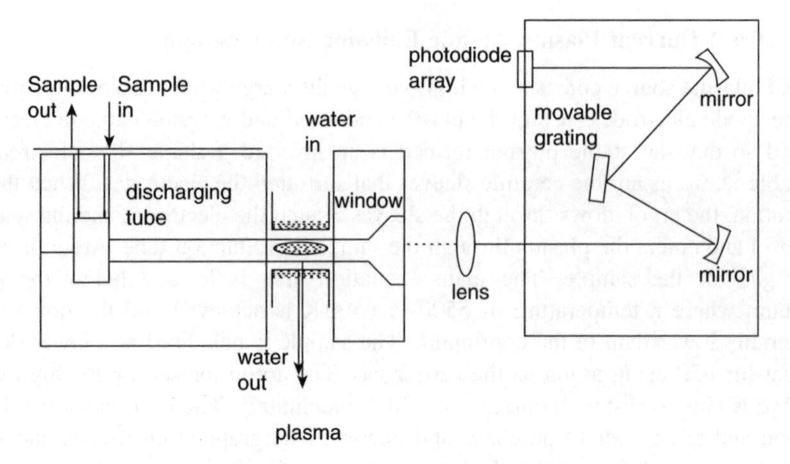

Figure 2.4 *Basic MIP configuration*

A Tesla coil is used to create a spark to generate the argon or helium plasma. The electrons generated oscillate in the microwave field and gain sufficient kinetic energy to ionise either gas by rapid and violent collisions. This is achieved by using a microwave frequency of ~2500 MHz. Elements such as fluoride, chloride, bromide, iodide, sulphur, phosphorous, and nitrogen, which are not possible to measure by ICP-AES or DCP-AES, can be measured by MIP.

2.2.3 Optical Emission Spectroscopy

The basic principle of ICP-AES now used worldwide is described by Greenfield *et al.* [1]; in latter years major improvements have taken place, but the underlying principles are the same (Figure 2.5). The latest instruments are adorned with sophisticated microcomputers, packages suitable for control, measurement, signal display, results and reporting. There are two main types of plasma emission spectrometers: sequential and simultaneous (selected multi-channel) analysers. The simultaneous can also be described as a direct reading spectrometer. The sequential uses only one channel and scans for selected lines (wavelengths) of interest.

2.2.3.1 Sequential ICP-OES. The sequential ICP-OES is the most common and is slightly cheaper for accurate quantitative analytical work that measures selected wavelengths one after another. This type of ICP-OES uses a one channel monochromator that is rotated using a computer control led stepper motor to locate and select the wavelength positions. Each wavelength can be selected at different locations for each element under consideration across the spectral profile incorporating a slew scan monochromator between 160 and 700 nm. Slew scan monochromators use a two speed wavelength movement and can provide a significant saving in analytical time and sample consumption over a measurement scanning. The grating is rapidly moved or 'slewed' to a wavelength near the analyte line of interest and final wavelength position is achieved by slowly positioning itself on the wavelength in small steps until the true position is obtained. These instruments do not have restrictions to line selection of which several can

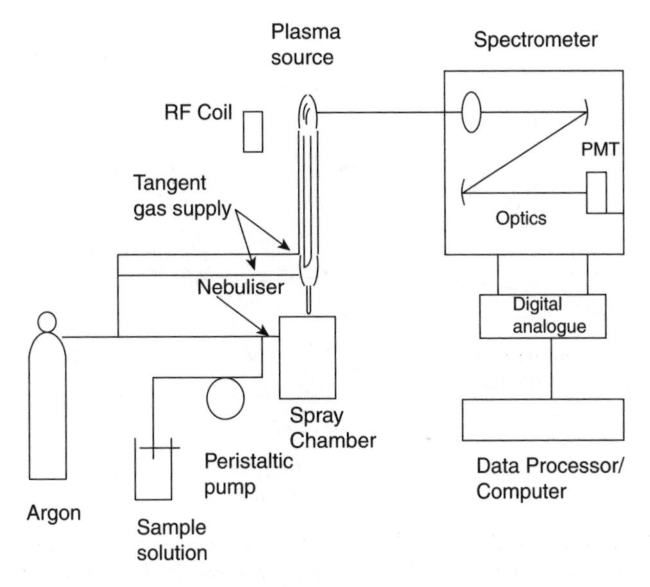

Figure 2.5 *Diagram of main components of ICP-AES*

be selected for the same element. The most common mount used for sequential instruments is the Czerny-Turner type that contains two separate mirrors, the colliminator and the separator as part of the optics design. The Ebert type mount contains both the colliminator and collector on the same mirror. In the latest design, instruments are turning towards the echelle grating in conjunction with a prism where excellent resolution can be achieved by precision ruled grating. These rulings are prepared using microwave etching to form the precise angles and faces and may be in the order of 100 gratings mm^{-1}. The signal identification and quantification programme can be used to measure wavelength and background intensities on one side or both sides of the analytical line. These instruments require very accurate presetting of the wavelength, of which half-width of a spectral line may be as narrow as 10^{-5} degrees. Several elements can be analysed using the sequential ICP-AES, but are slow when compared with simultaneous spectrometers. The torches where atomisation and excitation take place can be mounted either in axial or radial positions.

2.2.3.2 Simultaneous ICP-OES. The simultaneous ICP-OES measures all elements at the same time. A large number of samples can be analysed in a short period of time making it useful for rapid analysis. They are very expensive and are used where routine multiple sample analysis is required on a regular basis, usually on the same elements and samples. Most simultaneous instruments are custom designed for a selection of elements at specific wavelengths and some instruments can have between 10 and 100 slits and are factory fixed for selected elements. Typical applications would be in the water industries where analysis for metal content would be important for health reasons, in the water supply to power stations where analysis is usually carried out before and after treatment prior to use so as to avoid contamination of turbine blades, in the food industries, in mineral exploration or any other routine analysis where metal analysis requirements do

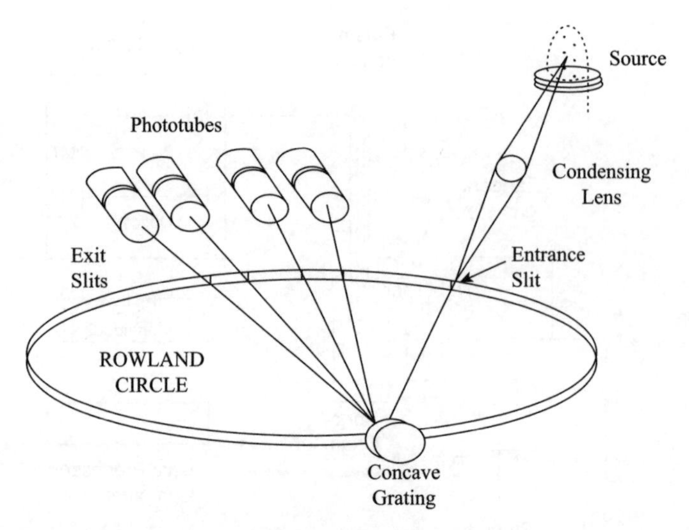

Figure 2.6 *Basic design of direct reading on a Paschen-Runge mount with fixed optics on a Rowland circle used in simultaneous ICP-AES. (Reproduced by kind permission: copyright © 1999–2008, all rights reserved, PerkinElmer, Inc.)*

not vary. These types of instruments commonly use the Paschen-Runge mount where the grating, entrance and exit slits are fixed on the Rowland circle (Figure 2.6). More sophisticated instruments are designed to allow the slits to be moved tangentially to the Rowland circle by means of a stepper motor, permitting scanning over peaks and providing a mechanism for background corrections. Several lines of the same elements can be included for confirmation purposes. The recent development of semiconductor type detectors offer the potential of multi-channel readout devices and include the introduction of photodiode-array systems, giving the possibility of detecting a wide spectral range from a single plate.

Trace analysis requiring background correction can usually be achieved by computer control displacement of the entrance slit or by rotation of a quartz refractor plate behind the entrance slit. Using this technique allows the application of slew scan methods which allow scans from different samples to be superimposed with solutions of samples, dissolution solvent(s), etc.

2.2.3.3 Dual View (Radial and Axial) ICP-OES Plasma. Since the development of ICP-OES, ICP torches were constructed in an upright position allowing the excess waste (sample/gases/toxins) to be extracted by an extractor above the torch. These early plasmas were viewed radially which means that only a small section of the light passed through the plasma perpendicularly through the analyte channel. Modern radial viewing ICP instruments allow the operator to select viewing heights which can be anywhere between 1–4 mm above the radio frequency induction coil restricting the observation zone to a segment of the analyte channel where the excitation is the highest and most stable for that wavelength. Some modern instruments facilitate scanning the excitation zone to reflect the behaviour of the entire excitation range and allow selection of the best viewing heights for that analysis.

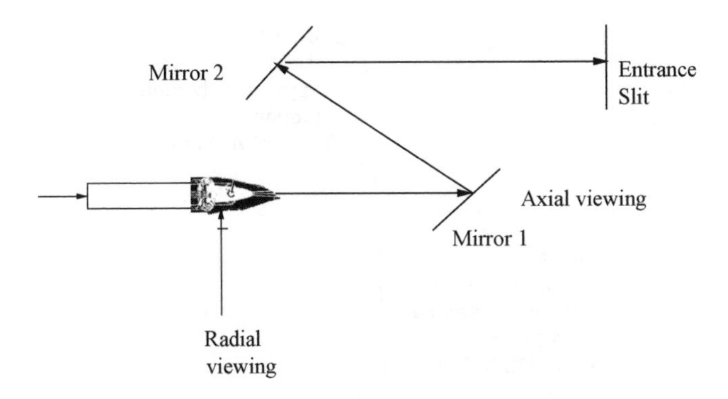

Figure 2.7 *Schematic diagram of a dual viewing ICP-OES*

A dual viewing ICP plasma can be observed in two directions: radial (sideways through the optical path) and/or axially (lengthways through the optical pathway) and these geometric views can be used individually or as a combined system depending on the range of and concentrations of analytes in a sample (Figure 2.7). The idea of axial viewing was considered at the beginning of ICP development but was prohibited due to design, technical, computing and analytical problems for around 25 years. The modern technique (introduced in the mid 1990s) makes the analyses more concentration friendly in terms of samples containing high and low concentrations of metals, i.e. samples of high concentrations can be analysed using the radial viewing while the low concentrations can be analysed using the axial viewing in a single run.

Axial viewing (end on) along the entire length of the plasma is a symmetry axis of rotation and is found only with instruments that have horizontal plasmas and will give improved sensitivity of several orders of magnitude. The limit of detection and signal to noise background ratio using the ICP axial viewing method is also improved by an order of a factor when compared with the radial viewing method and this is due the 'visibility' of a greater number of excited atoms and ions in the sample. The peaks in Figure 2.8 illustrate the differences in height for the same concentration of metal using identical conditions.

The linear range for axial and radial viewing is different. In axial viewing the optical pathway is longer thus increasing the risk of self-absorption and self-reversal. The latter is more likely when the analyte has to pass through a longer but cooler zone. However, this can be reduced or removed by inserting skimmers at the tip of the plasma zone or using a shear gas of argon, nitrogen or air perpendicular to the tip of the plasma torch and removing the wasted analytes and tail from the cooler zone of the plasma. The cooler zone includes the recombination fraction, which can act as an absorption zone for the excitation light, however, the wasted heat from the sample matrix could affect the signal measurement causing poor background, reducing the linearity range and distorting the signal structure. The disadvantage of using air instead of the inert gases nitrogen or argon as a shear gas is that the transparency in the vacuum-UV range is reduced due to the presence of oxygen in the air that absorbs the light below 200 nm.

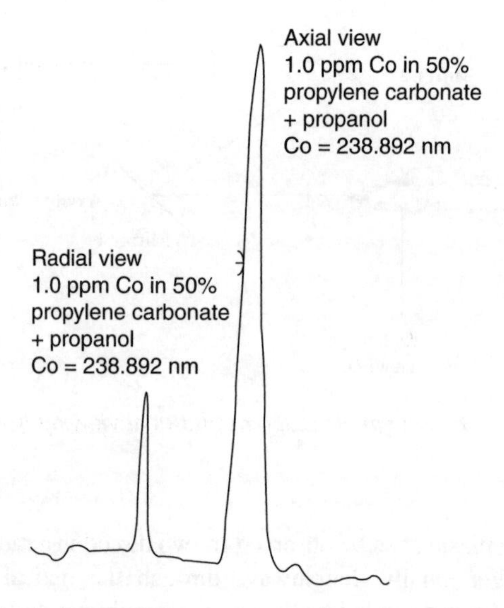

Figure 2.8 *Comparative peak height for 1.0 μg ml^{-1} (ppm) B using radial and axial viewing. Scans obtained using PerkinElmer dual view Optima 2100DV ICP-OES*

2.3 Sample Introduction Systems [4]

Sample introduction is probably one of the most important stages for reproducible measurements and is related to the efficiency of sample uptake to the plasma source. Normally samples are introduced in solution form and in latter years sample introduction as solids and gases directly or from GC columns is now commonly employed on a routine basis where applicable. Selection for the best sample introduction method needs careful consideration, keeping in mind that the properties of the atomiser will dictate its design and operation. For adequate thermal dissociation, volatilisation, excitation and atomisation of aerosol particles, the efficiency of nebulisation will determine the sensitivity and reproducibility of analyte response. The following requirements must be considered when analysing samples using atomic emission methods:

- type of sample i.e. solid, liquid or gas;
- levels and range for elements to be determined;
- accuracy required;
- precision required;
- number of determinations per sample;
- amount of material available;
- requirements for speciation studies;
- applicability of internal standard;
- linear range.

Sample introduction is an extension of sample preparation and therefore selection of a suitable introduction technique can also depend on the sample preparation technique.

To fully understand the limitations of practical sample introduction systems, it is necessary to reverse the normal train of thought which tends to flow in the direction of sample, i.e. solution-nebuliser-spray chamber-atomiser, and consider the sequence from the opposite direction. Looking at sample introduction from the viewpoint of the atomiser, the choice of procedure will cling on to what the atomiser can accept. Different properties of temperature, chemical composition, solvent(s), interferences, etc., and an introduction procedure must be selected that will result in rapid breakdown of species in the atomiser irrespective of the sample matrix.

2.3.1 Mechanical Transfer of Sample/Standards Using Peristaltic Pump, Pressure Valves, Motorised Syringes, etc.

The first and important part of sample introduction is a means of getting the sample to the plasma source in as constant, continuous, reproducible and non-interfering state as possible. Should any one of these requirements fail the signal reproducibility and precision response will be poor. The simplest method for getting the sample solution to the plasma is by use of a peristaltic pump and it must be designed to move segments of the liquid forward as constantly as possible. The pulsation rate is very much governed by the number of rollers built into the roller head (the more the better). It is important that the distance between each roller allows it to rotate freely, so as to maintain a constant and reproducible feed to the plasma. The pressure bearing on the tubing should be just sufficient to maintain flow and not be too tight so as wear the tubing out rapidly. Modern instruments have as part of instrument design programs for tubing speed/uptake rate to give the maximum signal response. Some instruments have uptake altering abilities to differentiate between sample uptake during measurements and sample washout time to reduce memory effects between each reading. The sample uptake rate should be such that the pressure of the argon flow should be efficient in forming as high a concentration of aerosols after the nebuliser as possible. Modern nebulisers can create aerosols of diameter in the 0.1–5µm range for a 3–5% efficiency using a gas flow of 1.5 L min^{-1} and a sample uptake of 2.0 ml min^{-1}). The selection of pump tubing is also important and must be of a high durability, flexibility, contamination free, of suitable diameter and wall thickness and compatible for the type of sample being analysed and reproducible for continuous use. It must also be chemically resistant to a wide range of solvents and samples and be readily available locally.

A controlled pressure valve to force the sample solution to the plasma has had some success, particularly where aggressive solvents are used and tubing could be avoided. Another idea investigated was a motorised syringe where samples are transported to the plasma by a constant rotating shaft pushing the plunger of the syringe continuously. Both the pressure method and motorised syringe have poor reproducibility and further work may need to be carried out in order to perfect these alternatives.

2.3.2 Nebulisers

The liquid sample introduced to the ICP-AES plasma torch must be nebulised prior to excitation and atomisation. The basic function of the nebuliser is to convert a solvent containing the analyte solution into an aerosol by action of a very high pressure carrier gas created by the nebuliser. In combination, the spray chamber in which the nebuliser is sited in a suitable position will further reduce the aerosol particle size to an ideal size by further surface collision. The ultimate design of the spray chamber is important because it must be such that condensation is reduced hence improving the efficiency of the nebuliser/spray chamber combination. It must be borne in mind that the efficiency is also affected by solution viscosity, surface tension, ease of evaporation, vapour pressure, etc., and will also affect the carrier gas uptake rate. Some of these problems are overcome by use of a peristaltic pump to aid transportation of the sample to the nebuliser. Unfortunately, a design to reduce condensation further has not altered since it was introduced in the early 1970s. The ease of use and acceptability of existing nebulisers, which normally offer problem free operation, makes it a less studied part of the plasma introduction operation. Therefore the precise microscopic process for which it operates is not totally understood. As mentioned earlier, the liquid jet is shattered by interaction with a high velocity gas jet and such a process leads to a type of solvent stripping mechanism so that successive thin surface films are removed by the violent gas flow. The carrier gas that causes the nebulisation forms a negative pressure zone that breaks up the solution into small droplets. These droplets spontaneously collapse under surface tension forces to produce the aerosol mist prior to plasma contact. In any event the entire process and mechanism result in an aerosol that has a very wide drop size distribution and only 2–5% actually arrive at the plasma torch. The drop sizes are normally very small for ICP-OES operation while in atomic absorption they can be larger.

There are several different types of nebulisers available from local instrument suppliers (Figure 2.9). They are expensive due to the inert material used and precise engineering required to make them. The size of the hole for the gas outlet must be big enough to sustain the very high pressure required to force the sample solution to move violently and rapidly throughout the spray chamber and small enough to create a very high pressure. The two most commonly used nebulisers are pneumatic and ultrasonic.

Most nebulisers, with the exception of the ultrasonic nebuliser, have low transport efficiencies. Ultrasonic nebulisers can have from 5 to 20% efficiency depending on the sample solution. Transport efficiency is defined as the amount of original sample solution converted to an aerosol before entering the plasma source. The remaining 95–98% for conventional nebulisers and 80–95% for ultrasonic nebulisers goes to waste. Care must be taken of the amount of mist/solvent getting into the plasma, bearing in mind the smaller the droplet size, the easier it is to dry the droplets and to achieve all the subsequent steps in the plasma. However, if too much sample is introduced into the plasma it can cause it to extinguish, as the energy achieved in the plasma design is only sufficiently high enough to maintain it for as much solvent free analyte reaching the torch as possible for analytical purposes. Evidence of incorrect mist volume to the plasma will cause it to flicker leading to poor stability, precision and reproducibility of the measurements. With this in mind, the supply of mist/sample to the plasma must be optimised so as to produce as high a sensitivity as possible while maintaining the stability of the plasma. Therefore good nebulisers which cause few problems to the plasma must be the ultimate choice of selection in the type of analyses being carried out.

2.3.2.1 Pneumatic Concentric Nebuliser. The current design of the pneumatic concentric nebuliser is probably the most common type used. It consists of a concentric glass through which a capillary tube is fitted. The sample is drawn up from the spray

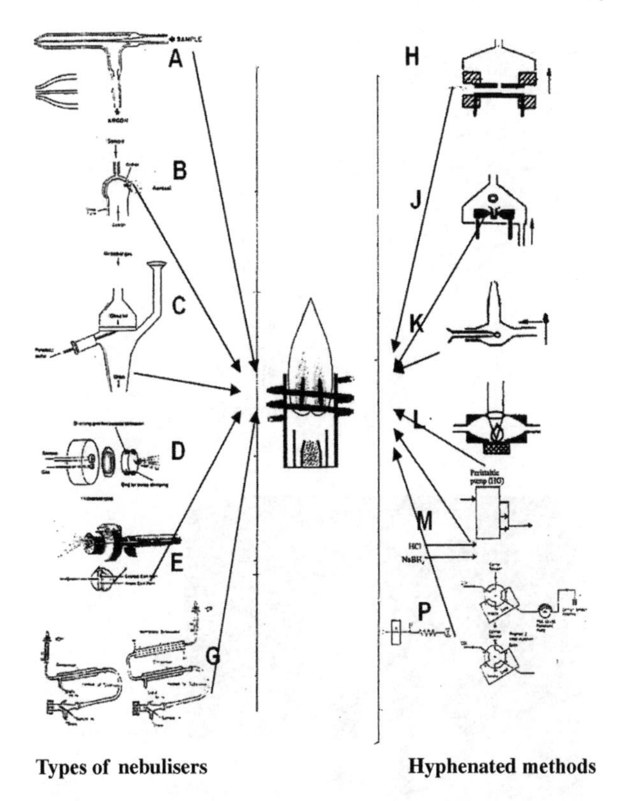

Figure 2.9 *Overview of sample introduction methods and hyphenated techniques used in ICP-AES. (A) Pneumatic concentric (sometimes called the Meinhard nebuliser); (B) Babington; (C) fritted disc; (D) Hildebrand nebuliser; (E) cross flow; (G) standard ultrasonic nebuliser for aqueous and non-aqueous solvents; (H) electro-thermal graphite; (J) electro-thermal carbon cup; (K) graphite tip filament; (L) laser ablation; (M) hydride generation; (P) flow injection*

chamber through the capillary by the argon carrier gas which can have pressure up to 300 psi escaping through the exit orifice that exists between the outside of the capillary tube and the inside of the concentric tube. The combination of speed, pressure and force caused by the small orifice avoids re-combination of the droplets once the mist is formed. The size of the toroid created by the two tubes is ~30 μm in diameter. The force of the escaping argon gas and liquid sample is capable of producing an aerosol from the Venturi effect. The solvent uptake can be between 0.2 and 5 ml min^{-1} and because of the Venturi effect it may not be necessary to install a peristaltic pump to get the liquid sample to the plasma.

The blockage problems associated with nebulisers occur where samples have high salt content and the concentric nebuliser is no exception. The maximum salt content that can be tolerated in these nebulisers is usually in the order of ~2.5% which depends on the aerosol gas flow and the type of salt solution which usually forms on the tip of the inner plasma torch and rarely at the orifice of the nebuliser. Figure 2.10(a) shows the Meinhard nebuliser, which by its simple design provides good stability, good endurance and is easy to maintain, but unfortunately suffers from blockage caused by higher than ~2.5% of salt

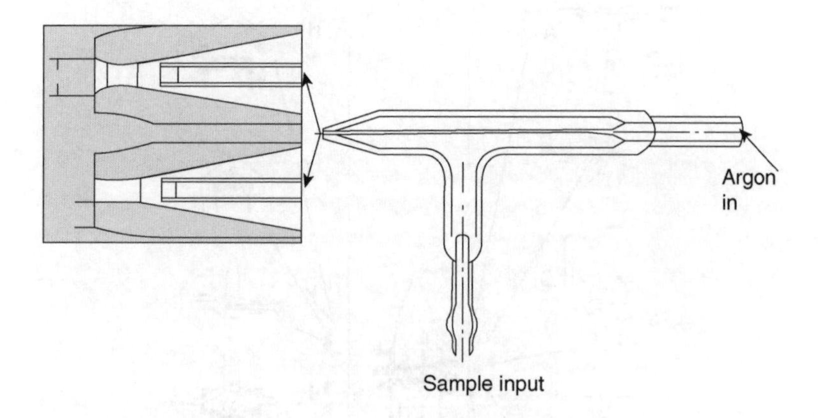

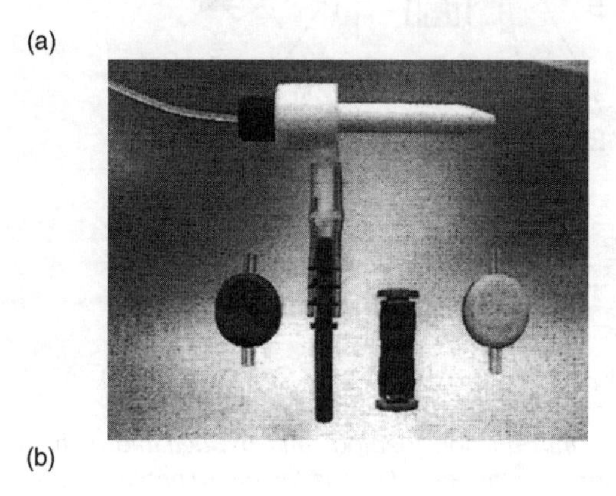

Figure 2.10 *Designs of (a) concentric glass nebuliser (Meinhard type) available showing enlarged fine bore tip to suit clear solutions and wide bore tip to suit solutions containing high salt content; (b) PTFE nebuliser suitable for HF acids with fittings. (Reproduced with kind permission from E-POND S.A., C.P. 389, CH-1800 Vevey, Switzerland)*

particles in solution. A unique Meinhard design is where the tip is wider so as to reduce blockage to some extent.

2.3.2.2 Babington Nebuliser. The design of a Babington nebuliser allows analysis of samples containing a high concentration of solids that have a tendency to precipitate out of solution (Figure 2.11). It functions by nebulising a high salt content liquid film by blowing it against a wall, causing coarse droplet formation. The main advantage of the Babington nebuliser is that the liquid need not be aspirated through a narrow orifice which avoids any blockage taking place during the course of analysis for solutions containing high solids content. A mixture of solids suspended in solution as slurry can be analysed using the Babington nebuliser and calibrations are carried out using a standard prepared in the same solvents used to suspend the solid particles. This has been described as a slurry analytical technique and has several useful applications. The most important parameter associated with slurry analysis is that the particle size must be suitably small, i.e. 10 μm or less. This technique has been used to determine trace to major levels of

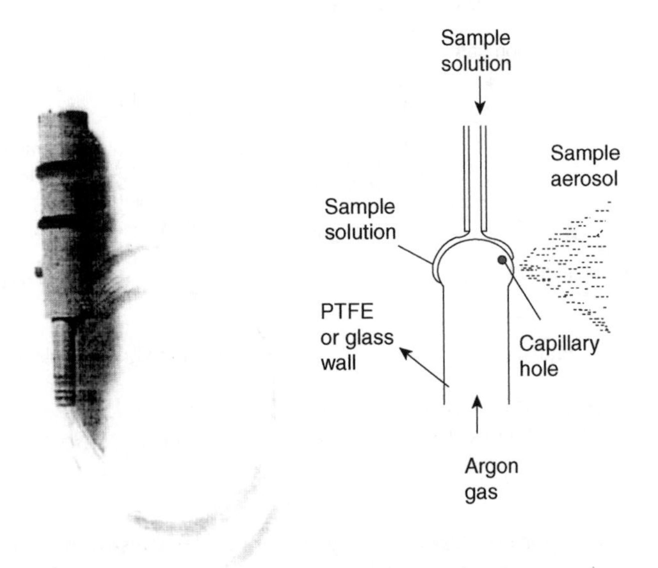

Figure 2.11 *Basic design of a Babington nebuliser showing the simple operation. (Reproduced with kind permission from E-POND S.A., C.P. 389, CH-1800 Vevey, Switzerzerland)*

elements in solution with good success. To achieve analysis of such solutions the following parameters are necessary:

- nebuliser suitable for high solids content is essential;
- injector set further from the plasma than usual so as to avoid solid/salt build-up at the tip;
- use a higher radio frequency;
- installation of an argon humidifier;
- rapid removal of sample after analysis and entire sample input cleaned with solvent between subsequent samples analysis;
- must be free of spectral interferences;
- free from solvent affect;
- slurry must be uniform throughout solvent;
- favourable internal standard comparison;
- free from agglomerates.

2.3.2.3 Fritted Discs [5]. The limitation of all conventional pneumatic nebuliser designs is that they produce aerosols with a very wide range in drop sizes. The net effect is that the transport efficiencies are reduced because of the high percentage of large droplets arriving at the plasma and, if they arrive at a higher level than the plasma can tolerate, they may quench it due to overloading. The balance between large droplets/mist is critical in maintaining plasma operation and signal reproducibility. Higher transport efficiencies can be achieved if the concentration of large droplets is reduced to finer mist and one such method is the fritted disc nebuliser.

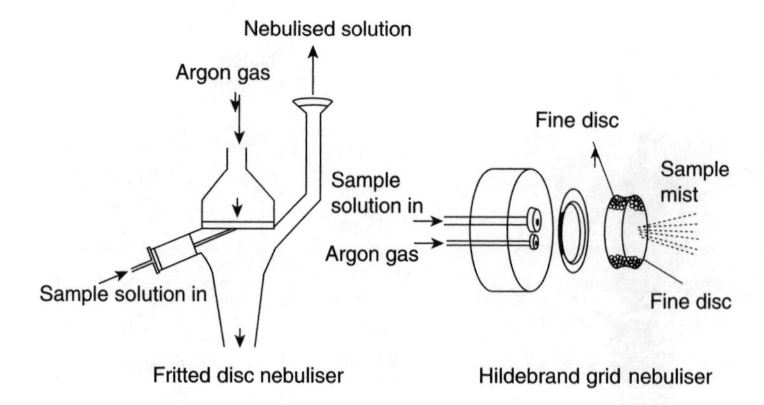

Figure 2.12 Diagram of a fritted disc nebuliser and a Hildebrand grid nebuliser. (Reproduced with kind permission from E-POND S.A., C.P. 389, CH-1800 Vevey, Switzerland)

This nebuliser uses a porous glass plate to give high efficiencies from ~2 to 5% for the conventional nebuliser to ~50% for the fritted disc nebuliser (Figure 2.12). This nebuliser requires very little sample for analysis making it a very powerful tool where only a minute sample is available. This device has been claimed to generate particles/mist of 1.0 μm or less and to achieve a high concentration of sample solution very low flow rates are required. This type of nebuliser is useful for the analysis of organic solutions and can be coupled with high performance liquid chromatography or gas–liquid chromatography. The main disadvantage of this type of nebuliser is that it suffers from memory effects with over usage hence it requires considerably longer periods of washout time between each measurement. It also tends to froth and block the pores when a high concentration of elements in solution is nebulised for a minute or more. This can occur for both organic and aqueous solutions. The need for blank checking needs to be carried out frequently before and after sample measurements.

2.3.2.4 The Hildebrand Grid Nebuliser. This nebuliser comprises a sample/gas inlet port with a shearing force grid attached (Figure 2.12). The grid vibrates at a very high frequency causing the sample solution to shear and causing the liquid to shatter into very tiny droplets. They have high efficiencies with low sample consumption and find excellent use with organic solvents or coupling to gas/liquid chromatographs. Unfortunately, they failed to gain popularity because they suffer from severe memory effects. The rinsing times are longer and potential clogging effect at the grid causes poor reproducibility between each sample measurement. Similar problems are associated with this nebuliser as with the fritted disc. These nebulisers use the same principle as the Babington except with a multitude of orifices. They can be used for high salt solutions since no constricting orifices are needed to produce the aerosol and are relatively blockage free.

2.3.2.5 Cross Flow Nebulisers [6]. The cross flow nebuliser design is based on a V-groove principle (Figure 2.13). This type of nebuliser is less sensitive to high salt content and can be used for aqueous and non-aqueous samples. It needs a peristaltic pump to transport the sample that must contain a sufficient number of precision rollers that do not

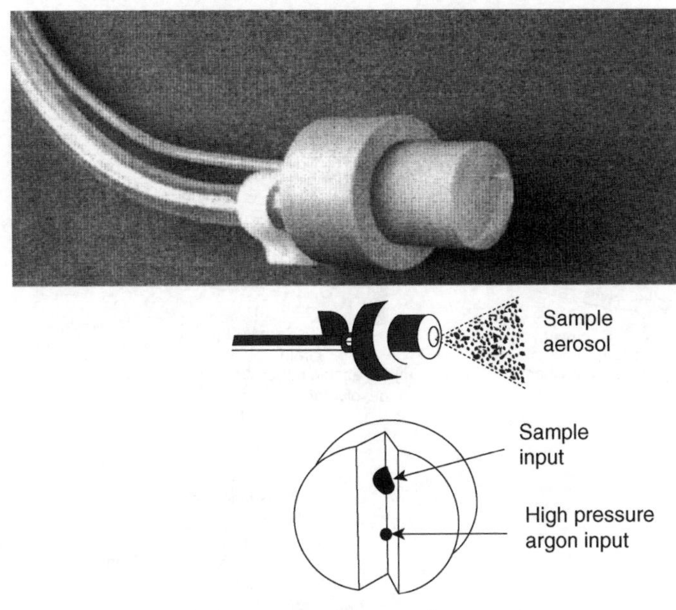

Figure 2.13 *Diagram of V-groove cross flow nebuliser suitable for high solids content. (Reproduced with kind permission from E-POND S.A., C.P. 389, CH-1800 Vevey, Switzerland)*

pulsate. The V-groove type nebuliser is where the sample is fed through a capillary and flows down the V-groove to a second smaller orifice capillary a few millimetres underneath to where a rapid flowing high-pressure argon gas can escape. This gas produces a mist of very fine particles that are eventually forced to the plasma for excitation. The diameter of the sample inlet capillary is usually large enough to allow samples containing high salt content to be nebulised and as expected the capillary for the gas inlet is considerably smaller.

2.3.2.6 Ultrasonic Nebulisers [7]. The ultrasonic nebuliser was first discussed in 1927 and developed in the 1960s to create smaller droplets than the conventional nebuliser (Figure 2.14). The sample is delivered using a peristaltic pump to a transducer that is vibrating at a very high frequency of approximately 1 MHz. The waves produced by the transducer are very efficient in turning the sample into a very fine aerosol. The fine aerosol is then carried by an argon stream through a heated tube and finally through a cooled tube to condense the solvent. The procedure ensures that most of the sample reaches the plasma as a cloud of fine, dry particles. This effect ensures that more atoms are available for excitation and atomisation. The signals for a range of elements are improved from 2 to 20 times which means detection limits are also improved. The principle of aerosol production is significantly different from pneumatic nebulisers. In the ultrasonic nebuliser, instead of the droplets being stripped from a liquid cylinder by a high velocity gas jet, surface instability is generated in a pool of liquid by a focused or unfocused ultrasonic beam. The beam is generated by the piezoelectric transducer that produces aerosol mean diameter ranging from $\sim$1.0 to 3.0 μm using a 0.5 ml min^{-1}

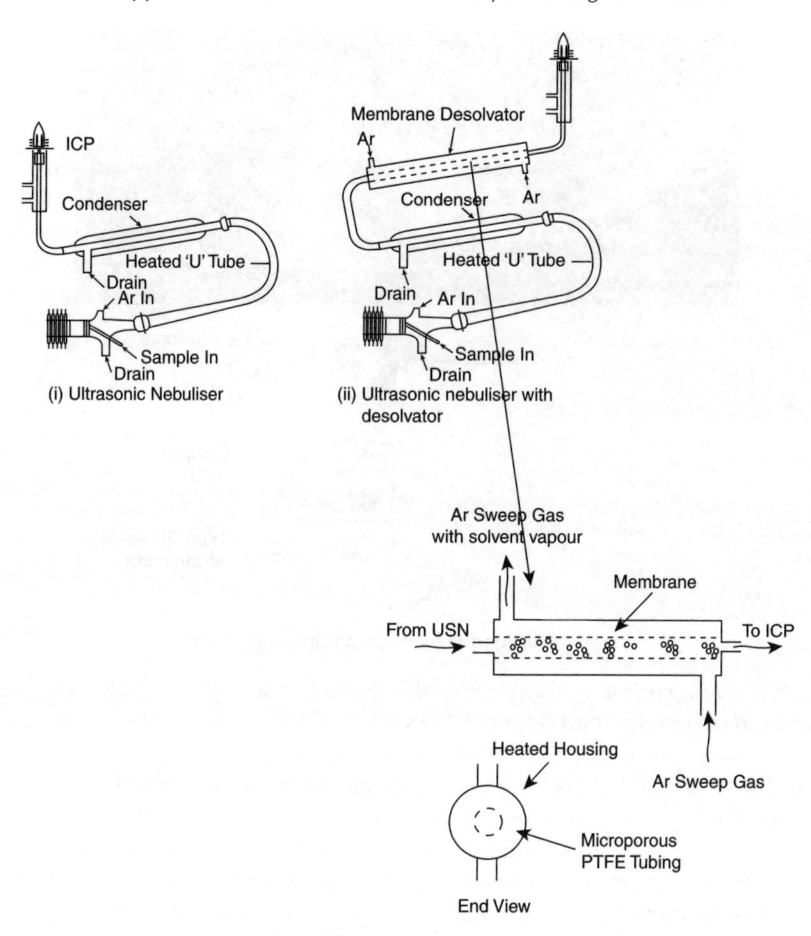

Figure 2.14 *Schematic diagram of Cetac Ultrasonic UT5000 and UT6000 complete with the membrane desolvator. (Reproduced with kind permission from CETAC Laboratories)*

solution flow rate. The ultrasonic nebulisers will improve efficiencies from ~2–5% for conventional nebulisers to 8–30% depending on the sample solution under investigation.

The desolvation step is carried out after vaporisation and condensation and has an enrichment factor of several orders depending on the solvent and element under test. The removal of organic solvents that could otherwise destabilise the plasma has to be beneficial so as to allow almost all solvents to be used as part of the analytical test and allow analysis of samples containing very volatile solvents.

Cross contamination encountered with desolvation systems has been greatly reduced by using a concentric sheath to prevent deposits on tube walls. It is important to note that nebulisers and spray chambers operate interactively and must be optimised as a unit rather than individually. There are, however, certain parameters that need to be considered in relation to the spray chamber:

- effective removal of aerosol droplets rather than the cut off diameter for interference free measurement;

- washout times must be sufficient to reduce memory effect;
- smooth drainage of waste aerosol from chamber to avoid pressure pulses;
- no build up of salt deposits or precipitation within the chamber;
- rapid clean out of the spray both at the entrance and drainage using faster pump flows;
- avoid using concentric nebulisers when working with samples containing a high concentration of elements in solution.

The membrane desolvator is very effective in reducing the solvent loading further when used in conjunction with the ultrasonic nebuliser. This will allow a range of solvents to be used for ICP-AES that would otherwise quench the plasma by almost totally removing the solvent from the sample and only allowing the dried particles containing the elements of interest to enter the source.

For ICP-OES-MS (inductively coupled plasma-optical emission spectroscopy-mass spectroscopy) work, the desolvator will remove oxide and hydride polyatomic ion interferences, i.e. $ArO+$ is reduced $\sim$100 fold, which allows for improved detection of Fe. The solvent loading reduction is caused by volatiles passing through the walls of a tubular microporous Teflon PTFE membrane. The argon gas removes the solvent vapour from the exterior of the membrane. Solvent-free analytes remain inside the membrane and are carried to the plasma for atomisation and excitation.

2.3.3 Brief Outline of Atomic Spectroscopy Hyphenated Systems

2.3.3.1 Electro-Thermal Methods [8]. Atomic spectroscopy of solid samples has been in existence since the mid 1960s as an arcs and sparks analytical technique. The sample is placed into an anode cup that is the anode electrode and a cathode electrode completes the circuit. The arc is struck and the sample vaporises into a discharge region where excitation and emission occur. As a large amount of sample is vaporised, detection limits are low but the unstable behaviour of the discharge leads to poor precision. This technique is used for semi-quantitative and qualitative analysis and used mainly to determine the metal profile of samples. In most samples elements would need to be present in a reasonable concentration to be detected, hence it is unsuitable for trace analysis. If light from the discharge is dispersed by a prism and a camera of long focal length is used, a photograph of the spectrum can be obtained. As the range of quantitative atomic spectroscopy instruments became available and affordable these instruments fell out of favour. Modern arc and spark instruments are used extensively in the steel industry for identification of various steels.

Introducing samples to the plasma via liquids reduces sensitivity because the concentration of the analyte is limited to the volume of solvent that the plasma can tolerate. An electro-thermal method seems an obvious choice to increase the detection limit as it will vaporise entirely most neat samples or using an increased concentration of sample in a suitable solvent. The sample is placed on a suitable open graphite rod in an enclosed compartment and heated rapidly (Figure 2.15). The electronics required for ICP-OES-ETV (inductively coupled plasma–optical emission spectroscopy–electro-thermal volatilisation) is similar to that for AAS and detection limits are better than ICP-AES.

The advantage of vapour introduction over liquid introduction is that ETV allows pre-concentration of the sample from a reasonably large sample size or volume of solution and with efficiency as high as 90–100% the detection limits would be considerably improved. This system finds many applications for the analysis of samples for arsenic,

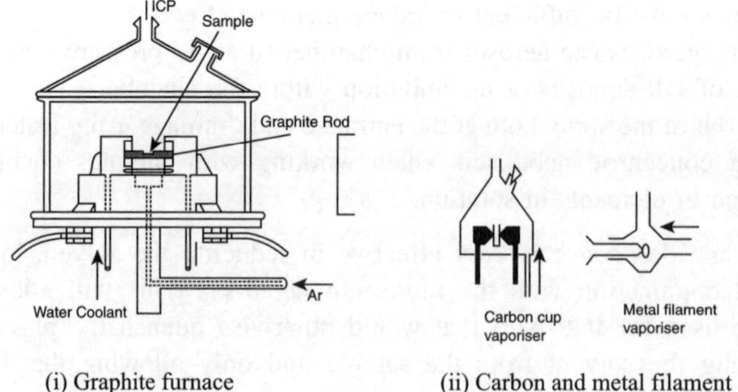

Figure 2.15 *Schematic diagram of electro-thermal vaporisation apparatus suitable for attachment to ICP-OES. (Reproduced by kind permission: copyright © 1999–2008, all rights reserved, Perkin Elmer, Inc.)*

tellurium, selenium, lead, antimony and mercury content, for which very low detection limits are needed to meet environmental and medical requirements.

Solid samples can be introduced via a graphite, tantalum or tungsten probe, which vaporises the sample after a rapid high voltage surge from an external supply. The probe heats up rapidly as does the sample. The vapour from the sample is carried to the plasma with argon gas. This operation takes 1–2 s which is sufficient time to record a signal output. The precision and reproducibility are poor unless the sample is added using precise automatic methods.

2.3.3.2 Laser Ablation [7]. The modern method for quantitative solid analyses is carried out using a laser ablation technique (Figure 2.16). The laser, usually in the form of Nd:YAG (Neodymium – Doped Yttrium Aluminium Garnet), is focused on to the surface of a sample which, by continuous pulsing, leads to vaporisation at that point and the vapour is transported directly to the plasma with argon for detection and quantification. Detection limits are

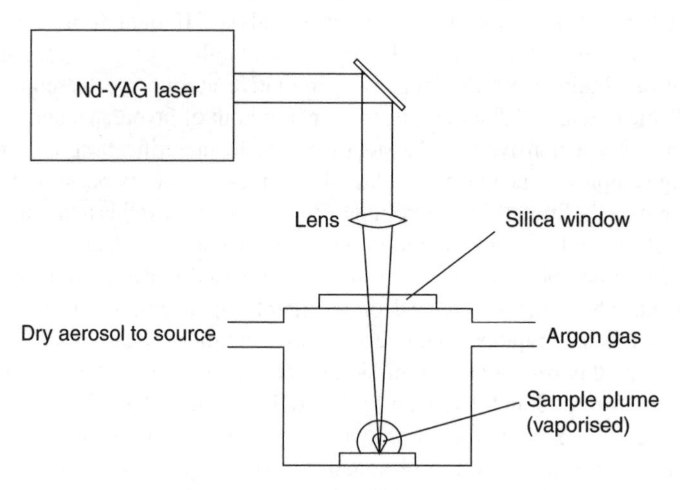

Figure 2.16 *Basic diagram of laser ablation suitable for attachment to ICP-OES. (Reproduced with kind permission from CETAC Laboratories)*

reasonably good making it possible to analyse very small sample sizes. Studies of solid samples can be carried out using this technique using a depth profiling method to determine the consistency of solid sample from the surface to a few micrometres below the surface.

A major problem with laser ablation is that it is very difficult to generate accurate calibration and requires close matrix matched standards that are not available for real sample analysis. A useful application is comparison between 'good' and 'bad' samples that could give information about matrix properties. This will be discussed in detail in Chapter 7.

2.3.3.3 Hydride Generation [9]. The hydride generation technique is probably the most sensitive for direct ICP-AES measurement/detection (Figure 2.17). The sensitivity of this procedure is 50 to 300 times greater than that by direct nebulisation. The method is relatively free from interferences, as it involves separation of the metals as hydride gases from the sample solution after reaction with sodium borohydride in the presence of acid. The technique is limited to the elements As, Bi, Ge, Pb, Se, Sb, Sn and Te, which are known to form readily volatile covalent hydrides. The hydrides are 'purged' directly into the plasma where they are atomised, excited and measured by ICP-AES in the normal way.

The reductant commonly used is sodium borohydride ($NaBH_4$) as it is slightly superior to zinc–hydrochloric acid reductant. The hydride from the $NaBH_4$ forms faster and a collection reservoir is not needed. An acid solution containing 0.5 to 2% acid is sufficient depending on the element and concentration of interest. The detection limits are greatly improved for these elements and can also be as low as those of GFAAS (Graphite Furnace Atomic Absorption Spectroscopy). It is worth noting that certain valence states of some elements are more sensitive than others. A known example is the valence of +3 for As or Sb is more sensitive than the +5 state. These elements may be reduced first using iodine prior to hydride generation. Similarly, selenium and tellurium in the +6 state must be reduced to the +4 state prior to hydride generation as the +6 state does not form a hydride. This can be achieved by heating with concentrated HCl. Some very stable compounds of these elements do not form hydrides and may need to be broken down by rigorous methods, e.g. microwave digestion, bomb combustion, etc. This method of analysis is not without its problems as the hydride forming stage may be suppressed by interferences in the sample. Elements such as copper, silver, nickel and

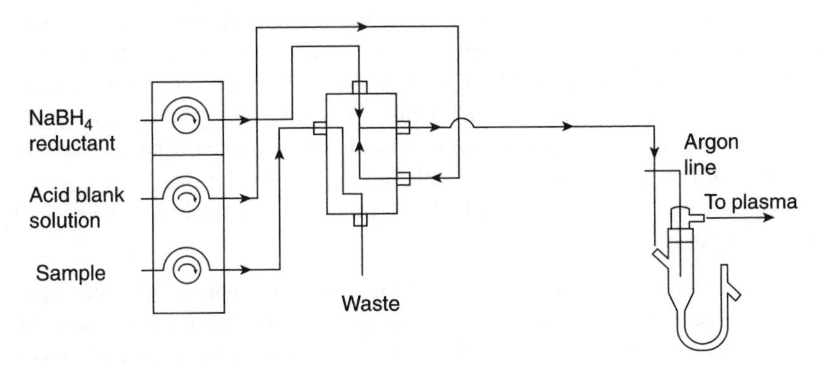

Figure 2.17 *Schematic diagram of continuous flow hydride generator. (Reproduced with kind permission from PSAnalytical, Orpington BR5 3HP, UK)*

Table 2.1 *Comparison of detection limits using pneumatic, graphite and hydride generation*

Element	Pneumatic	Graphite	Hydride
As	1500	0.25	0.005
Bi	1200	0.15	0.010
Sb	2000	0.20	0.015
Se	4000	0.20	0.050
Sn	1500	2.0	0.020
Te	3000	0.2	0.020

gold can have suppression effects; however, they can be removed by using masking agents.

The basic principle of hydride generation is the conversion into volatile hydrides by use of the sodium borohydride as a reducing agent. The hydride can then be readily dissociated into atomic vapour by ICP-OES.

The following is considered as a basic reaction with As(III):

$$3BH_4^- + 3H^- + 4H_3AsO_3 \rightarrow 3H_3BO_3 + 4AsH_3 + 3H_2O$$

In the presence of basic borohydride in acidic solution, excess hydrogen is formed as follows:

$$BH_4^- + 3H_2O + H^- \rightarrow H_3BO_3 + 4H_2$$

The hydride formed by chemical reaction is swept out of solution to the plasma by argon gas and the signal measured.

The diagram in Figure 2.17 is based on a continuous flow vapour system. During an analysis the solenoid valve switches from the acid blank to the sample at a certain time and automatically switches back to the blank after sample measurement. The resulting gases/liquid are separated by a gas separator with a constant head U-tube, which allows the liquid to drain to waste automatically. The chemically formed hydride of the element is swept to the plasma using argon gas for atomisation. The advantage of the continuous flow system over the gas bulb/purge method is that a steady signal is obtained which is reproducible and suitable for quantitative analysis. This design is also suitable for mercury analysis using tin (II) chloride/hydrochloric acid as the reductant. The sensitivity of the method is improved several fold i.e. from 10 to 100 depending on the element of interest. This method is now finding applications where very precise trace analyses of these metals are required, e.g. environmental, foods, medicine and where other safety and health specifications are required. The results in Table 2.1 are a comparison of detection limits using pneumatic, graphite and hydride generation methods for trace analysis of metals As, Bi, Sb, Se, Sn and Te.

2.3.3.4 Flow Injection Analysis [10,11]. Flow injection analysis involves injecting a known volume of sample solution into a continuous flowing liquid carrier stream usually of the same solvent that the sample is dissolved in (Figure 2.18). A loop of fixed volume is attached to a rotating valve which can be connected and disconnected manually or by computer to a flowing stream between sample analyses. As the loop is fixed the volume

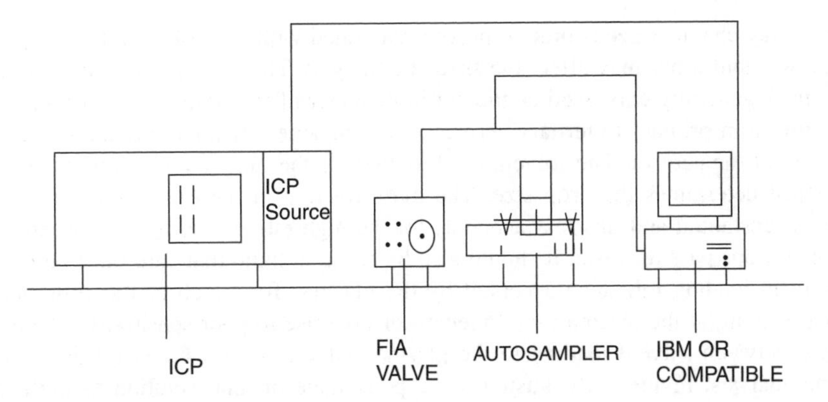

Figure 2.18 *Diagram of an automated FIA attached to ICP-AES. (Reproduced with kind permission from PSAnalytical, Orpington, BR5 3HP, UK)*

will be constant during the course of standard and sample analysis. The nebuliser, spray chamber and torch are immediately cleaned with the carrier stream after the sample has passed. These and other hyphenated attachments will be discussed in detail in Chapter 7.

2.4 Spray Chambers [12]

Nebulisers are usually mounted in a spray chamber and become the second part of sample aerosol preparation prior to plasma penetration. The function of spray chambers is to separate the larger droplets and discharge them as waste. Most chambers are designed to have maximum impact surfaces to assist the nebuliser in further fractionating the droplets and increasing the sample efficiency to the nebuliser. The earliest and most commonly used spray chamber was designed by Scott *et al.*, but others have followed with different designs all having their characteristic and beneficial effects. They must be carefully constructed so that the larger droplets are effectively removed and the noise, reproducibility and precision are reduced. The separation of large droplets in the spray chamber results mainly from the following criteria:

- gravitational fall;
- inertia deposition;
- sonic binding;
- electrostatic deposition;
- agitation;
- diffusion deposition;
- solvent flow.

The chamber can be regarded as a filter tolerating aerosol of diameter 0.1–10 µM for standard nebulisers through which the aerosol is passed; droplets larger than 0.1–3 µM are drained to waste, while the droplets smaller than this are passed to the plasma torch. The percentage of drops going to the plasma is in the order of 1–2% and any value above this tends to quench it. Studies have been carried out to show that agitation of the aerosol within the chamber plays a very important role in the separation process followed by gravitational settling of the drops. The final separation is due to the inertia deposition of droplets.

The spray chamber needs time to become saturated with aerosol; therefore, the build-up and washout times may affect the speed of analysis. The effectiveness in the removal of drops is generally expressed as the cut in diameter of the drop size distribution. The transition from primary to tertiary aerosol is accompanied by a large reduction in mean drop size of the aerosol. The solvent used to dissolve the sample containing the element of interest determines the drop size. The mass fraction of aerosol contained as larger drops is substantial and this loss accounts for the high rate of wastage as essentially all the large drops pass to waste. It should also be borne in mind that care must be taken of the solvent loading tolerance accepted by the plasma. Too much or too little aerosol/vapour arriving at the plasma may 'quench' or give rise to poor sensitivity. The correct balance between solvent loading to the plasma and waste must be established prior to sample analysis. Figure 2.19 illustrates the percentage of total solution as particles of varying sizes successfully reaching the plasma or going to waste.

Figure 2.19 shows the approximate percentage of droplets reaching the plasma torch for pneumatic, cross flow and V-groove nebulisers in the region below (a). The region below (b) is the percentage reaching the plasma torch with ultrasonic, fritted disc and grid nebulisers. The remaining solution beyond (a) and (b), depending on the nebuliser used, goes to waste.

The washout times can determine the speed of analysis, therefore rinsing out times are critical in removing traces from previous analysis and must be as short as possible without sacrificing analytical precision. This is particularly important when trace analysis needs to be carried out. In estimating the washout times of the ICP-OES sample contact components, a standard containing 10 μg ml^{-1} Ca is nebulised for 2 min. and then washed out with the same solvent used to prepare the Ca solution. The time taken to reach a level baseline is the time required to achieve a total metal free ICP-OES, and this washout time is used for subsequent analysis.

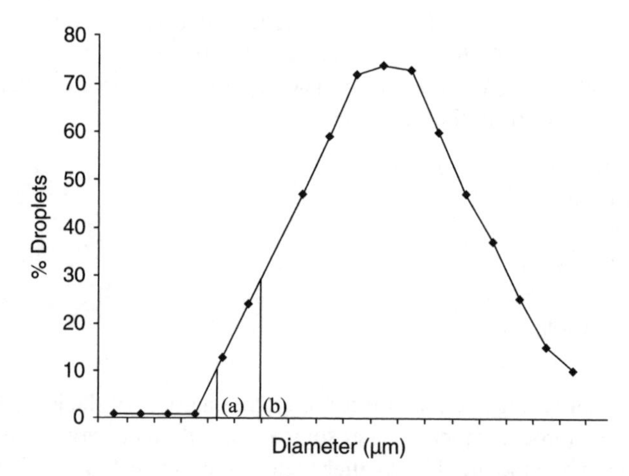

Figure 2.19 *Schematic diagram of the percentage of droplet mist successfully going to the plasma source. Line (a) is ~2% for standard nebulisers and line (b) is ~20% for ultrasonic nebulisers*

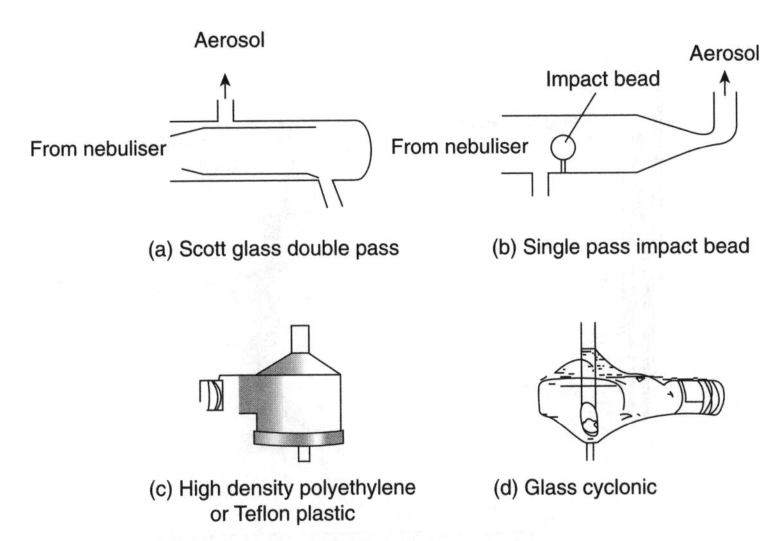

Figure 2.20 *List of common spray chambers used in ICP-OES torches. (Reproduced with kind permission from E-POND S.A., C.P. 389, CH-1800 Vevey, Switzerland)*

There are several designs of spray chamber available; the four shown in Figure 2.20 are the most commonly used.

2.5 ICP-OES Torches [13]

The torch design and construction is critical to the success of ICP-OES analysis (Figure 2.21). The inductively coupled plasma torch is made up of three concentric tubes of low flow and sealed together as one piece each of which is opened at the top with inner tubes specifically designed to allow the inert gas to flow at different rates. The inner tube has a relatively slow flow rate and is used to direct the nebulised liquid sample to the torch. The outer tube is used to allow argon gas to flow tangentially and acts as a coolant. The coolant, argon, also isolates the plasma from the outer tube and prevents overheating and melting. A water-cooled induction coil connected to a high frequency generator operating at ~40 MHz surrounds the torch at approximately 15 to 20 mm from the top. The coil operates as an intense oscillating magnetic field that induces electrons and ion currents and the energy is transferred from the coil to the plasma. A spark from the Telsa coil initiates the ionisation of the flowing argon gas to form the plasma. The plasma forms above the auxiliary and injector tubes but within the outer tube. The observation is just above the plasma fireball that is the optimum measurement position. The induced current flows in a circular path that heats the gas through violent collisions to a high ionising temperature that sustains plasma.

ICP torches are designed with appropriate settings of gas flow rates to give toroidal (or doughnut) or annular plasma at 40 MHz frequency which allows reproducible

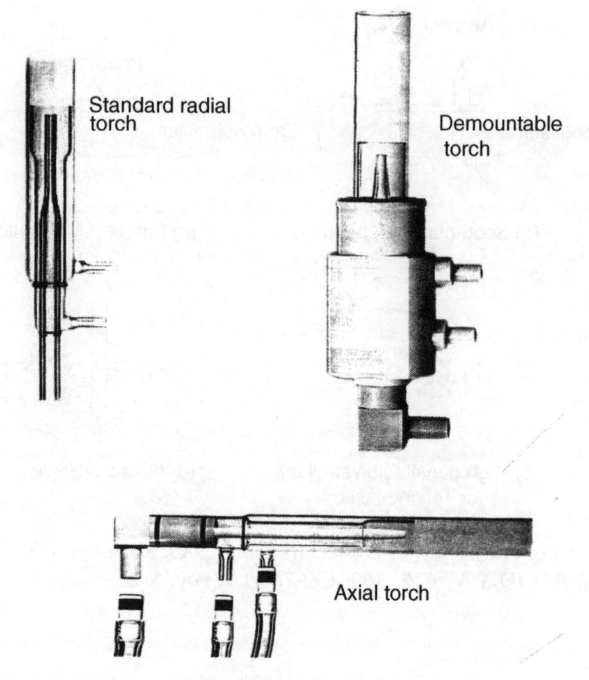

Figure 2.21 *Common types of ICP-AES torches available. (Reproduced with kind permission from E-POND S.A., C.P. 389, CH-1800 Vevey, Switzerland)*

introduction of samples as aerosols. The skin effect of the radio frequency heating, the axial zones of the plasma, are relatively cool compared with the surrounding zones. A stream of gas with a relatively small cross section bores a hole in the torus of the torch without upsetting its stability. The aerosol from the sample goes into the high temperature core to give rise to rapid and efficient desolvation, volatisation, ionisation, atomisation and excitation of the sample solution. The core of the plasma is made up of continuums of argon ions with electrons and superimposed on this is the atomic spectrum for argon. Spectral observations are normally made at a height of 5–25 mm above the induction coil where the background radiation is free from argon lines. This area is known as the normal analytical zone (NAZ). It is in this region that all the analyte atoms and ions are in their excited states (Figure 2.22).

The radial viewed torch, which is vertical, is viewed by the optics perpendicular to the axis of the plasma. The axial view is horizontal and is viewed along its axis. Different manufacturers of axial ICP-OES instruments have their own way of maintaining a cooled tip so as to avoid damage to the cone. Some use cooled cone interface, while others blow cooled argon across the tip that deflects the plasma tail outward. An axial viewed plasma can achieve detection ~6–10 times higher (depending on the analyte) than radial due to the fact it can see ~6–10 times more of the analyte (see Figure 2.7). The limitation is that the sample solution must not have more than 2.5% dissolved solids and is not organic solvent friendly except with the use of a desolvator attached to an ultrasonic nebuliser, otherwise such samples are easier to analyse with the radial design. The light emitted

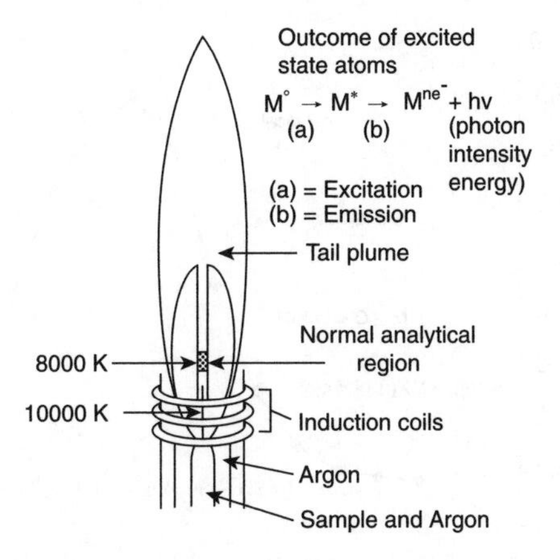

Figure 2.22 *ICP plasma torch showing the main parts and excitation regions*

from the plasma is very intense and the analytical measurements are made in the cooler tail of the plasma.

For torches to provide maximum excitation and atomisation, the following parameters must be known for each analyte-matrix-atomiser combination:

- acceptable drop size;
- optimum solvent loading (aerosol and vapour);
- acceptable analyte loading;
- correct gas flow for effective plasma penetration;
- suitable observation height;
- accurate wavelength peaking selection;
- suitable power range for plasma;
- stable plasma;
- presence or absence of particles in solution, e.g. slurry analysis;
- ancillary attachments, e.g. FIA, internal standards, etc.

Demountable torches are also available where individual parts can be easily removed for cleaning or replacement. The major disadvantage of the demountable torch is that reassembling precisely to ensure maximum sensitivity requires special skills. The sample/gas flow from the inner tube is designed to give the correct power to penetrate the bottom of the plasma and to push itself through without extinguishing it.

The first torches were designed by S. Greenfield and V. Fassel. The outer diameter is slightly wider with the Greenfield torch and requires higher radio frequency and gas power than the narrower Fassel design. The plasma formed takes the shape of a toroidal cone and is an important parameter which gives rise to a plume (tail-flame) to achieve excellent detection and sensitivity when used as a spectroscopic source. This tail-flame emits very little continuum radiation as the spectral background is low. Little mixing takes place between the central gas stream carrying the sample for analysis and the surrounding gas flows. As expected, the cooler gases surrounding the tail-flame contain very few atoms which means that the tail-flame behaves as an optical source that exhibits

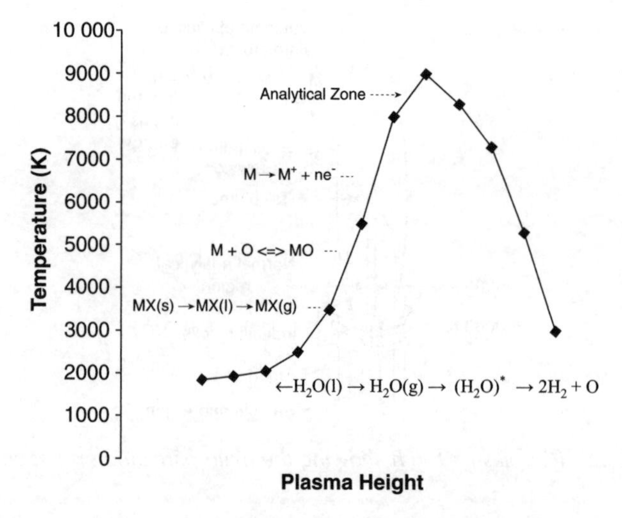

Figure 2.23 *Profile of plasma torch height and temperature*

little self absorption (Figure 2.23). The outer layer of the plasma torch takes most of the plasma power while not interfering with the atoms in the channel hence t is independent of the sample. This feature provides good stability and makes this technique a good spectroscopic source.

The outer channel conducts argon gas at $\sim 15 - 17\,\mathrm{L\,min}^1$ to the plasma to sustain the plasma and to isolate the quartz tube from high temperatures. The middle channel conducts the auxiliary argon at $\sim 1.0\,\mathrm{L\ min}^{-1}$ and is used for initiating the plasma or organic solvents. The ICP-AES has an annular or doughnut shape viewed from above. The sample hole has a lower temperature than the doughnut body and offers less resistance to the sample injection. The argon gas is initially ionised using the copper coil connected to a radio frequency generator. The radio frequency generator can achieve up to 2.0 kW forward powers at a frequency of 40 MHz. The high frequency current flowing in the copper coil generates oscillating magnetic fields whose lines of force are axially oriented inside the quartz tube and follow elliptical closed paths outside the coil. Electrons and ions passing through the oscillating electromagnetic field flow at high acceleration rates in closed annular paths inside the quartz tube space. The induced magnetic fields, direction and strength vary with time resulting in electron acceleration on each half cycle. Collisions between accelerated electrons and ions and unionised argon gas cause further ionisation. The collisions cause ohmic heating and, when measured spectroscopically, give thermal temperatures from 5000 to 10 000 K.

Correct settings of gas flows will form a toroidal or an annual plasma aided with a 40 MHz radio frequency coil which allows reproducible introduction of mist or aerosols. The power in the plasma is dissipated in the outer layers whose electrical conductivity is unaltered by the presence of sample atoms and ions in the tunnel and is independent of the nature and concentration of the sample. This parameter makes the toroidal plasma maintain good stability and an excellent spectroscopic source. The task of aerosol penetration is made easier through the toroidal plasma because all gases are flowing in a single direction, i.e. outwards. In order to penetrate the plasma, the sample/gas flow must overcome the 'magneto-hydrodynamic thrust velocity'. To achieve this, the aerosol injector tube must be of a narrow diameter so that the force will be great enough. The tip of the injector tube must be as close as possible to the base of the plasma but not too close to avoid melting it.

2.6 Optics

The excellent capabilities of plasma are not without problems. Due to the high temperature and high excitation power that can be achieved, their spectra are rich with ions and atomic lines. The richness of lines created by these excitation sources can lead to possible spectral interferences from nearby lines and give rise to stray spectral overlap that can affect the intensity of the spectral line of interest. Therefore the practicality, selectivity, qualitative and quantitative application rests on the ability to isolate the line(s) of interest.

The detection of emission lines is achieved using high-resolution techniques. The methods successfully developed to isolate these lines include Fabry-Perot interferometry, echelle and grating and latterly Fourier spectrometry. The Fabry offers excellent spectral resolution but poor spectral selectivity. The echelle developed by Harrison [14] is a coarsely ruled grating in high orders and, in conjunction with a prism, this unique combination of optical devices is able to have high resolving powers. Echelle grating monochromators possess significant advantages in comparison with other types of monochromators in that the resolving power and dispersion are both increased by a factor of 10 to 100. The combined prism/grating design is compact and easy to use when compared with conventional diffraction grating units of similar resolution. The echelle grating is ruled plane diffraction which has few rulings and is a precisely controlled shape and appears to be coarse. Typically, echelle gratings have less than 300 grooves mm^{-1} Echelles are used effectively at high angles of incidence which may exceed 45° or greater and have high orders of interference, usually 10 to 1000. Their application at high angles is the reason that ruling precision must be high and is much better when compared with conventional grating.

A grating is a reflective optical plate with a series of closely ruled lines and when light is reflected or transmitted through the grating each line generated behaves as a separated source of radiation. The bending of light by gratings is called diffraction and its rotation allows different wavelengths to pass through the exit slit towards the detector. Diffraction by a reflection grating is shown in Figure 2.24 which is ruled with close parallel grooves with a distance α between them. The grating is coated with aluminium to aid reflection and behaves as a source of radiation.

A typical Littrow echelle arrangement appears as a coarse grating that has a high angle of incidence showing that the steep side of the groove is used. In most cases echelles have ~35, 80 and 316 grooves mm^{-1} and the angle of incidence and angle of diffraction are almost equal and the same as the blaze angle: 63° 26'.

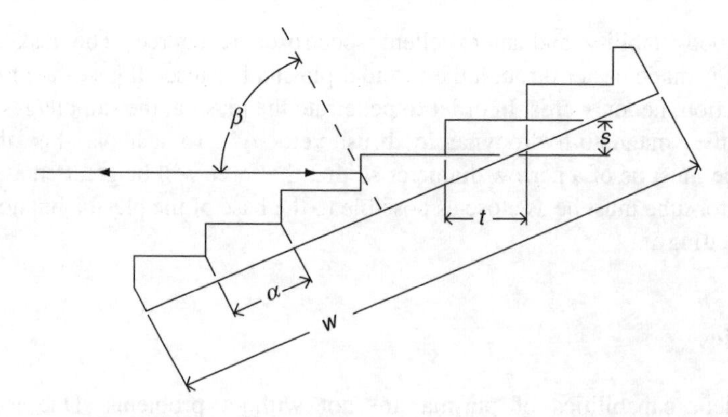

Figure 2.24 *Cross section of an echelle grating mirror*

The cross section of an echelle diffraction grating is given by:

$$\text{Tan}\,\beta = t/s \tag{1}$$

The grating equation for the Littrow mode is given as:

$$m\lambda = 2a\sin\beta \text{ or } m\lambda = 2t \tag{2}$$

where t is the width of one echelle step.

Therefore, the angular dispersion for the Littrow configuration is:

$$\frac{\sigma\beta}{\sigma\lambda} = \frac{2\text{Tan}\,\beta}{\lambda} \tag{3}$$

As shown by the above equation dispersion at a given wavelength is strictly a function of Tanβ and high angles of diffraction will lead directly to increased dispersion and hence high resolution. The angle Tanβ is the ratio of the groove width to the groove height, *t/s* and is sometimes called the 'r/number' control echelle dispersion. Increased values of Tanβ or r/number correspond to increased dispersion. Groove ratios of 2:1, 4:1 and 5:1 correspond to angles of 63° 26', 75° 56' and 78° 41', respectively, and are the reason these have been used as echelle blaze angles.

The linear dispersion of an echelle is just the focal length (L) of the lens used times the angular dispersion. This is expressed mathematically as follows:

$$\frac{\sigma l}{\sigma\lambda} = \frac{2f\,\text{Tan}\,\beta}{\lambda} \tag{4}$$

The theoretical resolution of grating (R) of an echelle is derived as for conventional diffraction gratings as follows:

$$R = \frac{\lambda}{\Delta\lambda} = mN \tag{5}$$

where m is the order of diffraction and N is the total number of rulings illuminated. Solving the grating equation for m and knowing that the ruled width w of a grating is Na, hence,

$$R = \frac{w(\sin\alpha + \sin\beta)}{\lambda} \tag{6}$$

where α is the angle of incidence with respect to normal grating, β is the angle of diffraction with respect to normal grating and λ is the diffracted wavelength.

2.6.1 Grating Orders [14]

It is important to note that for a given set of angles and grooves spacing (a), the grating equation is satisfied by more than one wavelength. In fact, subject to certain restrictions, there may be hundreds of discrete wavelengths that multiplied by successive integers (m) satisfy the interference conditions. The physical significance is that reinforcing of light coming from successive grooves merely requires that each ray be restarted in phases by a whole number of wavelengths (λ). This happens first when the retardation of one wavelength is of first order ($m = 1$) or any other multiple or order. The grating equation indicates that it is possible to have negative and positive values of m; negative orders are when the angle of diffraction both exceeds the angle of incidence in magnitude and is on the opposite side of the normal grating. This is simply due to sign convention with no physical significance. In any grating instrument configuration, the spectral slit image similar to wavelength λ will coincide with the second order of $\lambda/2$ and the third order image $\lambda/3$. Hence the simplest systems are those with first order grating. The main problem of multiple order of grating is that successive orders can overlap. To alleviate overlapping problems in higher orders a prism is placed at right angles to the grating, as shown in Figure 2.25.

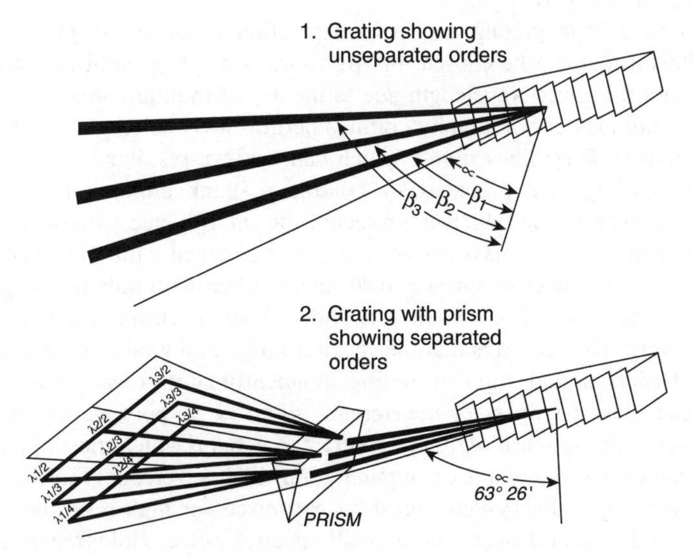

Figure 2.25 Schematic diagram showing separated and unseparated orders with and without the prism

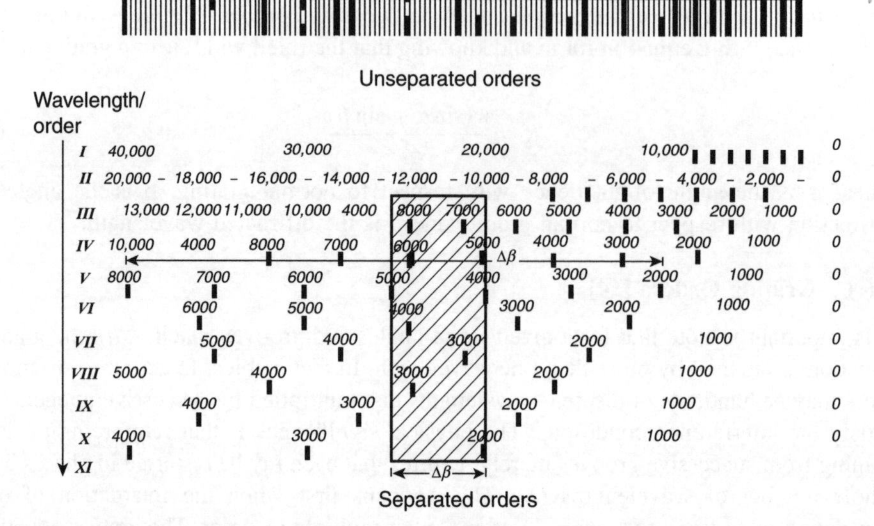

Figure 2.26 *Schematic diagram showing wavelength/orders separation*

The basic function of the grating/prism is when light coming from the main collimating mirror passes through the prism, it is diffracted by the grating and passes back through the prism to the focusing mirror. The diffracted light passes through the prism placed so that dispersion is at right angles with the echelle so that orders are separated. The final result is a compact two-dimensional spectrum with the orders stacked in the vertical position. The wavelengths are dispersed horizontally within the orders as shown in Figure 2.26.

This type of echelle grating and order separation gives rise to the most sensitive analytical line(s) that can be chosen for any element. High spectral efficiencies can be obtained across the entire wavelength due to the use of multiple orders.

Modern techniques of holography gratings permit finely ruled gratings with grooves density as high as 5000 lines mm^{-1} which can achieve resolving powers >500 000. Holographic gratings are prepared by coating a blank aluminium surface with a photosensitive material on which is projected the interference pattern of two uniting lasers to form the lines. A glass surface can also be etched with a layer of aluminium deposit to form the reflective surface. Instruments fitted with holography grating have several advantages relative to echelle and Fabry-Perot spectrometers and interferometers, respectively. The reason is that the spectral range of a grating used in low order is very large. Orders organisation in grating monochromators is very easily controlled with a simple glass bandpass or interference filters. It is possible to operate in one order over the entire spectral region in contrast to echelle which must operate in many orders and requires very complex computer controlled wavelength scanning programs. The mirror coating of Fabry-Perot must be optimised for high reflectivity to achieve high finesse and is useful over only a small spectral range. Holographic gratings are almost free of 'ghosting' because the line pattern is uniform, however it is less efficient.

Table 2.2 *Number of excited known lines for some elements*

Element	No. of lines
Li	38
Ar	1301
Ca	663
Pb	466
Au	330
Fe	5760
Total	8558

The higher resolving power of the echelle when used in high order relative to the diffraction grating used in the first order allows instruments to achieve high resolving power within a compact echelle instrument. The echelle is very efficient when used close to the optimum blaze angle and detection limits are similar to those obtained by grating instruments.

The selection of lines (wavelengths) of the analyte elements must be accessible, appropriate to the concentration of the analyte under test and must be within the working range. If outside the range a different line may be used. However, if working close to the detection limits, the most sensitive line is used. The nature of the origin of the emission must also be considered, i.e. whether the ion or an atom transition is useful in terms of interferences from EIE effect. When using an internal standard approach, the internal standard much be as close as possible to the analyte, i.e. excitation energy, oxidation state, etc. The selected wavelength(s) must be free from interferences as most elements emit several lines and the challenge of modern optics is to ensure that all the lines are readily isolated without interferences from other lines of the same element or lines from other elements. Table 2.2 illustrates the number of possible lines available for some elements in solution.

Inductively coupled plasma Fourier transform (ICP-FT) can be useful for correcting unexpected spectral interference. However, such a technique has limited applications and development is only in its infancy.

2.7 Signal Detectors

2.7.1 Photomultiplier Tubes [15]

After the important resolution of wavelengths, detection is the subsequent stage of signal monitoring. The earliest successful quantitative detection was by photomultiplier tube (PMT) that translates photon flux into electron pulses that are amplified through dynodes. The PMT consists of partially evacuated transparent envelope containing a photocathode which ejects electrons when struck by electromagnetic radiation (Figure 2.27). Photomultipliers can have from 6 to 12 dynode stages and convert light signals from the beam intensity by the free analyte atoms into electrical signals which can be displayed on a suitable calibrated voltameter or other display screen through the use of microprocessors.

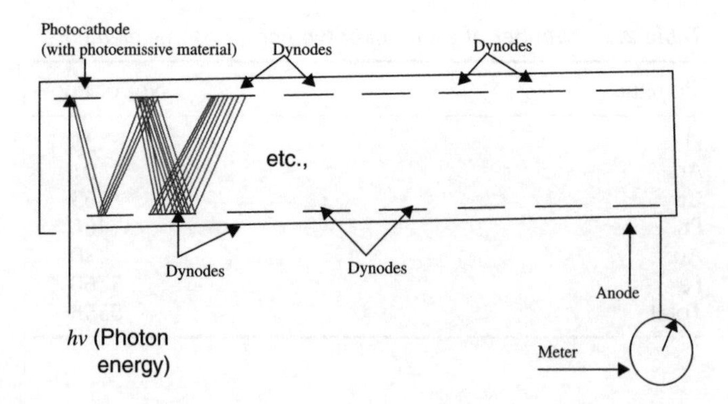

Figure 2.27 *Schematic diagram of a standard PMT used in ICP-OES*

The cathode and anode are separated by the dynodes which provide electron multiplication. A photoelectron ejected by the cathode upon incidence of a photon strikes the first dynode and produces up to five secondary electrons. Each secondary electron is accelerated by the field between the first and second dynode and strikes the next dynode with sufficient energy to release another five electrons. An electron avalanche takes place which results in an amplification of $\sim 10^7$. The cathode is at 450 V to 2500 V with respect to the anode.

The spectral sensitivity of PMTs depends primarily upon the photosensitive material used to coat the cathode. These materials are usually alloys of alkali metals with antimony, bismuth and silver. Most of these materials provide adequate output at wavelengths down to 160 nm (under vacuum) provided that the envelope material has adequate transmission. Different composition of metal for the coating offers detection at the upper end of the spectrum (~700 nm); and the caesium-antimony cathodes operate well up to 500 nm but are not suitable above this. Elements such as potassium rubidium will lose out for detection. Trialkali cathodes of antimony-sodium-potassium-caesium respond well up to 850 nm taking potassium and caesium into the detection zone. Gallium arsenide cathodes are the latest types to become available which respond well at high wavelengths (Figure 2.28).

The PMT is designed to act as a detector and measure light intensity for a given series of wavelengths and must be sensitive with good spatial resolution. When a photon of sufficient energy is incident on the photocathode, a single electron may be ejected (a photoelectron) and its probability is called the quantum efficiency and can be as low as 25%. The photoelectron is accelerated through a potential difference, and hits an electrode called a dynode. The photoelectron energy from the excited atoms attacks the first dynode and frees other electrons, and this second group of electrons is further accelerated and focused on the second dynode, and so on. The multiplication process takes place for 10–12 dynodes with a multiplication factor of 10^{7-10}. The electrons are guided by an electric field between the dynodes and the entire process takes place with an avalanche effect. The output signal's current pulses correspond to a detected photon which is converted to voltages by a voltage converter. The PMT is usually operated in a cooled environment so as to reduce the background thermal

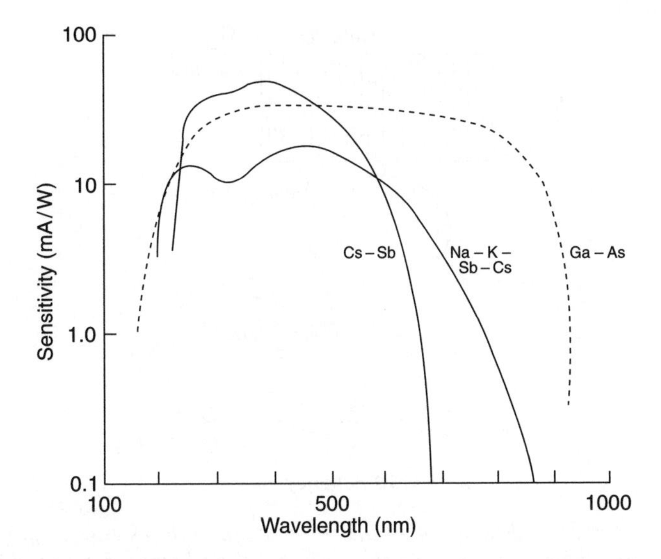

Figure 2.28 *Diagram showing typical spectral sensitivity ranges for dynode materials used in a PMT courtery of PerkinElmer*

noise. Dark currents also give rise to noise due to the darkness of the cell in which the PMT is mounted. All noise must be as low as possible so as to be able to achieve low detection limits.

2.7.2 Charge Coupled Devices [16]

All solid state detectors are called charge transfer devices (CTDs) and consist of doped pure silicon. The light sensitive device produces charges when struck by electrons (photons) even though the charge is positive. The charge injection device (CID), charge coupled device (CCD) and segmented charge device (SCD) [17] offer continuous wavelength coverage with high sensitivity (after sufficient accumulation time). These detectors consist of an array of closely spaced metal-insulator-semiconductor diodes in which the incident light is converted into a signal. The advantage of such a detector is that the optimum wavelength can be selected for each element in all types of samples. They can monitor large portions of the spectrum in multiple orders by taking 'electronic photographs'. These detectors are reasonably sensitive and store photogenerated charge in a two-dimensional array. The CID offers high sensitivity and continuous wavelength coverage. The CCD contains linear photodetector arrays on a silicon chip and detects several analytical lines of high sensitivity and large dynamic range that are free from spectral interferences (Figure 2.29). The array segments detect three to four analytical lines of high sensitivity and dynamic range and are reasonably free from spectral interferences. Each subarray is illuminated by over 6000 emission lines made up of pixels which are photosensitive areas of silicon and are positioned on the detector at x, y locations that correspond to the locations of the selected emission lines resolved by the echelle grating spectrometer. The emission lines are detected by means of their location

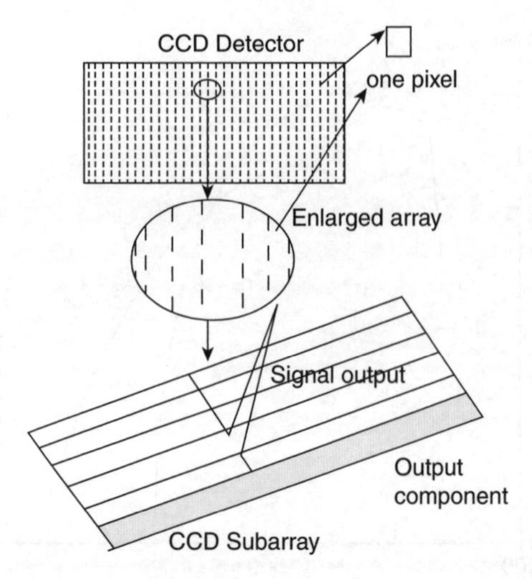

Figure 2.29 *A schematic diagram of a subarray with adjacent pixels of 30 μm × 30 μm on a solid state detector. These detectors have the ability to detect before, during and after the signal so that background corrections are carried out. (Copyright © 1999–2008, all rights reserved, PerkinElmer, Inc.)*

on the chip and several lines can be measured simultaneously. On completion of analysis the detector can be cleaned and used for the next sample to be analysed. Lines that are affected by interferences can be eliminated from the analyses. After the desired observation time, electrons are stored in each pixel until the maximum is reached and read. Charge transfer from each pixel is efficient with low loss of electrons. CCD detectors can view 5–50 times more lines than PMT detectors and offer an improvement in quantum efficiency and a reduced dark current effect.

Each detector (pixel) element on the CCD/CID (pixels can be arranged in an array 512 × 512 to 2400 × 2400) measures the intensity of light. The higher the pixels the better the resolution and higher the signal. The photoelectric effect produces a charge on its capacitor and converts to a voltage. The quantum efficiency is as high as 70%. The signal scan (see Figure 2.30) illustrates approximately the differences between PMT and CID/CCD detectors. There are no hard rules to state that this is true for all elements but generally it is true for the common elements. The main advantages of CCD detectors are the optimum wavelength that can be selected for each element in every type of sample. Some disadvantages of the CCD are that limitations may occur with readout noise because the detector is read many times so as to accumulate the charge during the numerous subsequent measurement cycles that are required to achieve a large linear range.

The sensitivity of the CCD is determined from electrons generated per incident photon, low background electrical noise and low readout noise. The most sensitive element/line detection is governed by the following criteria:

- *Plasma nebuliser gas flow* will influence the aerosol drop size, efficiency, stability and temperature.

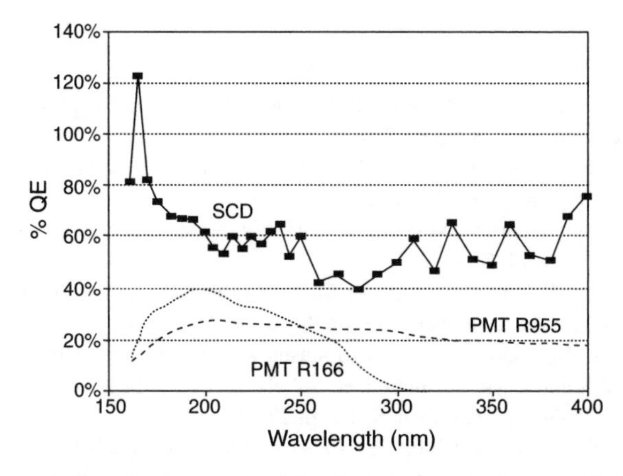

Figure 2.30 *Comparison of spectra of PMTs (two types) and SCD showing the quantum efficiencies of different detectors. (Reproduced by kind permission: copyright © 1999–2008, all rights reserved, PerkinElmer, Inc.)*

- *Power of the plasma* controls the plasma volume and is optimum for soft lines (e.g. Na, Mg, Ca) and for hard lines (e.g. B, W, P) and metals in organic solutions, etc.
- *Correct x, y plasma position* (i.e. horizontal and height) is a compromise between the analyte number densities which are highest in the lower zones and the completeness of atomisation and excitation which may be positioned 2–4 mm above the radio frequecy coil.
- *Wavelength* must be stable, constant, free from electronic noise and frequent atmospheric temperature change. These are important parameters to be optimised so as to obtain the maximum signal for that element. The alkali and alkaline metals are usually observed in the tail flame while elements such as Cr, Nb, Ta and Be are observed in the body of the plasma. A compromise set of conditions may necessitate the analysis of several elements.

The advantages of coupled detectors (CCD, CID, etc.) are that they can detect and measure a wide range of wavelengths, hence elements. They do not require high voltages like PMTs, can detect unknown elements and carry out simultaneous analysis with background corrections.

The disadvantages of coupled detectors are that they can give smaller signals due to smaller surface area of the light sensitive region and have higher background noise which is responsible for the poorer signal to background ratios. They have low time resolution and are not suitable to detect weak photon energies without the aid of a light amplifier. A photon producing one electron in the detector may not be detected as it could get lost during transportation or be buried in the noise. Good spectral resolutions require complicated mathematical corrections and detection limits are poorer. Blooming by close pixels of other elements at high intensities may lead to crossover intensities giving incorrect readings. The SCD is designed to avoid blooming and is subdivided into small photosensitive subarrays during manufacture. The subarrays prevent any crossover of charges to adjoining subarrays and are controlled by computer software. High quantum energies are not necessarily a good measure of how good a detector is.

References

[1] Greenfield, S., Jones, I.L. and Berry, C.T. (1964) High pressure plasmas as spectroscopic emission sources, *Analyst*, **89**, pp713–720.

[2] Wendth, R.H. and Fassel, V.A. (1965) Induction coupled plasma spectroscopic excitation source, *Analytical Chemistry*, **37**, pp920–922.

[3] Thompson, M. and Ramsey, M.H. (1985) Matrix effect due to calcium using ICP–AES, *Analyst*, **110**, pp1413–1422.

[4] Browner, R.F. and Boorn, A.W. (1984) Sample introduction: the Achilles heel of atomic spectrometry, *Analytical Chemistry*, **56**, pp786A–798A.

[5] Liu, H., Clifford, R.H., Dolan, S.P. and Monaser, A. (1996) Investigation of a high-efficiency nebuliser and thimble glass frit nebuliser for elemental analysis of biological materials by ICP-AES, *Spectochimica Acta, Part B*, **51**, pp27–40.

[6] Fuishiro, M., Kubota, M. and Ishida, R. (1984) A study of designs of cross flow nebulisers for ICP atomic emission spectrometry, *Spectrochimica Acta, Part B,* **39**, pp617–620.

[7] CETAC Technologies. *Ultrasonic Nebulisation of Liquid Samples for Analytical ICP Atomic Spectroscopy*, South Shields: CETAC.

[8] Aziz, A., Broekaert, J.A.C. and Leis, F. (1982) Analysis of microamounts of biological samples by evaporation in a graphite furnace and ICP atomic emission spectroscopy, *Spectrochimica Acta, Part B*, **37**, pp369–379.

[9] Thompson, M., Pahlavanpour, B., Walton, J. and Kirkbright, G.F. (1979) Simultaneous determination of As, Sb, Se, Bi and Te in aqueous solutions by introduction of gaseous hydrides into ICP-AES source of emission spectrometry, *Analyst*, **103**, pp568–579.

[10] Ruzicka, J. and Hansen, E.H. (1978) Flow Injection Analysis Part X, Theory, technique and trends, *Analytica Chimica Acta*, (1978) **99** 37.

[11] Reijn, J.M., Linden, W.E. and Poppe, H. (1981) Transport phenomena in flow injection analysis without chemical reaction, *Anal. chim. Acta*, **126** (1).

[12] Routh, M. W. (1986) Characterisation of ICP nebuliser aerosols using Fraunhofer diffraction, *Spectrochimica Acta*, **41B**, pp39–48.

[13] Allemand, D. and Barnes, R.M. (1977) A study of inductively coupled plasma torch configurations, *Applied Spectroscopy*, **31**, pp434–443.

[14] Harrison, G.R. (1949) The production of diffraction gratings: II. The design of echelle grating and spectrographs, *Journal of the Optical Society of America*, **39**, pp522–528.

[15] Matthee, K. and Visser, K. (1995) Background correction in atomic emission spectrometry using repetitive harmonic wavelength scanning and applying Fourier analysis theory, *Spectrochimica Acta*, **50B**, pp823–835.

[16] Kirk, R.E. and Othmer, D.F. (1982) *Encyclopedia of Chemical Technology*, 3rd Edn.Vol. 17, Chichester: John Wiley & Sons, Ltd. p664.

[17] Mermet, J.M. and Ivaldi, J.C. (1993) Real time internal standardisation for ICP–AES using custom segmented array charge coupled device detector, *Journal of Analytical Atomic Spectrometry*, **8**, pp795–801.

3

Methodologies of Metal Analysis of Organic Matrices Using ICP-OES

3.1 Sample Preparation Techniques and Methods of Analysis

During the past decade analytical science has gone through a revolution brought about by the efforts of thousands of scientists. The growth of modern technology has confronted the analytical scientist with a host of new and increasingly complex materials and has called upon the analyst to provide information about constituents previously not known or ignored because of lack of scientific knowledge. The modern scientist has been given the task of providing greater sensitivity, reliability and a more rapid turnaround time in analysis. The demand for information has also necessitated that research develop newer, more advanced analytical techniques, instrumentation, procedures and purer reagents. It is only by experience that an analyst can gain a perception of the scope, advantage and limitation of new instrumentation and methodology at his disposal.

The availability of more sophisticated instruments with the capability of analysing smaller sample sizes and better sensitivity has complicated the analyst's quest of searching for the best procedure for analysing new or unusual samples. The data and conclusion on which the selections are based is scattered through dozens of journals and monographs that may be tedious to collect and difficult to assimilate. This often causes a loss of a simpler or shorter and better method for obtaining the desired results. The application of an inferior procedure must be avoided merely because it is easier to access, while a better procedure could be hidden in the maze of analytical literature available in libraries. A major challenge for the analyst is to suggest an analytical approach about which no specific practical solution has been previously devised and to be successful in obtaining the required information about the unknown sample and avoid reporting false and inaccurate results. This challenge may be the development of a new method or be included as supplement to an existing method.

A Practical Approach to Quantitative Metal Analysis of Organic Matrices Martin Brennan

The availability of a good analytical method is essential if precise, accurate and reliable results are to be confidently reported. The imperative for cleanliness and meticulous attention to detail at each step in the analytical process cannot be overemphasised. Good analytical results depend on careful sample preparation along with meaningful interpretation. It is important that an analyst acquires an appreciation of the fundamentals of sampling, statistics and assessment of quality criteria, together with an understanding of the significance of the results obtained. Analytical science is a powerful aid used as a part of problem-solving through proof of reliable measurements for interpretations and support in decision-making processes. Problems may be associated with the environment, health, drinking-water supply, analysis of engine oil as part of engine-wear study, refined oil quality, pharmaceuticals, forensic examinations, manufacturing and quality assurance. The definition of chemical analysis is the application of multiple-step processes used to identify, and in most cases quantify, one or more substances present in sample(s) and/or the determination of structures of chemical compounds. Therefore, the application of analytical chemistry is very broad and requires a wide range of manual, chemical and instrumental techniques. In everyday life everybody benefits from analytical chemistry in some way or other, e.g. smelling of perfume or aftershave, cooking different types of food, smell or taste of rancid or fresh foods, gas or solvent leaks, sweet or sour tastes, and so on. The application of science to obtain advanced understanding requires the dedication of trained scientists to unearth the more complex nature and, in most instances, requires the application of analytical laboratories equipped with sophisticated instrumentation to support their quest for qualitative or quantitative information.

3.2 Defining Goals

When a sample is presented to the analyst, the first requirement is to ascertain what the sample is and the substances present for reasons of safe handling, and to find out whether it has been contaminated, accidentally or maliciously. An important part of the analyst's task is the determination of how much of a particular component is present and such a requirement offers a greater challenge to the analyst. This part of the task falls into the realm of quantitative analysis and requires the application of sophisticated techniques used by intelligent and well-trained personnel. With increasing demands for higher standards in the quality of raw materials and finished products – be it foods, pharmaceuticals, industrial, forensics, or whatever – analytical science plays a very important role in ensuring that these standards are maintained.

Manufacturing industries rely on both qualitative and quantitative chemical analysis to ensure that all stages in the process meet the specifications for that product and supports cost-saving beneficiaries. The development of new products that are usually mixtures of reacted and unreacted raw materials may also require the analytical chemist to ensure that the product(s) formulations are correct and meet the customer's standards. Many industrial processes give rise to pollutants that can present health problems and, with the support of analytical chemistry, as much chemical information as possible is made known about the pollutants. Analysis of air, water, and soil samples as a result of industrial pollutants must be monitored to establish safe limits after removal and/or disposal.

Similarly, in hospitals, chemical analysis can also assist in the diagnosis of illness and can be used to monitor the conditions of patients and assist medical personnel. In agriculture, monitoring the level of fertilisers through their elements for benefiting or non-benefiting effects, e.g. phosphorous, potassium, transition elements, etc., is also important and analytical science plays a decisive role here.

All of the above can be related to metal analysis as well as analysis for other components. Therefore, in order to analyse samples for metals or other unknown components, the analyst must have available the necessary information on the samples, suitable instruments, and procedures/methods for measuring the chemical and physical properties, all of which are an essential part of the analytical protocol. That reporting of measured results should include the support of statistical data is of paramount importance, and an inadequate knowledge of the same hinders confidence in the reported results.

3.3 Steps in Chemical Analytical Protocol [1]

All samples submitted for metal analysis, particularly trace levels, must be taken through a series of steps and procedures to remove as much as possible of contamination, matrix, chemical, physical and other problematic interferences. It is important that each step is carried out so as to minimise errors and obtain meaningful results and, if the analysis is carried out with care, it will increase confidence in the results. However, the obvious strategy that the analyst must bear in mind is that the more tedious the sample preparation the more care will be needed to ensure that the elements being tested are totally associated with the sample and not from contamination during sampling, handling or from reagents used in the preparation; this is particularly important for trace analysis. Most modern instrumental techniques are designed for speed of analysis, and greater sensitivity and, because of this, sample preparation becomes even more important. However, analysis using slurry techniques which may involve little or no sample preparation can be used for limited numbers of samples provided the particles are suitably small and do not block the sample transport lines or nebulisers. The ideal situation would be to automate the sample preparation, analytical techniques and the reporting/presentation of data as used in most clinical laboratories. Unfortunately, this level of automated and speed analysis has not arrived for non-clinical samples. Clinical samples are a constant matrix for which it is relatively easy to design automated analytical systems. Table 3.1 illustrates steps that are usually applied for most non-clinical samples.

The application of statistics to support analytical results is usually the final step in reporting. Statistics can reveal much information about the determined result and ensure confidence in results. It can be applied in several ways and one of its most effective uses is the generation of the control charts to monitor the routine analysis of samples to determine whether the preparation of standards and instrument parameters are correct and no contamination has crept into the sample, reagents and instrument or during sample preparation. A control chart is generated from a control standard and is a visual display of confidence in the method. It can warn the operator if the sample/instrument parameters are in, or out of, control and whether corrections are necessary before proceeding with the analysis.

Table 3.1 Common stages of chemical analysis

No.	Steps	Preparation of liquids	Preparation of powder/solids
1	Sampling	Samples must be homogenous and traceable	Powder or solid samples must be made representative using subdividing methods. May require more than one sample per bulk material. Liquid samples must be homogenous
2	Preparation of sample	Determination of sample weight/volume for density, etc.	Reduction of particles size, drying, test for dissolution in solvents compatible with plasma sources
3	Dissolution of sample	Extent of dissolution to fit calibration range, or limit of detection	Complex formation, extraction, wet digestion, dry ashing, fusion, combustion
4	Removal of interferences	Filtration, extraction, ion exchange, chromatography	Same for all samples
5	Measurement	Emission response, calibration curve, sample/standard matching, etc.	Same for all samples
6	Results	Calculation, statistical data to support results	Same for all samples
7	Presentation	Printout, LIMS report, etc.	Same for all samples

It is useful to note that statistical handling for a relatively small number of systematically planned measurements may yield more information than a large number of repeated identical measurements. Consider triplicate analysis of the same sample using different weights or volumes. This may reveal errors that would not be detected if repeated similar sample sizes were taken. A report entitled 'Principles of Environmental Analysis' [2] which can be applied to all types of analyses, states:

> 'the single most characteristic of any result obtained from one or more analytical measurements is an adequate statement of its uncertainty interval.'

The objective purpose of chemical analysis has to be sensibly assessed prior to selecting the appropriate procedure.

3.4 Sampling and its Importance

The first and most important part of sample analysis is that the sample must be representative. The analytical result is wasted if considerable time is spent preparing the sample, using expensive equipment with state-of-the-art sensors and submitting an extensive report if the sample is not representative. Care must be taken with the sampling procedure, particularly with powder samples, as variations can occur and can give poor repeatability for analysis of several samples from the same bulk product. Analysing solid samples may require more than one sample be taken from different parts of the bulk product for analysis. This would give dual information of the uniformity of the bulk product and scatter of results throughout the bulk. For homogeneous liquids it is usually easier to obtain a representative sample.

3.5 Sample Preparation Methods

In most trace analysis a considerable amount of time is spent preparing samples for accurate and precise measurements. Some sample preparation techniques used today are almost 100–150 years old and may involve time-consuming efforts. The continued tediousness and dangerous nature of traditional sample preparation techniques prompted the need for modern alternatives.

Fortunately, in recent years, improved methods involving chemical isolation of metals from difficult matrices through organometallic complexes, microwave acid digestion, chromatography, etc., are now the norm. Some of the newer sample preparation methods will speed up the analysis time but most are aimed at detecting lower levels, achieving close to ~100% recovery from samples and allowing greater choices in dissolving them in suitable solvents for accurate measurement and instrument compatibility. Sample preparation usually consumes the large share of analytical time and the process has economic significance. The bar chart in Figure 3.1 is an approximate breakdown of the spread of time for complete sample analysis.

In the application of atomic emission spectroscopy for quantitative analysis, samples must be prepared in liquid form of a suitable solvent unless it is already presented in that form. The exceptions are solids where samples can be analysed as received using rapid heating electro-thermal excitation sources, such as graphite furnace heating or laser ablation methods. Aqueous samples, e.g. domestic water, boiler water, natural spring, wines, beers and urines, can be analysed for toxic and non-toxic metals as received with

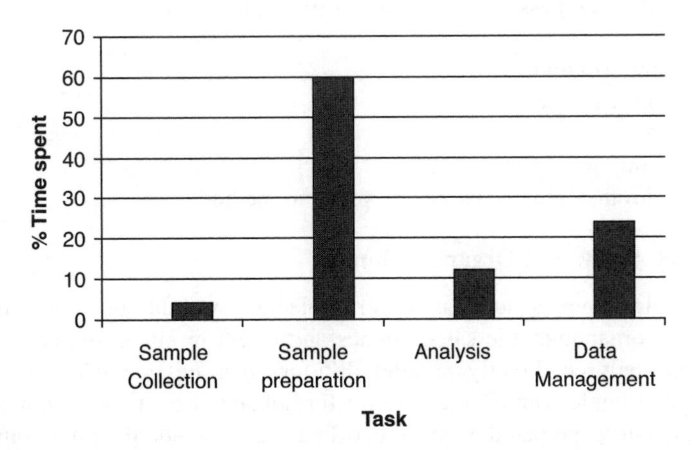

Figure 3.1 *Breakdown of approximate time spent during a typical analysis*

little or no further treatment. Domestic drinking water must be analysed for trace metals content and must meet rigid specifications before local authorities will release it into the domestic mains supply.

Accurate sample preparation prior to chemical measurement is often a limiting step; not only is it an important source of uncertainty in final results consideration but is also lengthy and labour-intensive. The preparation step is intermediate between the correct sampling procedure of the bulk matrix and actual measurement. Sample preparation for modern chemical analysis involves a considerable amount of analytical time (see Figure 3.1) and is considered as the main bulk in costing.

It is important to ensure the sample is representative, free from contamination and stored in a suitable unreactive and contamination-free container. The most sophisticated technique cannot rectify the problems generated by poor sampling procedures.

Serious consideration and time must be given to whether the component of interest is in exceptionally high concentration which, in such cases, may have to be diluted to fit calibration standards or, if low, may require pre-concentration. In some cases it may be necessary to carry out a trial-and-error to ascertain the approximate concentration of metals in samples. The low value must at least be at quantitative limits (i.e. ten times the standard deviation of baseline noise, to be confident of results; see Section 3.8.1.7). If it is lower than ten times standard deviation of the baseline the sample may have to be pre-concentrated prior to analysis to a level that can be comfortably detected and is suitable for reproducible measurements. The following is a list of common methods of sample preparation techniques. Choosing the correct method is of primary importance and poses a challenge to most analysts, particularly for unknown samples:

1 Direct
2 Sample dissolution
3 Extraction using acids/bases, organometallic complexing agents
4 Dry ashing without aid
5 Dry ashing with aid
6 Acid digestion using microwave
7 Oxygen bomb flask combustion (low pressure)
8 Oxygen bomb stainless steel combustion (high pressure)
9 Fusion
10 Slurry sample solution
11 Acid/solvent leaching
12 UV digester
13 Plasma ashing
14 Spinning chromatography sample preparation method

3.5.1 Direct Analysis of Organic Solutions

Certain organic samples such as lower molecular weight alcohols, polyalcohols, organic acids, organometallic salts, amines and esters in either aliphatic or aromatic form can be analysed directly or after dilution in a suitable ICP-AES compatible solvent. These samples can be analysed for formulated or contaminated metals against a calibration curve prepared from a certified stock standard in the same solvent. Organic samples with varying viscosities, particularly for trace metals content, may be analysed using the method of standard addition or internal standard after careful preparation.

3.5.2 Sample Dissolution

Sample dissolution is probably one of the most common operations in analytical chemistry and is carried out by dissolving in a suitable solvent to a suitable concentration that the analyte of interest can be reproducibly measured. If the composition of the non-aqueous solution is amenable to combustion in a flame or plasma, direct aspiration is possible. Unfortunately, ICP-AES instruments do not have the same solvent tolerance as AAS and require that the solvent selected be stable, non-quenching and non-interfering. Calibration standards are usually prepared in the same metal-free solvent, keeping in mind the effect of sample in the solvent. If the nebulisation efficiency of sample/solvent mixture is different to standards prepared in the same solvent only, then corrective actions must be taken so this anomaly can be taken into consideration.

3.5.3 Chemical Extraction of Metals from Organic Matrices

Chemical separation for preparing samples for ICP-AES analysis is an excellent technique to use for samples containing major and trace concentrations, and there are several reasons for doing this. Three of the most important are:

(a) concentration of element is too low for detection using normal preparation methods;
(b) separation from interfering components in sample;
(c) sample contains solids that cannot be handled by the nebuliser.

Extraction of metals using acids, bases or organometallic complexes are the most common techniques, as they can be reduced to their simplest form and have efficiencies as high as 95–100%. The use of chelating or complexing agents as an aid in trace analysis is a powerful technique and a number are available. A good example is ammonium pyrrolidine dithiocarbamate (APDC) developed by Malissa and Schoffmann [3] that is used to complex metals over a wide pH range.

APDC can complex up to 25 elements at pH range from 0 to 14 and its main advantage is the ease with which a single extraction can achieve almost 100% recovery. Malissa and Schoffmann [3] devised a list of elements that can be extracted and their best pH, which is shown in Table 3.2. The structure of APDC is:

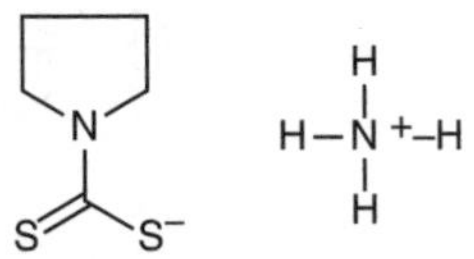

The advantage of using APDC is the metal salts are readily soluble in most organic solvents and will separate them from high concentrations of other solutes that could cause difficulties in nebulisation and atomisation. Large bulk of aqueous sample may be extracted efficiently into a smaller volume of an organic solvent and this can be further

Table 3.2 *List of metals forming complexes with APDC and their respective suitable pH for extraction*

Metal	Nominal pH	pH range (approx.)	Metal	Nominal pH	pH range (approx.)
V	5.0	4–6	Se	3.0	2–7
Cr	5.0	3–7	Ru	7.0	4–10
Mn	5.0	4–6	Rh	7.0	4–10
Fe	5.0	1–10	Ag	7.0	4–10
Co	5.0	1–10	Cd	7.0	3–10
Ni	5.0	1–10	In	7.0	3–10
Cu	5.0	0–14	Sn	5.0	4–6
Ga	5.0	3–8	Sb	7.0	5–9
As	3.0	0–4	Te	3.0	2–5
Th	5.0	4–6	I	7.0	2–12
Bi	5.0	1–10	Ti	7.0	2–9
Pb	5.0	0–9			

pre-concentrated by evaporation to yield low quantitative metal analysis. APDC complexes are soluble in a number of ketones especially methyl isobutyl ketone or chloroform. These solvents have a very low solubility factor in aqueous solutions and have high partition coefficient that is suitable for extraction but they are not suitable on ICP-AES due to their noisy and quenching effect. However, such solvents can be evaporated to a low volume and the concentrate re-dissolved in a solvent that is suitable for ICP-AES analysis, e.g. kerosene, IPA, glacial acetic acid, etc.

Other extraction agents such as dithizone, diethyl dithiocarbamates, and 8-hydroxy-quinoline (oxine) are also useful as chelating agents and extract several metals including transition metals, alkali and alkaline metals. The dithizone can extract up to 20 elements and these complexes behave the same and there is no additional advantage of using one over the other. The compound 8-hydroxyquinoline will form stable complexes with Al, Ca, Sr and Mg at pH 8.0; these metals form unstable complexes with APDC. At carefully controlled pH the oxidation states of metals may also separated, e.g. Fe[II] and Fe[III]. The structure of 8-hydroxyquinoline is:

M^{n+} + n [8-hydroxyquinolinate: O⁻, N, H⁺] = M[O, N chelate]$_n$

3.5.4 Dry Ashing without Retaining Aids [4]

Ashing of samples may be defined as heating a substance to leave an ash that is non-combustible and that is soluble in most dilute acid or base solutions. The resulting solution is analysed for elemental composition against certified standards prepared in the same acid or base solvents. The advantages of ashing is that a large sample size can be prepared at one time requiring little or no reagents and it is safe to use. Disadvantages could be loss of powder dust and certain volatile elements during the heating cycle. It is possible to add more sample to the same dish and repeat the ashing procedure increasing the concentration of elements of interest for analysis; this would be useful for trace analysis. In assessing any method of sample preparation of organic matrices two requirements need to be considered. First, the method must destroy all the organic matter completely and effectively and within an acceptable time, and, secondly, the elements of interest must be retained in their original quantity. The first requirement can be recognized immediately by visual observation and the time required for complete destruction can be estimated. The time required for ashing may be different for different compounds and is generally accepted that if a 1.0 to 2.0 g sample is subjected to temperatures of 550–650°C for 3–4 h it should be sufficient to completely burn the sample using a muffle furnace. Microwave ovens may shorten this time by a factor of a half because of the more efficient heating of the sample (Figure 3.2).

The second consideration for quantitative analysis is that if the elements are present in trace quantities the problems of multiple sample preparation by ashing increase. A vicious circle is endured when the quantity and percentage recovery of the element cannot be determined until the quantity of the element present is known for dilution limits and range of standard calibration curve required. If the quantity is known, the percentage recovery may be determined by calibration curve or by standard addition. The percentage recovery results should be $\sim$100% $\pm$ 2% to allowing for errors and loss. Drawbacks to this method are:

(a) the chemical form of the elements in the sample can influence their behaviour during the ashing process so that there is no guarantee that they will behave in a similar manner. Therefore, 'spiked' elements may not necessarily parallel the behaviour of the elements in the sample under test;

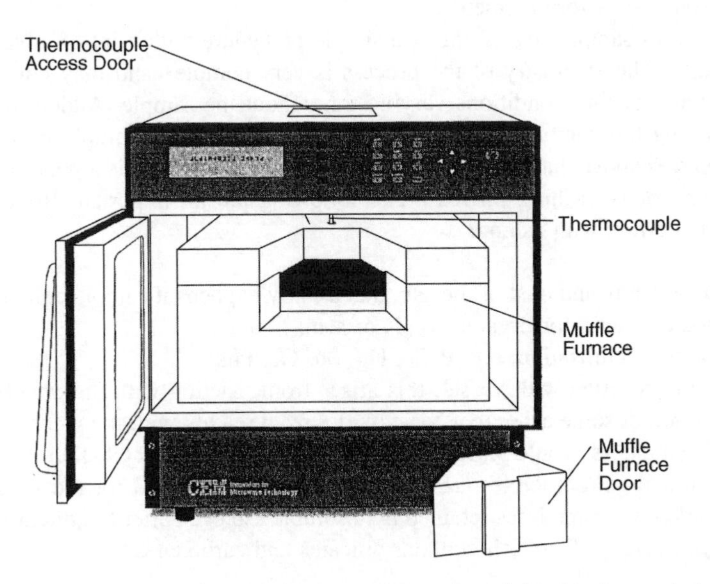

Figure 3.2 *Dry ashing microwave furnace. (Reproduced by kind permission of CEM Corporation)*

(b) it is best not to have any further steps in the sample preparation other than dissolving the ash in a suitable solvent. Extra steps may give rise to reduced recovery and errors;
(c) avoidance of contamination and/or loss of sample or 'spike' during the ashing process is paramount in a recovery study.

The methodology involves heating a known weight of organic sample in an open silica, platinum or gold vessel at a series of time-controlled ramping temperatures in either a standard muffle furnace or microwave oven to a maximum of 550–650°C. The ashing should be efficient and destroys the unwanted matrix without spitting or catching fire. The ashing temperature should be kept as low as possible (to avoid loss of volatile elements) but must be high enough to ensure complete combustion. Usually, the above temperature range is sufficient for a large number of samples of organic compounds. A microwave asher can ash a sample in less than half the time of a muffle furnace due to its heating efficiency. In both cases it is important that the heating cycle is controlled and sample sizes from 1.0 to 5.0 g are sufficient for most analyses. To increase detection of certain elements at trace levels it may be possible to continue burning two or three lots of 5.0 g consecutively. Depending on the size of the muffle furnace up to 10 to 20 samples or 5 to 10 duplicates may be ashed at the same time. The final ash is dissolved in 1.0–2.0 M acid or base solutions and the elements measured against a standard calibration curve prepared from certified standards in the same acid or base solutions.

The ashing of a sample is usually easy to perform but not all resulting ashes are soluble in acid, bases, etc. It may be possible to analyse these samples for elements in the ash that will dissolve in 1.0–2.0 M HCl and separate the unwanted ash by filtering. The insoluble products in these cases usually indicate mainly sand, talc and silica. If analysis is required for the complete ash a stronger acid or acid mixtures may be required, e.g. H_2SO_4, HCl, HNO_3, HF, etc. This method works well for most alkali, alkaline and transition elements including some refractory elements.

Preparation of samples by ashing is a simple procedure with a lot of advantages and disadvantages. The chemistry of the process is very complex and may consist of both oxidising and reducing conditions varying throughout the sample. Added to this is the fact that during the combustion process the temperature in the sample may be several hundred degrees above that set by the furnace, particularly if there is a good flow of air at the beginning of the ashing process. The following are an important list of potential problems and errors with ashing:

(a) loss due to spray and dust of the ash; this usually happens if sample containing water is heated too rapid for organic solids or semi-liquids;
(b) loss due to volatilisation, e.g. P, Se, Hg, Sb, Ge, PbI;
(c) loss due to reaction with vessel; this arises from reaction with the crucible and the composition of some ashes, e.g basic oxides react readily with the glaze on porcelain or with silica, as would nitrates, sulphates or carbonates at higher temperatures. Platinum and gold crucibles are relatively inert and are good for most samples;
(d) trace metals are sometimes retained by insoluble ash even after treatment with strong acids particularly if the ash contains silicates and carbonates.

The vessels used for ashing, particularly for trace elements, must be carefully cleaned (usually in 1–2 M boiling HCl for 10 or 15 min), rinsed with de-ionised water and dried in an oven prior to use. It would be important to do this immediately after use so as to avoid acidic, basic other residues damaging these expensive vessels.

3.5.5 Dry Ashing with Retaining Aids

Dry ashing methods are more efficient when carried out in the presence of a metal retaining additive. These additives may have several advantages, i.e. may accelerate oxidation, prevent volatilisation of certain elements and prevent reactions with the vessel used to ash the sample. Common retaining aids such as magnesium nitrate, sulphuric acid and *para*-toluene sulphonic acid (PTSA) are applicable for organic samples where volatile metals may be present and are converted to nitrate or sulphate salts in order to be retained at elevated temperatures. Nitric acid is also a good oxidant but could not be added to platinum and gold dishes as it would react with them. High molecular weight organic samples such as oils and monomers may be treated with powdered cellulose so that the sample then burns easily and smoothly without the temperature going out of control.

Samples of heavy duty and crude oils, lubricating oils, worn oils, organic polymers, plastics, grease, etc., may be ashed with the retaining acid PTSA using the dry-ashing method up to temperatures of 650°C. This compound will retain most mono- and divalent states of elements in samples. The structure of PTSA is:

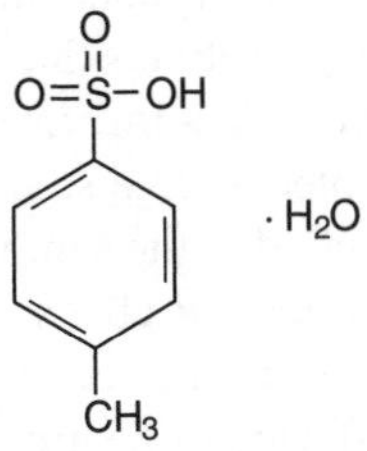

O
O=S–OH
· H_2O
CH_3

It is important to note that the same rules and observations need to be applied when preparing samples without additives/aids.

3.5.6 Acid Digestion Using Microwave Oven [5]

Sample preparation using microwave digestion is a relatively modern addition to the list of analytical sample preparation techniques for analysis by ICP-AES, ICP-MS, AAS or polarography (Figure 3.3). The technique of using microwave energy was invented by accident by Dr Percy Spenser during his work on improving radar technology shortly after the Second World War. He went on to develop it for many uses in domestic and industrial applications and in the laboratory. The aim of every sample preparation technique is to achieve a clear and complete decomposition of the sample and it must also be time-saving, avoiding analyte loss and contamination and be easy to work with. Fortunately, microwave digestion meets most of these criteria and is applied to a wide range of samples for major and trace analysis.

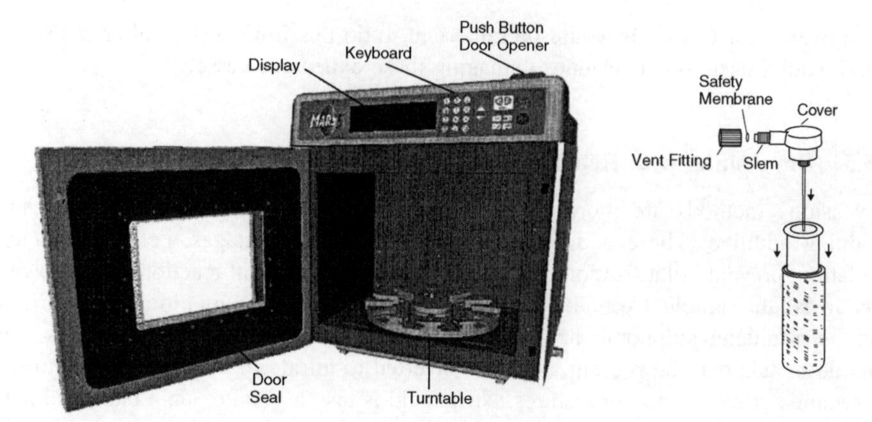

Figure 3.3 *Microwave acid digester with temperature controller, safety membrane and vessel with fittings. (Reproduced by kind permission of CEM Corporation)*

The theory of microwave energy used for acid digestion is based on electromagnetic energy that causes molecular motion by movement of ions and rotation of molecular and atomic dipoles and does not affect the structure of the molecules. The heat caused by reduction in current due to resistance in the sample is associated with ionic conduction and is affected by the relative concentration and mobility in the sample medium. When microwave radiation penetrates the sample, it is absorbed by an amount dependent on the sample. Microwave heating involves ionic conduction and dipole rotation. The electric field of the molecule in a sample gives rise to dipole rotation and as the electric field in the oven is applied to the sample the molecules becomes polarised and aligned and when the field is removed chaos returns to the molecule. This jump from chaos to order occurs 10 billion times a second and in doing so causes the sample to heat very rapidly; however, its efficacy of heating depends on the sample matrix and viscosity.

The time taken to dissolve a sample in concentrated acids on an opened hotplate can vary from 1 to 16 h and using a closed microwave oven the sample could be prepared in 15 to 30 min. The reason for this is that the efficiency of heat transfer by the microwave oven is greater than that of a hotplate. The dual contribution of ionic conduction and dipole rotation relaxation times to heating depends on the nature of the sample and one may predominate over the other. If the ionic mobility and the concentration of ions are low, then the heating will be dominated by dipole rotation. On the other hand, if the mobility and concentration of ions are high the heating will be dominated by ionic conduction and heating time will be dependent on the dielectric relaxation times of the sample solution. As the ionic concentration increases the dissipation factor increases and heating times will decrease. It is worth noting that heating times also depend on the microwave design and sample size and not on the dielectric absorption of the sample. The digesting concentrated acids containing the metals of interest can be reduced by attaching the vessels to a micro-vap accessory, and by heat and vacuum the acids can be reduced to a fraction of the original volume. This makes it easier to dilute the digested sample and reduces the concentration of acids in the analytical solution (Figure 3.4).

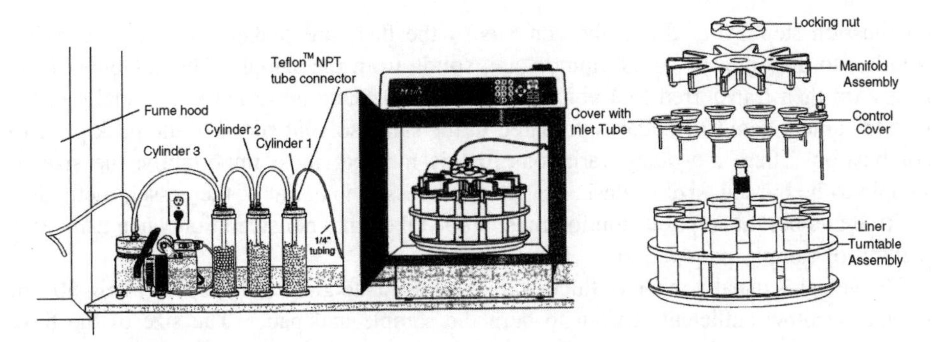

Figure 3.4 Microwave acid digester and micro-vap acid/solvent reducer. (Reproduced by kind permission of CEM Corporation)

3.5.7 Oxygen Bomb Flask Combustion (Low Pressure)

Hempel and Angew [6] proposed as early as 1892 that some organic samples could be combusted in a closed 10 l vessel filled with oxygen at atmospheric pressure and the resulting material re-dissolved in a suitable aqueous solution already in the vessel. Schöniger [7], who reduced the vessel to a micro-scale size, improved this cumbersome technique. The increased sensitivity of modern instruments enabled reduction in sample and vessel size to be used for preparation by this method and enhance the safety aspects in its use. The technique simply involves wrapping a known weight of sample in an ignitable filter paper, as shown in Figure 3.5, which is cut so as to have a protruding paper tab. The sample is held by a wire support that is capable of rapid heat and ignites with the application of an electrical current. An absorption solution (depending on the sample) is poured into the flask that is then flushed with oxygen gas at approximately atmospheric pressure and closed immediately. An electric current is applied while the flask is held firmly and carefully behind a viewing screen, making sure that it does not open during the

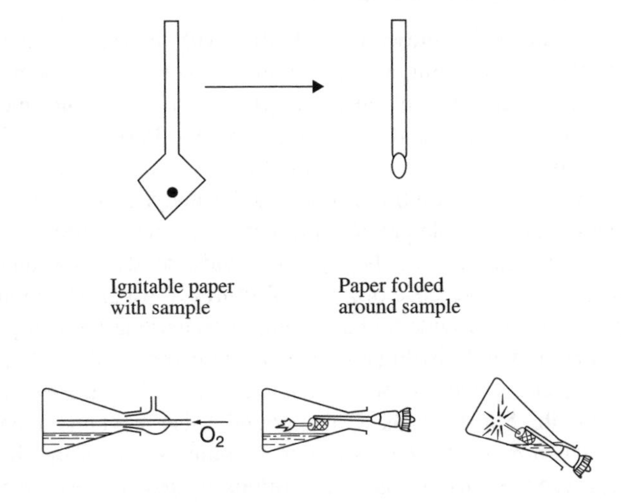

Figure 3.5 Diagram of the low pressure oxygen flask and the sample insert

combustion step. After firing, the contents of the flask are shaken with the absorption solution to dissolve the gases, liquids, and solids from the sample. The contents of the flasks are then transferred to a volumetric flask and made up to mark and analysed for metals content against standards prepared in the same solvent added to the flask prior to combustion. There are many variations to this method; most importantly, the size of sample to be burned, size of flask, shape of the vessel, material of the vessel, method of ignition and best absorption solution must all be taken into consideration when using this method of sample preparation.

The vessel is usually made of thick-walled borosilicate glass and is usually 250–500 ml in size to allow sufficient oxygen to burn the sample and paper. The size of the flask depends on the sample to be burned and it is fitted with a ground glass stopper containing electric wires as a means of connecting to an electric current. High-density plastic flasks have been used with success but have a tendency to be attacked by some absorption solutions and by heat of combustion. Combustion aids such as sugars, sodium carbonate and cellulose may also be added to the flask prior to ignition. They can be mixed intimately with the sample or impregnated in the filter paper in which the sample is wrapped. These aids usually supply more hydrogen and in doing so combine with the added oxygen to produce more water in the combustion process. The added oxygen has a dual purpose – it ignites readily causing a rapid heating process, and it reacts with the compound itself. The pressure of the final solution and gases will be reduced to a safe level after standing for a short time at room temperature so as to allow opening of the flask but care will still have to be taken and the process must still be carried out behind a safe screen.

This sample preparation technique is widely used in the determination of halogens, sulphur, phosphorous in organic compounds as well as for the determination of Hg, Zn, Mn, Ni, Co, Fe, Cu, etc. An excellent application is the separation of compounds by paper chromatography in which the spot of interest is cut out of the paper and burned, as described above.

3.5.8 High Pressure Oxygen Combustion

Sample preparation using a high pressure bomb filled with oxygen at an elevated pressure is an excellent method of obtaining a clean solution containing the elements of interest (Figure 3.6). The technique was previously described as a bomb calorimeter filled with oxygen as the combustion aid and was developed by Berthelot [8] in 1885. This technique is also used as a method for preparing samples because of its ability to convert complex matrices into a simple solution containing the analytes of interest and has a wide range of applications. This sample preparation method is popular because practically all organic materials, including most foods, organic liquids, solids and a range of fuels can be burned completely in the presence of a few millilitres of water when ignited in excess oxygen under pressure and provide a clean solution containing the analytes of interest. These materials contain mainly hydrogen, oxygen and carbon and react readily with the oxygen to form carbon dioxide and water that are removed by a special release valve attached to the bomb. Some of these compounds also contain sulphur, halogens, phosphorus, etc., which must also be taken into account as part of the final analysis.

High pressure oxygen bombs are made from thick walled stainless steel vessels with capacities from 100 to 300 ml for general use. The sample to be burned is weighed into a

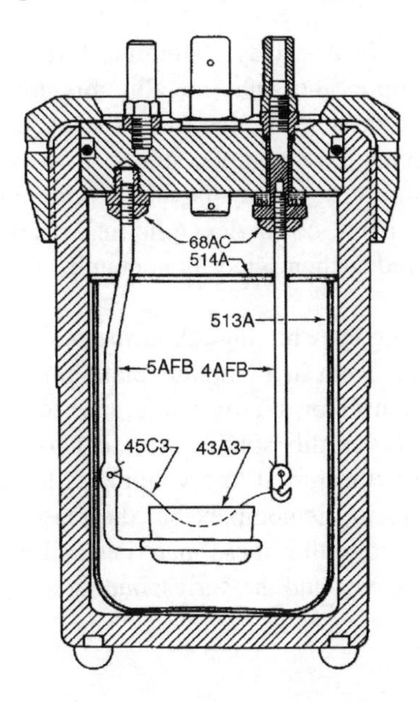

Figure 3.6 *High pressure oxygen bomb fitted with Quartz liner. (Reproduced by kind permission of Parr Instruments and Medical and Scientific Products)*

carbon-coated stainless steel cup or platinum or gold dish of suitable size to fit the special holder supported by the lid of the bomb. Attached to this are electrical contacts to which an ignition wire can be fitted in such a way that it is close to the sample. The bomb is assembled as per manufacturer's instructions and filled with oxygen to approximately 30 atm. The filled bomb is the placed under water to check for gas leaks and if there are none, the bomb is fired by passing a current through the ignition wire. Care must be taken that the possible errors from the ignition wire, absorption solution and aids are not added in when calculating the final results. Blank must also be prepared in exactly the same way but without the sample.

Liquid and solid samples are placed in an opened cup while volatile samples are placed in a gelatine capsule and placed in the cup as for a normal sample. A suitable absorption solution is placed at the bottom of the bomb along with the prepared sample. It may be necessary to include an aid with combustion resistant samples as a support to combustion and such additives are low molecular weight paraffin oils, ammonium nitrates, benzoic acid, decalin, etc.

The advantages of this technique are that it is easy to use and solutions obtained are easy to analyse. The solutions contain virtually no contamination and there is little or no loss of analytes by volatilisation. The disadvantages are that extreme care must be taken during use and it can only handle one sample at a time.

3.5.9 Sample Preparation Using Fusion Methods [9]

A number of materials requiring analysis that are difficult to dissolve by conventional methods are best fused with a fusion reagent to form a 'fused' mixture that dissolves easily. Samples containing silicates, clays, alumina, lavas, slags, refractory, ceramics, cements, etc. and other complex inorganic metallic mixtures that are insoluble even in strong acids can be made to dissolve in these acids after fusing. The mixture of fusion reagent and samples containing these complex minerals convert the minerals into soluble versions by reacting and forming soluble salts. Some of these materials are used as fillers in organic compounds, e.g. paint, cosmetics, toiletries, pharmaceuticals, adhesives, etc. and may need to be monitored as themselves or contaminants as part of quality control of these products.

The technique of fusion involves mixing a known weight of sample with about three times its weight of a fusion reagent in a graphite, platinum or gold crucible. The mixture is heated to 900°C using a muffle or microwave furnace for 2–4 h to form a 'melt'. This 'melt' is allowed to cool and is readily soluble in a single or mixture of strong acid (HCl, HNO_3, HF, etc.) that is then transferred to a volumetric flask and diluted to mark. The chemistry associated with fusion is complex but the reason it is used is the ability to break the strong bonds formed within these materials and converting them into soluble salts. The popular fusion reagents and the corresponding acids required to dissolve them are listed in Table 3.3.

Table 3.3 *List of common fusion reagents used in sample preparation*

Reagent	Comment
Sodium hydroxide	Rarely used, 500°C, attacks platinum and porcelain, Ni or graphite crucibles, dissolves in 10 M HCl
Sodium carbonate	Good technique, 800°C, dissolves in HCl
Lithium metaborate	Good technique, 900°C, dissolves in HCl
Lithium tetraborate	Good technique, 900°C, dissolves in HCl
Potassium hydroxide	Rarely used, 500°C, use Ni or graphite
Potassium persulphate	Good technique, 900°C, dissolves in H_2SO_4
Sodium nitrate	Good technique, 500°C, dissolves in HNO_3

Disadvantages of the fusion method are that some elements may be volatile at 900°C, the fusion reagent may cause contamination, and the presence of high amounts of dissolved solid content may not be suitable for trace analysis. Blanks of fusion reagents must also be prepared alongside samples. The fusion fluxes are expensive and give rise to spectral interferences and must be considered a last resort.

3.5.10 Analysis Using Slurry Solution Method

The development of the Babington [10] cross flow nebuliser allowed samples containing high salt content and slurries (max. ~20 μm) to be analysed with considerable ease.

Slurry solutions can be nebulised and introduced as an aerosol to the plasma source similar to that for clear solutions and the solvent containing the particles also assists in transporting them for elemental analysis. Care must be taken in avoiding the tendency to form agglomerates and samples must be kept stirred or shaken using an ultrasonic bath prior to nebulisation. Samples of products containing very small particles, e.g. fillers in paints, crayons, adhesives, etc. can be analysed by dissolving the organic components in an organic solution compatible with the ICP-AES torch and the insoluble component stirred as slurry and aspirated in the normal way against standards prepared in the same solvent. The method of standard addition is the best approach for these samples. In most cases stabilisation using wetting agents is unnecessary providing agglomeration is avoided.

The application of slurry analysis technique is successful provided that the particles are suitably small. It is a very useful technique for samples that contain small particles and avoiding the tedious procedures of dissolving them. A range of samples containing these fillers can be analysed with a high degree of accuracy using this method. Some products are formulated with fillers to improve their thixotropic properties for certain applications e.g. non-drip paints, toothpastes, lipsticks, toiletries, adhesives, etc. It is important not to overload the slurry solution so as to avoid blockage in the sample supply line.

3.5.11 Sample Preparation Using Leaching Method

Some samples such as clays, silicates, refractory slags, quartz, etc. may be contaminated with free and unbound elements that will dissolve readily by leaching into an acid solution with stirring and filtering. The procedure simply involves stirring a known weight of powder sample for a known period in an acid solution and centrifuging or filtering through a fine bore filter paper. The supernatant liquid containing the leached metals is analysed for their metal content against standards prepared in the same acid solution. This procedure is often governed by the difficulties in dissolving these samples, clays, etc. and is a useful technique for analysing free and unbound metals providing that analysis of the main element constituents of the fillers are not required. An interesting application is the analysis of fine powder of clays and sand for trace levels of Hg, Cd, Se, Sb and Pb content that are readily leached into an acid solution giving $\sim$100% recovery as part of environmental monitoring.

3.5.12 Sample Preparation Using a UV Digester

Digestion is relatively decisive and reliable for ultra-trace analyses in natural samples. The UV digester is suitable for water samples containing low to moderate concentrations of organic material insoluble or soluble in a range of aqueous samples. Samples that contain low or high levels of metals and contaminated with organic matter could cause spectral interferences in ICP-AES, particularly where trace analysis needs to be carried out. In such samples it is possible to use UV photolysis with a minute level of hydrogen peroxide to decompose the organic material achieving a total aqueous solution for analysis. A blank with the same level of hydrogen peroxide in deionised water should also be prepared under exactly the same conditions as a blank.

Digestion using this procedure involves the generation of OH radicals from the hydrogen peroxide by photolytic effect of the UV lamp and these radicals attack the organic matter in the water and degrade them to such an extent that they are made soluble and harmless or are removed by boiling. The OH radicals generated by the UV radiation act as initiators of radical reactions. This is an excellent method for eliminating organic matter in water producing a clean homogenous sample. Temperature of 80 to 100°C is achieved which accelerates the digestion and which is high enough to break down the organic compounds but low enough to avoid loss of volatile elements of interest. To validate the method it would be wise to halve the sample and 'spike' one half with metals of interest and compare percentage recovery. It is possible to prepare several samples at the same time.

3.6 Non-Spectral Corrections Using ICP-OES

3.6.1 Effect of Solvents on ICP-OES

The analysis of organic matrices dissolved in solvents using ICP-OES is finding an increased number of analytical applications in laboratories worldwide. These methods are important in terms of rapid sample preparation, reduction in contamination, loss of elements through sample preparation, etc. A considerable number of organic-based metal solutions are used in industrial, medical and pharmaceutical applications as initiators, activators, colorants, chemical catalysts, pharmaceutical preparations, etc. and need to be quantified as part of contamination monitoring or quality control.

Only a limited number of solvents can be used for direct analysis by ICP-OES. There are several methods available to overcome using problematic solvents and the best way is to evaporate to low volume and re-dissolve the sample in a solvent compatible with ICP-OES. In the latter years CETAC Technologies have developed an ultrasonic nebuliser with a solvent desolvator attached to it as described in Chapter 2. The 6000 AT^{+} solvent desolvator is used in conjunction with 5000 AT^{+} and is designed to remove most solvents prior to excitation and atomisation. The combined 5000 and 6000 AT^{+} apparatus has proven to be an excellent accessory to ICP-OES and has the advantage of improving detection limits by 10 to 50 times when compared with a conventional pneumatic nebuliser. This ultrasonic nebuliser efficiently converts liquid samples into aerosols suitable for ICP-OES analysis. They are more efficient than standard pneumatic nebulisers and do not block as readily. The 6000 AT^{+} system strips the solvent from the sample aerosol and enriches it prior to entering the ICP-OES torch which maintains the plasma energy required for excitation. This will reduce matrix effects caused by solvent loading and solvent derived interferences. Both are available from CETAC Technologies, Michigan, USA who have worldwide agencies.

Table 3.4 illustrates some of the physical properties of solvents directly nebulised and the effects that they have on the torch.

Table 3.4 *Properties of solvents used with ICP-OES*

Solvent	Run 1	Run 2	Run 3	Mean	% efficiency	Remarks
GAC	46.5	45.5	46.0	46.0	~8.0	Stable
Ethanol	44.0	44.5	44.0	44.2	~11.0	Noisy
IPA	45.0	44.0	43.5	44.2	~11.0	Noisy
Xylene	44.0	43.0	43.0	43.3	~13.0	Noisy
MIBK	42.0	42.5	41.5	42.0	~16.0	Noisy
Kerosene	44.5	45.5	46.0	45.3	~9.0	Noisy
Toluene	43.0	44.0	43.0	43.3	~12.0	Noisy
Water	48.0	49.0	48.5	48.5	~3.0	Stable
Petroleum spirit	45.8	46.9	46.4	46.3	~12.0	Noisy

GAC, glacial acetic acid; MIBK, methyl isobutyl ketone.

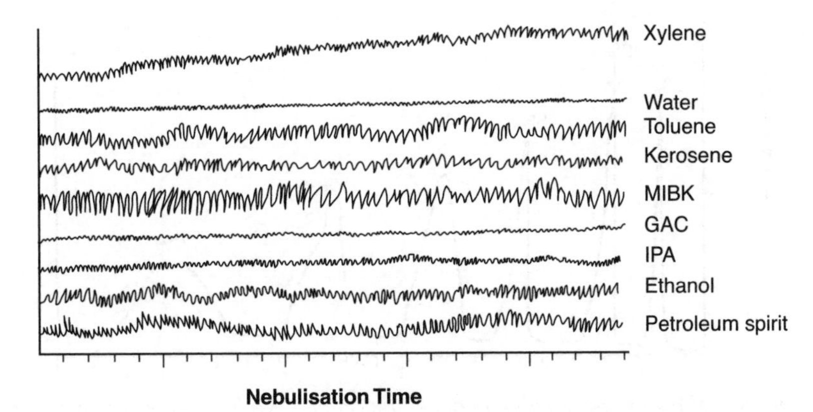

Figure 3.7 *Effect of solvents with ICP-OES plasma torch showing baseline noise of listed solvents nebulising for 10 min under identical conditions using a chart recorder [15]*

Figure 3.7 shows the results of a study of nebulisation efficiencies of solvents compatible with the ICP-AES plasma source under normal conditions (as listed in Table 3.4). An accurate volume of 50.0 ml of each solvent was nebulised through the plasma under normal conditions. The resultant waste volume was collected into a measuring cylinder in triplicate to give the values listed in Table 3.4. The efficiency was calculated from the average. The behaviour of each solvent was also noted under 'Remarks' [15].

The behaviour of solvents listed in Figure 3.7 on the ICP-OES is common and the best are water and glacial acetic acid which are almost identical in terms of sensitivity, stability, excitation, solubility, effect on the pump tubing, etc. Unfortunately, acetic acid has two main drawbacks, namely its odour and corrosive properties. The solvent kerosene finds many applications in the analysis of a wide range of oil and petroleum products, and is also stable.

3.6.2 Effect of Viscosity on Signal Response

The viscosity of the sample/solvent solution can influence the nebulisation efficiency and the number of atoms entering the plasma torch are reduced. Solutions with different viscosities can affect the rate of uptake of a sample by a given nebuliser and the surface tension affects the size distribution of the droplets formed hence the nebulisation efficiency. The solvent vapour pressure affects the drop size distribution during transport and nebulisation. The different particle size distribution which is governed by these properties is also important as the larger the particles the less the vaporisation efficiency. It was shown above that the introduction of an ultrasonic nebuliser reduces the interference effect caused by poor particle size distribution. The signal and viscosity effects are illustrated in Figures 3.8 and 3.9 showing the effect of adding increasing percentage of a viscous mineral oil to glacial acetic acid solvent to the same concentration of Fe metal.

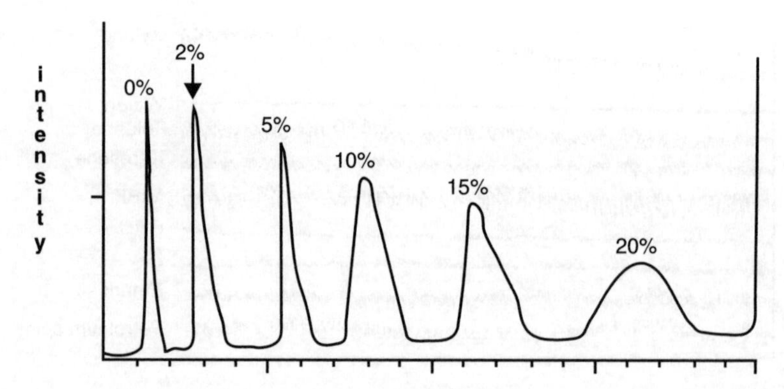

Figure 3.8 *Signal responses of known concentration of a synthetic 1000 cps lubricating oil containing 5.0 ppm Fe standard dissolved in glacial acetic acid. The samples were measured against a calibration curve of 0.0, 5.0 and 10.0 ppm Fe in glacial acetic acid only*

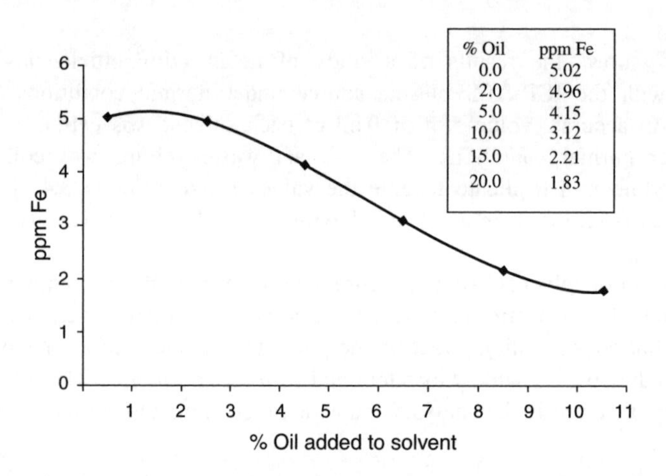

Figure 3.9 *Graphical illustration of signal responses obtained in Figure 3.8*

Samples dissolved in the same solvent as the standards that have different viscosities could present problems in terms of response, i.e. metal standards prepared in a solvent could differ from a sample prepared in the same solvent containing exactly the same concentration of metal as the standards. Therefore, if this anomaly is not taken into consideration accurate quantification would not be possible.

3.6.3 Comparison of Nebulisation Efficiency of Solvents Using ICP-OES

The sample transport system, nebuliser and spray chamber are designed to ensure the maximum amount of sample reaches the atomisation source without quenching it. Only a few solvents can be used that are compatible with direct injection to ICP-OES (see Table 3.5) and these solvents have been studied as part of nebulisation efficiency.

The behaviour of solvents for the analysis of metal ions is important because the determination of the correct concentration is paramount to whether the ICP-OES can handle a solvent or not. The journey from liquid to nebulisation, evaporation, desolvation, atomisation, and excitation is governed by the physical nature of the sample/solvent mixture. The formation of the droplet size is critical and must be similar for standards and sample. The solution emerging from the inlet tubing is shredded and contracted by the action of surface tension into small droplets which are further dispersed into even smaller droplets by the action of the nebuliser and spray chamber which is specially designed to assist this process. The drop size encountered by this process must be suitably small in order to achieve rapid evaporation of solvent from each droplet and the size depends on the solvent used. Recombination of droplets is possible and is avoided by rapid transfer of the sample droplets/mist to the plasma torch. The degree of reformation depends on the travel time of the solution in the nebuliser and spray chamber. For accurate analysis the behaviour must be the same for standards and samples.

The nebulisation efficiency is subsequently related to atomisation efficiency and depends mainly on the properties of the nebuliser, a significant fraction of the mist is usually lost in the expansion chamber and the plasma. The efficiency never achieves 100% and can be as low as 2% and as high as 6–10% depending on the solvent. The combined physical properties of organic solvents are governed by surface tension, density, viscosity, boiling point, vapour pressure and combustibility; these affect the sensitivity of plasma measurement. Logic would have us believe that the higher the efficiency the higher the signal counts for a comparable concentration of metal in each solvent and would result in a higher response. It has also been shown that a heated spray chamber can improve the efficiency and increase gas flow while maintaining a constant aspiration rate. The application of heat can alter drop size which produces a minimal effect until the temperature is raised sufficiently but not to boiling point. However, this is not a good practice because controlling the heat in a flowing system is difficult to maintain and the nebulisation rate is not constant. The concentration of the analyte is not necessarily proportional to the aspiration rate as the fraction of sample lost is not constant and increases with increasing aspiration rate, owing to the deterioration of the nebulisation action. High aspiration rate is accompanied by a decrease in atomisation efficiency, which in turn affects the desolvation, dissociation and excitation of the atomic elements. Solvent evaporation and volatility of solid particles determine the number of free atoms available for excitation and partial desolvation of the spray affects the size of the droplets reaching the excitation source.

When an analyte solution is sprayed into the plasma, the spray droplets are rapidly heated to the boiling point of the solvent. The solvent then starts evaporating at a rate depending on the rate of heat transfer from ambient plasma to the boiling point of the droplet. This process of solvent evaporation and vapour heating consumes some of the plasma heat and may affect the excitation efficiency of some elements under test. The vapour pressure varies from one atmosphere at the surface of the droplets to almost zero at a distance far from the droplet. The low surface tension and low viscosity of many organic solvents can lead to an increase of flow rate and favour dispersion, consequently increasing the concentration of the atom in the plasma source.

It must be stressed that the efficiency, as determined in Table 3.4, may not be entirely due to efficiency of the nebuliser because consideration must be given to loss due to

Table 3.5 *List of the physical properties of solvents considered for ICP-OES analysis. In all cases silicone tubing of 1.02 mm internal diameter was used*

Solvent	Density at 20°C (g cm^{-3})	Viscosity at 20°C (cps)	Boiling pt (°C)	% Carbon	Remarks
Carbon tetrachloride	1.6	0.99	76	7.9	Large plume, quenches plasma
Chloroform	1.7	0.57	61	23.8	Large plume, quenches plasma
Methanol	0.75	0.95	76	50.0	Large plume, quenches plasma
Ethanol	0.79	1.19	78	52.5	Large plume, jumpy plasma
Propanol	0.86	1.22	87	60.0	Large plume, jumpy plasma
Methyl isobutyl ketone	0.8		116	72	Large plume, jumpy plasma
Kerosene	0.85	0.65	108	?	Plume, jumpy plasma
Toluene	0.87	0.59	110	90.6	Large plume, quenches plasma
Xylene	0.86	0.45	137	91.4	Large plume, quenches plasma
Water	1.0	1.01	100	0	Stable plasma
Glacial acetic acid	1.06	1.21	118	40	Plume, stable plasma

evaporation, leakage, wetting characteristics, etc. However, there is some relationship and this is confirmed by the varying response of the same concentration of the same element in different solvents. The wear and tear on the tubing by the peristaltic pump by solvents such as methyl isobutyl ketone, xylene and toluene must be noted.

3.6.4 Choice of Carrier Liquid

The behaviour of solvents can be observed by monitoring the signal response of the same concentration of an element continuously over an extended period of time through a common nebuliser/spray chamber combination. The following examples were studied for the list of solvents in Table 3.5 using rapid measurements over 20 readings. The optimum parameters applied to ICP-OES for analysis using these solvents had to be altered to suit the solvent under test and this study was initiated to observe the effects of solvents using the same concentration of the same element.

The solvents listed in Table 3.5 are studied because they are used extensively in AAS analysis and were applied to ICP-OES. Demonstrating the effects of different solvents on signal response and overall behaviour on the ICP-OES were studied here; the long term use of these solvents would not be viable in a routine situation because of the unstable plasma, etc. The major problem with some solvents is that the rapid softening effect can shorten the peristaltic tubing life considerably increasing the cost of consumables, and frequent replacement is inconvenient. The deterioration of tubing was evident by the slow response and broadening of peaks and loss of sensitivity. Silicone tubing was found to be stable for a longer period of time and used in all cases as it is considered the best for its resistance to organic solvents. The best solvents are water, alcohols and glacial acetic acid.

The wavy curves illustrated in Figure 3.7 are obtained by allowing each solvent to pass through the plasma source as blanks for sufficient time before wearing of tubing and in some cases 'quenching' the plasma and producing stable signal response. If too much heat was generated around the torch it eventually quenched the torch using a safety thermal switch built in to the instrument to avoid mechanical damage to the optics and plasma jet source. Some solvents give rise to shaky plasma and most form visible 'plumes'. The unavoidable build-up of carbon at the tip of the sample injection tube is evident in most cases and this can cause distortion of the plasma with loss of sensitivity. Therefore, the combination of the shape of the plasma and carbon build-up on plasma stability is of great importance in ICP-OES analysis. The response for glacial acetic acid in Figure 3.7 shows that its behaviour is almost similar to water and it can be used as a suitable solvent for most metal analysis. A second advantage is that it will dissolve most organic and aqueous solvents. This solvent contains $\sim$54% oxygen to assist in the combustion process and is cheaply available. The disadvantages are that it is corrosive and has irritant odour effects.

3.7 Methodology of Measurement

Measurement in analytical chemistry is based on qualitative and quantitative analysis to determine one or more constituents of a compound or mixture. The methods employed are usually well tried and tested methods for a typical range of samples but new and unknown samples may need 'trial and error' tests before deciding which one is suitable for that sample matrix. The method employed should distinguish between the absolute and comparative methods and depend wholly on the nature of the substance being analysed and the importance attached to the required results. Qualitative analysis preceding precise analysis will also allow the analyst to gain knowledge of the approximate concentration of analyte present in an unknown sample before resorting to the more tedious quantitative analytical procedure. Quantitative analysis differs from qualitative analysis in that it is more tedious and time-consuming, and extreme care needs to be applied in preparing samples for analysis as well as measurements.

An absolute method is based on stoichiometric chemical reactions such as titrations (acid/base, redox, precipitation and chelometry, coulometry, voltammetry). Methods that are accepted or developed by official laboratories are usually accurate, precise and used by other laboratories throughout the world. A significant number of methods for atomic spectroscopy also fall into these categories and are readily available from the appropriate literature. Developed and accepted methods give confidence in reporting of results because all the teething problems and pitfalls associated with that method would have been observed and noted by other users. Standards must be as close as possible to the analyte in the sample and creating sample-standard similarity requires the skill of a competent analyst.

3.7.1 Choice of Standard Materials

In quantitative metal analysis, high purity metals are the best for preparing standards. The use of pure metals instead of compound removes stoichiometry as a factor that needs to be included in calculating the true concentration of the standard solution. These factors are difficult to establish with extreme accuracy for most compounds because of factors such as the stability, number of water molecules, dryness, contamination and reactivity which must taken into account before use. However, if pure metals are not available, metal compounds are used which are usually checked carefully against certified standards. Many metal standard solutions of various concentrations are available through commercial suppliers in solution form but can be very expensive. The common concentration supplied is $1.0\,g\,L^{-1}$ to $10.0\,g\,L^{-1}$ (or $1.0\,mg\,ml^{-1}$, $10.0\,mg\,ml^{-1}$). These values are usually quoted as ppm for convenience and used widely in most laboratories. Commercial standards are normally checked by other methods (e.g. nuclear activation, titration, etc.) and supplied with a Certificate of Analysis to meet most accreditation requirements. Other special concentrations are also available or especially prepared on request and they are also supplied with a Certificate of Analysis.

3.7.2 Quantitative Analysis Using Calibration Graph Method

Preparation of standards for ICP-OES is carried out by dissolving high purity metals, salts or certified standard solutions in high purity solvents to the desired concentrations to generate a working calibration curve. A calibration curve is defined as a plot or equation that describes the relationship between the concentration of an analyte and the response variable that is measured to indicate the presence and concentration of the analyte. The 'best practice' is to prepare standards in such a way that the analyte concentration in the sample will be as near the centre of the calibration curve as possible. In most cases this may not be possible because the concentration from one sample to the next may vary considerably to give results at some distance from the centre point but this will not deviate from the true result of analysis if the calibration curve is linear. Calibration curves are prepared to satisfy the Beer–Lambert Law of absorption for which the curve must be linear with zero intercept. A non-zero intercept is a function of non-analyte signals or impurities present in the sample or solvent and the slope is the result of increasing instrumental response with increasing concentration of standards.

Calibration graphs containing too high a concentration of elements may start to bend towards the concentration axis. The type of deviation from linearity is pronounced where higher concentration begins to saturate the detector. The ICP-OES has linear curves 50 to 200 times greater than that for AAS depending on the element. The non-linearity in AAS is caused by a decrease in the degree of dissociation with increasing concentration that results in a lower proportion of free atoms being available at higher concentration at a constant atomisation temperature. These effects are reduced or removed with ICP-OES as the signals obtained are based on emission rather than absorption and depend on the intensity of signals from the excited elements and they do, however, reach a maximum. Modern computers controlling the ICP-OES can generate a wealth of mathematical and statistical information about calibration curves and warn the operator when deviation from linearity is detected.

In all measurements used to generate calibration curves the following mathematical relationship is followed:

$$y = f(x) \tag{1}$$

where y is the measured analyte, x is the concentration and f is the proportionality constant function. The relationship more applicable to calibration curves generated by atomic spectroscopy is as follows:

$$y = a + b(x) \tag{2}$$

The equation follows a linear straight line if response y increases with increasing concentration of the analyte x. The value a is the intercept or blank and b is the proportionality function. The normal method for plotting curves using this equation is to plot the y values (response) on the vertical axis and the x values (concentration) on the horizontal axis. The intercept a is on the vertical axis and may be zero, or as near it as possible. It is included in signal response for standard/sample analysis and must be subtracted from each measurement.

Statistical inference is concerned with drawing conclusions from a number of observations in accordance with formalised assumptions and objective computational rules. Through statistics, trends in data may be sought and tests performed to track down non-random sources of error. Statistics can, with properly designed experiments, determine the experimental variables more efficiently than through traditional methods of holding all variables constant but one and investigating each variable in turn. Control charts are useful in evaluating day-to-day performance and identifying variations over long-term trends.

Generally, there are two classes of errors, the first is *determinate errors*, which are attributed to definite causes and are characterised by being unidirectional, e.g. positive error caused by weighing a hydroscopic sample, and it increases with sample size due to moisture absorption and varies according to time spent weighing, the humidity and the temperature. Negative errors occur when very volatile samples lose weight during the weighing in a short time. The second type is *indeterminate or random errors* caused by uncontrolled variables which are the summation of a series of small errors that can be ignored. There are also errors caused by uncertainties in the measurements which will produce a scatter of results for replicate measurements that can only be assessed by statistical analysis. A typical example is that the correction for solubility lost due to precipitation can be made but others would be introduced by changing the volume, temperature, etc.

Not all calibration curves generate a perfect straight line due to indeterminate or random errors. Most scattered points can be corrected using linear regression analysis to give an acceptable 'best fit' linear calibration curve. To establish a calibration curve (line) a set of solutions containing increasing levels of analyte are carefully prepared and usually span the concentration range of samples under test. A blank is also prepared minus the analyte. A calibration curve can be poor because of many factors but the most common are errors in preparation of standards, contamination, poor instrument calibrations, etc. Standards must be prepared fresh and must avoid storage particularly for low concentration standards. The reason is that the analyte concentration may decompose, volatilise, decrease or increase by absorption on the container walls or be contaminated by solvents used to prepare standards.

3.7.2.1 Linear Regression Analysis [11]. All atomic spectroscopy instruments use calibration curves for the majority of analysis for metals by ICP-AES and usually involve three or more standards. It is possible through the use of statistical linear regression analysis to correct for random errors and calculate the correlation coefficient of the slope, intercept, standard deviation and confidence limits of the curve generated. Random errors are associated with y values while x values are assumed to be correct because standards are certified and are prepared with a high degree of accuracy using calibrated balances, pipettes and volumetric flasks, in contamination-free solvents. Instruments are prone to a high degree of uncertainty and can give rise to errors in generating calibration curves.

Wavelength drift during analysis can also occur after prolonged use and must be corrected. If a straight line exhibits a correlation coefficient of $r = 0.999$ then the line is deemed acceptable. The following formula is generally used to calculate the r value of the calibration line:

$$r = \sum \{(x - \bar{x})(y - \bar{y})\} \quad (3)$$

$$\left\{\left[\sum (x - \bar{x})^2\right]\left[\sum (y - y)^2\right]\right\}^{1/2} \quad (4)$$

3.7.2.2 Interferences. Composition differences between standards, blanks and samples can give rise to interferences. The sample matrix can be very different from standards and must be carefully checked so that it does not deviate considerably from standards prepared separately. In practice it is very difficult to prepare all standards the same as the sample, but if the sample acts as a suppressant or enhancing effect, selecting the correct method to analyse such samples is paramount. The analyte in the sample must behave the same as the analyte in the standards or vice versa. Spectral interferences are typical in ICP-AES analysis if the matrix gives an interfering signal in the vicinity of the analyte that is very noticeable at low concentrations. To overcome this, a blank containing exactly the same amount of analyte in the sample but which does not contain the elements of interest may be used. However, in most cases this is not possible because not all samples can provide a blank matrix without the analyte. An ideal analytical method is used so that even in the presence of concomitants the measured signal is not affected. Unfortunately, this is not the case for complex matrices where the analyte is present at very low concentrations.

3.7.2.3 Quality Control of Calibration Graphs. The main objective of any analytical method is to report reliable and accurate results that are informative in supporting the quality of products, be it as a part of problem-solving, detecting anomalies in samples, contamination, forensic support in crime, etc. As part of development of a new method, stringent tests are carried out to check for trueness and accuracy of the method before it can be accepted for continuous use. The obvious procedure used to check a new method is to validate it against an established alternative method if it is available and use the statistical F- and t-test. Results of two methods are plotted using the y axis for one method and the x for the new method and the correlation coefficient plotted and the regression line are then studied. For a new method, which cannot be compared with an established method, to be accepted it must be validated using comprehensive statistical analysis. Details of a statistical approach to validating a method will be discussed later in Section 3.8.

The different graphs may be summarised as follows:

The graph in Figure 3.10(a) shows that if methods are identical then, $r = \sim +0.9998$ and $a = 0$ and $b = 1$ for the equation of the linear regression line. These results are rare because of systematic and random errors in the method. The straight line graph in Figure 3.10(b) has an $r = +0.9998$ but the intercept does not pass through the origin which indicates that one method gives higher values than the other due to the lack of background corrections. (*A possible correction of the background is by subtracting it from all measurements.*) This may be attributed to viscosity effect, poor standards preparation, poor precision of multiple measurements, etc. In Figure 3.10(c) if the regression line is less than 0.99, it means that an unacceptable error(s) has occurred.

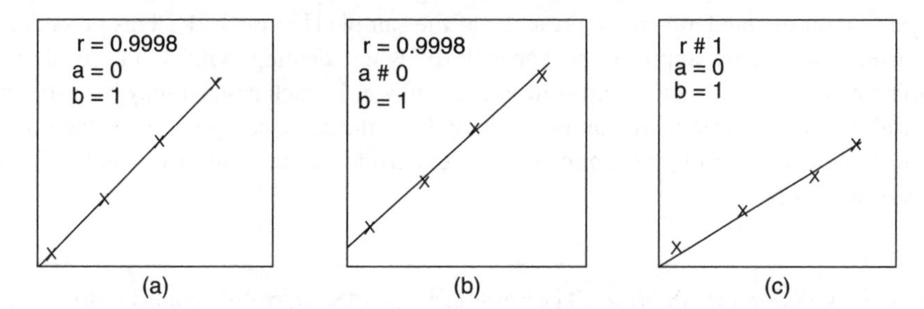

Figure 3.10 *Comparative study of types of linear regression graphs*

3.7.3 Quantitative Analysis Using Standard Addition Method

As is already illustrated in Figure 3.10, the signal given by the analyte can either be enhanced or suppressed by the presence of other sample components. Carrying out analysis within the sample matrix itself and correcting for the differences between addition and sample may solve this problem. The change in physical properties of the sample solution is noticeable when solutions containing viscous organic substances such as heavy crude oil, and thickening agents used in most chemical, industrial and household formulations give lower readings for the same concentration of metal prepared in the same dissolution solvent without the sample. It is fortunate that even when viscous or thickening agents are present very few other practical problems are encountered in the analysis other than the reduction or increase in signal response. The use of the standard

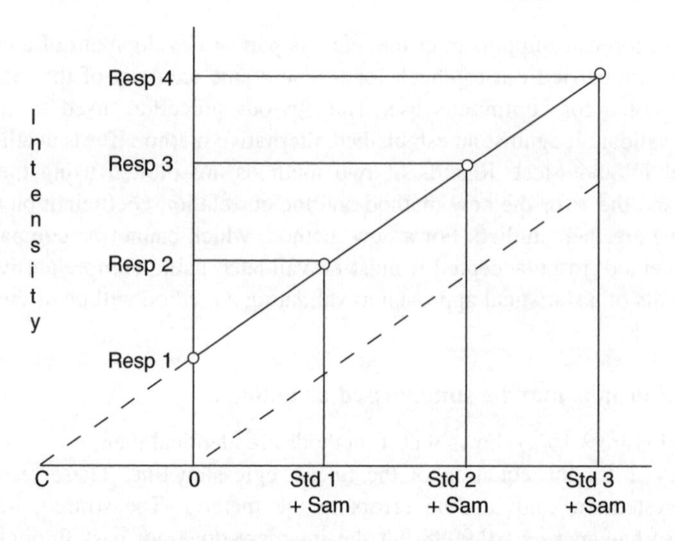

Figure 3.11 *Typical calibration curves for analysis using standard addition method. The full line represents results of sample with known added spiked standard. The dotted line represents the same curve without the sample. The extrapolating negative point for the sample can be used to determine the concentration of analyte in the sample*

addition approach is a good method for matching the sample with standards and involves analysing a set of standards in the presence of the sample (Figure 3.11). This procedure is very useful for many applications particularly when dealing with solutions that are complex and where the exact composition is unknown. In such cases it may be difficult or impossible to prepare standard solutions which have the same composition as the sample. The following is a brief description of the experimental procedure of analysis using standard addition.

Method for Standard Addition. The sample is divided into four equal aliquots into 100 ml volumetric flasks, all but one is 'spiked' with volumes of standards of increasing concentrations and diluted to the mark with solvent. Under these conditions, all the solutions differ in the analyte concentration but have the same matrix composition ensuring that the influence of the matrix will be the same for total analysis. A similar series of standards at exactly the same concentrations are prepared in the solvent only without the sample matrix and the intensity for each set of samples is measured against their concentrations. The concentration of the analyte in the sample is determined by extrapolating the plot back to the negative x axis where the concentration in the sample can be determined.

Interferences due to matrix effects can be detected by comparing the slopes of the curves for the spike sample and the pure standard solutions. In the absence of interferences both slopes should be parallel. In effect, the method is equivalent to preparing the standard calibration curve with exact matrix matching. To apply this approach to analysis sufficient sample must be available. The accuracy of the extrapolation method is never as good as interpolation methods but, if sufficient care is applied to sample/standard preparation, it can be very close to the true results. The method is rapid, relatively contamination-free and sometimes it is the only method available for some samples without resorting to other slow and tedious sample preparation techniques. See Figure 3.11 for a typical standard addition calibration curve with the corresponding curve without the sample. The following conditions apply when using this technique:

(i) There must be a linear relationship between response and concentration.
(ii) The response factor b must be constant over the concentration range.
(iii) The blank value is subtracted from each measurement.
(iv) It must not have spectral interferences.
(v) There must be no loss or gain of analyte due to container absorption or leaching.

3.7.4 Quantitative Analysis Using Internal Standard Method [12]

An internal standard is an element added at a known concentration to both standards and sample and corrects for random fluctuations of the signal as well as variations of the analyte signals due to matrix effects. The signal for the internal standard should be influenced the same way as that for the analyte in the sample. The correction of suppression or enhancement effects by the internal standard depends on the mass number, as they should be as close as feasible to that of the analyte element.

The method is based on the addition of a standard reference (internal standard) that is detected at a different wavelength from the analyte. The reference standard is added at the same concentration to samples and standards and diluted to mark in a volumetric flask. This technique uses the signal from the internal standard to correct for matrix interferences and is used with respect to precision and accuracy as well as eliminating the viscosity and matrix effects of the sample.

Consider an analytical technique for which the measured parameter y of a single analyte obeys the relationship:

$$y = kbx$$

where k is variable, e.g. temperature, volume, etc., b is detector response (constant value) and x is analyte concentration. Therefore,

$$y_1 = k_1 b_1 x_1 \quad \text{for analyte}$$

and

$$y_2 = k_2 b_2 x_2 \quad \text{for internal standard}$$

If both measurements y_1 and y_2 are obtained in the same sample, then k_1 and k_2 are equal. Therefore,

$$\frac{y_1}{y_2} = \frac{b_1 x_1}{b_2 x_2} \tag{5}$$

$$\frac{y_1}{y_2} = \frac{R x_1}{x_2} \quad \text{where } R = \frac{b_1}{b_2} \quad (R \text{ is response factor}) \tag{6}$$

When using the internal standard method, usually, in practice, a series of calibration solutions are prepared containing different concentrations of the analyte x_i, together with a constant known concentration of the internal standard x_2; the equation can then be rewritten as:

$$\frac{y_1}{y_2} = R\frac{x_1}{x_2} = R'x_1 \tag{7}$$

where R' is the quotient of the response factor R.

Calibration curves are generated using the same concentration of internal standard in different concentrations of standards (x_1).

The internal standard method can compensate for several types of errors that can be caused by sample matrix. Systematic errors due to matrix effects can sometimes be avoided. The internal standard method can also correct for fluctuations in experimental conditions: amount of sample analysed, sample introduction, emission source temperature assuming that the signal analyte and internal standard are influenced to the same extent. The main advantage of the internal method over usual calibration methods is that it can provide excellent accuracy and precision and at the same time correct for variable viscosity affects. The method is limited by the availability of a suitable reference element that behaves almost as close to the analyte under test in terms of ionisation energy, solubility, low memory effects, etc.

Method. Sample, standards and blank solutions are individually 'spiked' with a known and constant concentration of an internal standard. The internal standard corrects for most sample effects e.g. viscosity, change in nebulisation efficiency, etc., and the results are reported as if the sample behaves exactly similar to the standards. This will be expanded in Chapter 7 using an automated system.

(a) The internal standard is absent from the sample.
(b) The internal standard is soluble in sample and standards.
(c) It is pure and does not contain interfering analytes.
(d) It does not cause spectral interferences.
(e) It is as close as possible to excitation energies.
(f) Internal standard yields intensity with good reproducibility.

The purpose of using internal standard is to provide for matrix matching so that the sample and standards are measured simultaneously. Most computer software will calculate the ratios of the intensities of standards and samples and quantify the level of analyte in the sample.

3.7.5 Quantitative Analysis Using Matrix Matching Method

Analysis of complex samples by matrix matching is rare but the method can be applied if a sample with and without the metals of interest is available. If complex products are prepared for industrial, food, pharmaceutical or medicinal use and formulated with a critical concentration of metal salt it may be possible to monitor that metal by matrix matching by requesting a sample containing all additives without the metal addition. This sample can be used to prepare standards and blank, and the calibration curve generated is used to monitor the metal(s) content of the sample. This will ensure that the nebulisation efficiency for sample solution and standards solution are the same giving precise and accurate results. For complex matrices with unknown composition, complete matrix matching is not possible. This method has severe limitations and can only be applied in very special cases.

3.7.6 Quantitative Analysis Using Flow Injection Technique [13]

An automated flow injection (FI) system attached to the ICP-OES can determine major and trace elements as part of quantitative analysis. The emission signal is continuously recorded, and, after digitalization of peak height or peak area the concentration of analyte is recorded. Computer software is available for recording peaks, enhancing signal sensitivity, generating calibration curves, and printing results. Continuous nebulisation of solutions into the plasma is feasible using a FI technique and can be automated using an auto-sampler for batch operations. The first attempt to combine FI with ICP-OES was made as early as 1981 [14]; since then mathematical models have been created for using the standard addition principles involving FI-ICP-AES. Problems associated with nebulisation efficiency using FI attachment have been examined [14] and the same parameters as applied to direct nebulisation are also introduced to FI-ICP-OES. This method is equally capable of being quantitatively acceptable as direct injection.

Results for the determination of boron, molybdenum, tungsten, and zinc in non-aqueous solutions have been published [15]. Since then the principle has been extended to include other elements using a range of carrier streams and sample solution as 'plugs'. Correct selection of the carrier solvent is important with this technique as for direct analysis because the same rules apply. The combination of FI and ICP-OES means that a constant nebulisation is maintained over a longer period of time so that the plasma mechanics are not greatly affected going from one sample to another. The introduction of the sample solution as a 'plug' into the carrier stream causes a transient signal (peak) in the response, which soon decreases to the background level caused by the carrier liquid. The resulting signal compares well with the continuous nebulisation of the same sample using direct injection. This will be discussed in detail in Chapter 7.

The advantage of the combined FI-ICP-OES system allows a continuous carrier stream to transport a sample as a 'plug' to the nebuliser and to the plasma for atomisation and excitation. On its way the sample plug is partially dispersed and the degree of dispersion depends on the distance the injection point is from the nebuliser, the volume of sample, flow rate and inner diameter of the tubing. These parameters must be optimised for carrier liquids and viscosity affects as they play a major part in determining the shape of the signal response. In a given procedure experimental conditions must be kept constant for both the sample and standards and providing the viscosities, etc., are similar the concentration of analyte in the sample is easily determined. Analytical signals are usually obtained between 2 and 4 s that can lead to high standard/sample throughput so that several samples can be analysed in a short time. The disadvantage of the technique is that it will only analyse one analyte at a time using a sequential ICP-OES.

The complete system can be controlled by a computer with special software designed for this type of analysis and is available from PS Analytical (UK) as a 'Touchstone Package' (PSA 30.0). The carrier liquid and sample 'plug' is transported using a multi-roller peristaltic pump at a rate predetermined to suit the analysis of interest. Silicone tubing of 0.8 to 1.0 mm internal diameter is normally used and is found to be best for most organic solvents. A sample loop used for injection is usually in the volume range of 150 to 500 µl with an internal diameter ~0.8–1.0 mm and is made of solvent resistant plastic. The 'best' volume of loop is predetermined for a particular sample and analyte concentration.

A precision-controlled flow injection valve is necessary to keep the dispersion to a minimum and to keep the volume low and constant. A Teflon six port valve (obtained from PS Analytical, Orpington, UK, Cat No. PSA 60.00) was found to be the best. The ports are used for the loop, carrier stream, plasma supply and waste. The valve is controlled by the 'Touchstone Package'.

The FI valve is connected after the pump and as near possible to the nebuliser to reduce as much of the dispersion of the sample as possible. In normal analysis 15–20 s washout times should be sufficient between each injection to minimise memory effects, particularly for elements such as Mo, Ag, B and W.

A computer with an RS 232 interface, completed with an enhanced graphic adapter, screen and printer, controls the method. The 'Touchstone Package' data-acquisition program is designed to be user-friendly and follows a step-by-step guide through its facilities. This program can collect and process data generated as line intensities in the spectrometer and which are transferred through the serial asynchronous communication interface continuously. When measurements for standards are completed a standard calibration curve will be generated automatically, using the direct calibration, standard addition or internal standard method and this curve will be generated automatically on the screen to be used for subsequent quantitative analysis. The program may be interrupted at any time for logging in additional changes or information. The data collector allows statistical analysis to be carried out on a series of different samples by calculating the mean, standard deviation, correlation coefficient and slope of the curve. An expanded study of this technique will carried out in Chapter 7.

3.8 Validation of an Analytical Method [16]

The most important part of a new or developed method is to demonstrate its fitness of use for its intended purpose. The main criterion of testing for acceptance of a method is through a validation report with statistical support. A method must be able to provide timely, accurate and reliable results and must be relatively easy to understand and use. Validation does not rule out all potential problems, but should correct or address the more obvious ones with warnings or notifications attached. Problems increase when additional personnel, other laboratories, or different equipment are employed to perform the method but once the initial teething problems are sorted out all laboratories should be getting almost similar results. If any adjustments to a developed method are made they must be carefully noted so that future users can be informed. If these adjustments are not reported the feasibility to change is lost once the method is transferred to other users or laboratories. Validating a method cannot deviate from the original development conditions, otherwise, the developer will not know if the new conditions are acceptable until validation is performed. However, if a method does need to be changed then revalidation is necessary.

3.8.1 Method Validation of Analysis of Organic Matrices

Method validation makes use of a series of tests to determine its performance characteristics and to establish the method's acceptance for general use. The following are a list of criteria associated with the validation of a method: selectivity and specificity, linearity and calibration, accuracy or trueness, range, precision, limit of detection, limit of quantification, ruggedness and application.

3.8.1.1 Selectivity and Specificity. Selectivity and specificity are the ability to which an analytical method can measure and quantify the analyte precisely with or without interferents. These can be checked quantitatively by measuring the selectivity index A_{an}/A_{int}, where A_{an} is the sensitivity of the method and A_{int} is the sensitivity of the analyte with or without interferents. Any major differences would be an indication of negative effect by interferents and must be checked against a sample without interferents. If the interfent is too great for accurate quantitative analysis an alternative method must be used.

3.8.1.2 Linearity and Calibration Curve [17]. The linearity of an analytical calibration curve is its ability to confirm test results that are directly or mathematically/statistically proportional to the concentration of analyte in a sample. Linearity is determined by a series of measurement of standards of increasing concentration spanning 50–150% and this range is suggested to test for curvature at higher concentrations. This test should be carried out three times and results averaged. Validating over a wide range provides confidence that routine analysis is well removed from non-linear response concentrations, that the method covers a wide enough range to incorporate the limits of content uniformity testing and allows quantification analysis of unknown samples. Accepting linearity is judged by examining the parameters of the line equation. A linear regression equation is applied to the results and should have an intercept not significantly different from zero and a correlation coefficient $r = \sim 1$ and the linear curve obtained should give accurate results when analyte in samples is measured against it.

In most cases response of known increasing concentrations of standards are plotted graphically and a direct visual evaluation of whether the curve is linear is based on signal height as a function of concentration. Deviations from linearity are sometimes difficult to detect using visual inspections but by using mathematical and statistical calculations the degree of curvature can be roughly estimated.

Linearity is tested by examination of a plot produced by linear regression of responses in a calibration set. Unless there are serious errors in preparation of calibration standards, calibration errors are usually a minor component of the total uncertainty. Random errors resulting from calculation are part of run bias which is considered as a whole; systematic errors usually from laboratory bias are also considered as a whole. There are some characteristics of a calibration that are useful to know at the outset of method validation as they affect the optimal development of the procedure. Questions that need to be addressed are: is the calibration linear, does it pass through the origin or is it affected by the sample matrix?

Standard solutions are prepared at five or more concentrations covering at least 0.5–1.5 times the concentration of the sample and each concentration is analysed in triplicate to detect any curvature in the plotted data. Acceptance of a linear slope is determined by calculating its correlation coefficient which should attain a value of $r = +0.998$ or greater. If two variables x and y are related, either of two different scenarios may be recognised (both variables are subjected to comparable experimental error): one variable may be regarded as being determinable to so high a degree of precision that its uncertainty can be ignored. The second is more frequently encountered in analytical chemistry; usually we are interested in determining whether a statistically significant trend in results exists with some variable (sample size, correct volume, and temperature) that exerts only a small effect and therefore can be fixed as accurately as necessary.

The following is an example of a mathematical/statistical calculation of a calibration curve to test for true slope, residual standard deviation, confidence interval and correlation coefficient of a curve for a *fixed* or *relative* bias. A fixed bias means that all measurements are exhibiting an error of constant value. A relative bias means that the systematic error is proportional to the concentration being measured i.e. a constant proportional increase with increasing concentration.

This test was carried out to test the validity of the use of ICP-AES in the analysis of an organometallic vanadium compound (V^{n+}) using organic solvents. Validation involves either a visual or mathematical interpretation to illustrate whether the analytical method is fit for its intended use. This practical example was to prove if a method is accurate, precise, and reproducible for acceptance of a method. (This is the first step, and other steps involve a second analyst and finally an inter-laboratory study. The latter two will not be considered here.)

Method

1. A certified Conostan* vanadium alkylaryl sulphonate standard containing 5000 μg g^{-1} metal dissolved in a white mineral oil (Cas#8042-47-5) available from Conocophilips (MSDSconco155) was used as a controlled standard.
2. A stock solution of 100 μg ml^{-1} was prepared by transferring 2.0 ml of the Conostan standard above to a 100 ml grade 'A' glass volumetric flask. The standard was first dissolved in ~20 ml methyl isobutyl ketone (MIBK) and made up to mark with glacial acetic acid (GAC) and shaken to form a homogenous solution.
3. Aliquots of stock solution in step 2 were diluted to prepare 0.5, 1.0, 2.5, 5.0 and 10.0 μg ml^{-1} as standard solutions in GAC. A blank was prepared by diluting 20.0 ml MIBK to 100 ml with GAC.
4. The standards were measured on the ICP-AES using the standard conditions for this element after initial instrument set-up according to manufacturer's instructions.

*Conostan is a trade name for standards prepared in oils supplied by Conocophilips Ltd.

Results. The following calculations are examples to determine whether a fixed or relative bias is found in a calibration curve and in an attempt to separate the random variation from any systematic variations the following lines were calculated:

(a) fitting the 'best fit' line;
(b) fitting the 'best fit' regression line through the origin.

(a) Fitting the 'Best Fit' Line

Fitting the 'best fit' straight line to a set of points involves calculating the values for intercept a and slope b in the following line equation:

$$y = a + bx \tag{8}$$

The 'method of least squares' is used in calculating the slope b and intercept a using the following statistical/mathematical equations:

$$\text{Slope } b = \sum (x - \bar{x})(y - \bar{y}) / \sum (x - \bar{x})^2 \tag{9}$$

$$\text{Intercept } a = y - b\bar{x} \tag{10}$$

Table 3.6 shows the results of calculations for the slope and intercept of a line for true concentration and measured concentration.

Table 3.6 *Results of calculations for the slope of the 'best fit' line*

	True conc. x	Meas. conc. y	$(x - \bar{x})$	$(y - \bar{y})$	$(x - \bar{x})(y - \bar{y})$	$(x - \bar{x})^2$
	0.5	0.489	−3.3	−3.325	10.9725	10.89
	1.0	1.012	−2.8	−2.802	7.8456	7.84
	2.5	2.479	−1.3	−1.335	1.7355	1.69
	5.0	4.980	1.2	1.166	1.3992	1.44
	10.0	10.110	6.2	6.296	39.0352	38.44
Total	19.0	19.07			$\sum = 60.988$	$\sum = 60.3$
Mean	$\bar{x} = 3.8$	$\bar{y} = 3.814$				

$$\text{Calculation of} \sum (x - \bar{x})(y - \bar{y}) \quad \text{and} \quad \sum (x - \bar{x})^{1/2} \tag{11}$$

$$\begin{aligned} \text{Slope } b &= \sum (x - \bar{x})(y - \bar{y}) / \sum (x - \bar{x})^2 \\ &= 60.988/60.30 \\ &= 1.0114 \end{aligned} \tag{12}$$

$$\begin{aligned} \text{Intercept } a &= \bar{y} - b\bar{x} \\ &= 3.814 - 1.0114(3.80) \\ &= -0.028 \end{aligned} \tag{13}$$

The equation for the 'best fit' straight line from these calculated values is $y = -0.028 + 1.0114x$. This equation suggests that a constant error of −0.028 is evident regardless of the true concentration and this is a fixed bias of −0.028 and a relative bias of 1.14%. Using these figures it is possible to calculate the bias at any particular concentration.

The relative bias gives an error of −0.0223% for 0.5 ppm vanadium and 0.086% for 10.0 ppm vanadium metal and this shows that the relative bias exerts a greater influence on the determination than does the fixed bias (see Table 3.7). The estimated error due to relative bias is calculated by taking differences between 1.0114 and 1.000 of the perfect line and correcting for each concentration is used to calculate each predicted concentration. Table 3.7 gives results obtained along with each residual.

***Table** 3.7 Results of bias, errors and residuals for 'best fit' line through the centroid*

True conc.	Estim. error 'fixed bias'	Estim. error 'rel. bias' *(0.0114x)*	Estim. total error due to bias	Predicted conc.	Measured conc.	Residual	Residual [2]
0.5	−0.028	0.0057	−0.0223	0.4777	0.4870	0.0093	0.00009
1.0	−0.028	0.0114	−0.0166	0.9834	0.992	0.0086	0.00007
2.5	−0.028	0.0285	0.0005	2.5005	2.4970	−0.0035	0
5.0	−0.028	0.0570	0.0290	5.0290	5.105	0.0760	0.0058
10.0	−0.028	0.1140	0.0860	10.086	10.125	0.0390	0.0015
Total							0.0075

The total error is obtained by adding the fixed bias error and the relative bias error and adding this error to the true concentration returns the predicted concentration. Each predicted concentration is close to the measured concentration and the calculated differences are residuals. The residuals are estimates of random errors in the determination and can be used to estimate the precision of the test method. The residual sum of squares indicates how well the line fits the points, and estimates of random errors. Calculation of estimates can be determined by first converting the residual sum of squares into variance and then to standard deviation. To obtain this, divide the sum of squares by degrees of freedom and then calculate the square root to obtain the residual standard deviation. Two degrees of freedom are lost because both the slope and intercept are estimated by dividing the sum of squares by $(n-2)$, which will give residual variance; calculation of its square root will give the residual standard deviation *(estimate of precision).*

$$\text{Residual variance} = (\text{residual sum of squares})/(n-2) \tag{14}$$
$$= 0.0075/3$$
$$= 0.0025$$

$$\text{Residual standard deviation} = (\text{residual variance})^{1/2} \tag{15}$$
$$(\text{estimate of precision}) = (0.0025)^{1/2}$$
$$= 0.05$$

These values can be used to calculate confidence intervals for the *true intercept* and *true slope.* Multiple repeats of this experiment would also report different values of slope (b) and intercept (a) but all results would be in close proximity to each other.

A 95% confidence interval (CI) for the *true intercept a* can be calculated as follows:

$$\begin{aligned}\text{CI of true intercept} &= a \pm t(\text{ESD})(1/n + \bar{x}^2)/(x - \bar{x})^2 \qquad (16)\\ &= -0.028 \pm 3.18(0.05)(1/5 + 3.8^2)/(60.3)\\ &= -0.028 \pm 0.07\\ &= -0.098 \Leftrightarrow 0.042\end{aligned}$$

where $a = -0.028$, $t = 3.18$ (3 degrees of freedom and 5% significance), RSD = ESD = 0.05, $n = 5$, $\bar{x} = 3.18$ and $(x - \bar{x})^2 = 60.3$ from Table 3.7.

The *true intercept* is 95% confidently between −0.098 and 0.042 and since this interval includes zero therefore it may be possible that no fixed bias is evident. The confidence interval, the *true slope b* of a regression line, is given by:

$$\begin{aligned}\text{CI of true slope} &= b \pm t(\text{ESD})/(x - \bar{x})^{1/2} \qquad (17)\\ &= 1.0114 \pm 3.18(0.05)/(60.3)\\ &= 1.0114 \pm 0.026\\ &= 1.014 \Leftrightarrow 1.0088(1.4 \rightarrow 0.9\%) \text{ from the perfect line}\end{aligned}$$

where $b = 1.0114$, $t = 3.18$ (3 degrees of freedom and 5% significance), RSD = ESD = 0.05, $n = 5$, $\bar{x} = 3.18$ and $(x - \bar{x})^2 = 60.3$ from Table 3.7.

The above calculation states that the true slope lies between 0.9% and 1.4% which shows that a very small systematic increase is observed. The error at 0% is not included, and based on these calculations a small relative bias may exist.

(b) Fitting the 'Best Fit' Regression Line Through the Origin

The calculation above was unable to prove whether a fixed bias exists and the best way to estimate this is to fit the equation $y = bx$ forcing a zero intercept (Table 3.8).

$$\text{Slope } b = \sum xy / \sum x^2 \qquad (18)$$

$$\begin{aligned}\text{Slope} &= \sum xy / \sum x^2 \qquad (19)\\ &= 133.45/132.5\\ &= 1.007\end{aligned}$$

The equation for the 'best fit' line through the origin is:

$$y = 1.007x \qquad (20)$$

Table 3.8 *Calculation of slope for relative bias*

True conc. x	Measurement y	xy	x^2
0.5	0.489	0.2445	0.25
1.0	1.012	1.012	1.0
2.5	2.479	6.1975	6.25
5.0	4.98	24.0	25.0
10.0	10.11	101.1	100.0
		Sum = 133.45	Sum = 132.5

Table 3.9 *Results of bias, errors and residuals for 'relative bias'*

True conc.	Estimated error due to rel. bias (0.007x)	Predicted measurement	Measured conc.	Residual	Residual [2]
0.5	0.0035	0.5035	0.489	−0.0145	0.0002
1.0	0.0070	1.0070	1.012	0.0050	0
2.5	0.0175	2.5175	2.479	−0.0385	0.0015
5.0	0.0350	5.0350	4.980	−0.0550	0.0030
10.0	0.0700	10.070	10.110	0.0400	0.0016
				−0.063	0.0063

This line indicates a relative bias of 0.7% which compares well with the 'best fit' line of 1.14% estimated by the best straight line through the centroid. To compare the differences it is necessary to calculate the residual for this line as determined in Table 3.9.

The residual standard deviation is calculated by dividing the sum-of-squares by its degree of freedom, followed by the square root. Four degrees of freedom are used here as only one is lost while fitting the line through the origin therefore it contains no intercept, $t = 2.78$. Hence,

$$\text{Residual standard deviation} = (0.0063/4)^{1/2} \tag{21}$$
$$= 0.04$$

Similar for 'best fit' line, RSD = ESD.

The confidence interval for the slope of the line can be calculated as follows:

$$\text{CI} = b \pm t(\text{ESD})/(x^2)^{1/2} \tag{22}$$

$$1.007 \pm 2.78(0.04)/(132.5)^{1/2}$$

$$= 1.007 \pm 0.01$$

$$= 1.017 \Leftrightarrow 0.997$$

A 95% confidence interval of the true slope is between 1.017 and 0.997, therefore the relative bias will lie between 1.7% and 0.7%. In this case the interval is wider than the line through the centroid even with smaller *t-value* and *residual standard deviations*, and the fact that $\sum x^2$ was used instead of $\sum(x - x)^2$, and because of this it is expected that the line through the origin would give a narrower confidence interval. There are many reasons for this and they are beyond the scope of this book.

For a set of data to which regression analysis can be used, the correlation coefficient (a measure of linear association between two variables) can also be calculated using the formula:

$$\text{Correlation coefficient} = \sum (x - \bar{x})(y - \bar{y})/[(x - \bar{x})^2(y - \bar{y})^2]^{1/2} \tag{23}$$

$$= 60.988/[(60.3)(61.71)]^{1/2}$$

$$= 0.9998$$

This correlation coefficient shows how well the regression line fits the points and hence percentage fit as follows:

$$\begin{aligned} \text{Percentage fit} &= 100(\text{correlation coefficient})^2 \\ &= 100(0.9998)^2 \\ &= 99.96\% \end{aligned} \tag{24}$$

The sample-to-sample variation of the true concentration is 99.96% of variation in measured concentration, hence errors are small.

3.8.1.3 Accuracy or Trueness. The accuracy or trueness is a measure of closeness of the determined value of analyte to the true or added known value. The accuracy may be compared with an alternative proven method or other internationally recognized and validated method. The accuracy is accepted once precision, linearity and specificity have been determined and results are within the control parameters and capability of the test. Samples with impurities or other variable matrix can affect accuracy and precision. Spiking with known amounts of same impurities (if known) to check response can check their effects in the sample. As part of method validation, accuracy should be determined across the specified range of samples with variable matrices. A minimum of three or four standards containing the sample/matrix under investigation should be compared with standards prepared without the sample. The measured results should be close to the added value and the percentage recovery calculated. The recovery should be close to 100% to determine whether the matrix has a detrimental effect on signal response or not.

3.8.1.4 Range. The validated range is defined as the interval between the upper and lower concentration of analyte and must be such that it can give acceptable accuracy, linearity and precision. The criterion to determine an accurate concentration is that if the measured value falls between 90% and 110% of the true value in sample it may be acceptable as an accurate result. The validation study should operate the range close to the value in which the validation is being carried out and bearing in mind that the estimated uncertainty holds true.

3.8.1.5 Precision. The precision of analysis is important and can change as a function of the analyte concentration. The graph in Figure 3.12 gives an indication of change of percentage RSD with concentration.

The percentage RSD imprecision increases significantly as the concentration decreases towards detection limits of analyte, as shown in Figure 3.12. Higher variability is expected as the analyte approaches the detection limits for the method because of its poor precision at this level, and for which a decision would have to be made as to what level the imprecision becomes too great for acceptance of results in the presence or absence of interference.

Precision is the closeness of results between a series of measurements using the same homogenous sample and may be subdivided into three categories:

1. Repeatability as determined under the same operating conditions in one day.
2. Repeatability as determined with the same laboratory over three different days, with a different analyst, and if possible different equipment.
3. Reproducibility as determined in different laboratories (collaborate study).

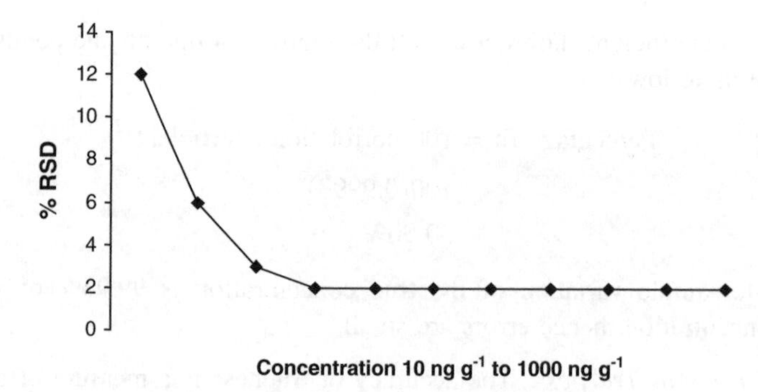

Figure 3.12 Percentage of RSD for precision of concentration range

3.8.1.6 Limit of Detection. Limit of detection is defined as the lowest amount of analyte that can be detected but not necessarily quantified in the sample. It is generally accepted by most international measurements standards that limit of detection is based on three times the standard deviation of the baseline noise, i.e.

$$DL = 3\sigma \tag{25}$$

where σ is the standard deviation of the baseline noise.

(i) The value σ may be determined from the standard deviation of the noise of the blank sample.
(ii) The value σ can also be determined by calculating the residual standard deviation of the regression line or standard deviation of the y intercepts of the regression lines.

The calculated detection limits may be further validated by the analysis of a suitable number of samples near this value.

3.8.1.7 Limit of Quantification. The limit of quantification is the lowest amount of analyte that can be quantitatively determined in a sample with a high degree of precision. The limit of quantification is usually calculated by multiplying the standard deviation of the baseline noise by ten, i.e.

$$QL = 10\sigma \tag{26}$$

where σ is the standard deviation of the baseline noise.

Similar to detection limits this value may be validated by several analyses near the limit of quantification of the analyte.

3.8.1.8 Ruggedness of the Method. The ruggedness of the method is the ability to maintain correct results when minor deviations from the method may occur during analysis. This can be studied by changing the parameters such as a different operator, different brand of solvent, different supplier of standards, concentration of standard, time between each measurement and slight change in carrier flow rate, change in gas flow, solution stability and any other deliberate variations in the method parameters. Other factors over which the operator has no control include wavelength drift, electronic noise and change in temperature, etc.

3.8.1.9 Application. The application of the method should indicate the analyte and reagent's purity, its state (oxidation, ligand, etc.) where appropriate, concentration range, effects, if any, with sample matrix, description of sample(s) and equipment, and procedures including permissible variation in specifications of samples under tests, etc.

3.9 Control and Range Charts

A control chart (see Figure 3.13 for an example) is a graphical illustration that measures a control sample at intervals and determines if corrective action needs to be taken to ensure that results are precise. These charts are sequential plots of some quality characteristic. It may be a day-to-day plot of the average concentration of a stable solution of a standard solution. The chart consists of upper and lower lines called control limits, and these lines are each side of the mean value and they are such that any measured value outside them indicates that the analysis is out of control. The 'out of control' could be associated with poor calibrations of the instrument at the outset of the analysis, the standards and control standards are prepared incorrectly or contamination occurs in either standards or control sample. The main purpose of control charts is the plotting of a sequence of points in order to make a continuous record of the quality characteristics. Plot trends of data or loss of precision can be visually evident so that the causes may be investigated. Control charts are plotted for many applications and the most common are for average and range observations. The control chart for the mean monitors the accuracy, with the target value being the known concentration of analyte in the standard. These charts are used to see if data are in statistical control, and may be regarded as random samples from a single population of data. The test for randomness with the control chart may be useful in seeking out sources of error in laboratory data, supporting the quality in manufacturing production or control of an analytical method.

The individual observations in sequential order are compared with control limits established from a past measurement to generate the control chart in the initial case. If the mean value $\bar{x}$ and standard deviation σ of a constant quantity have been established from 15 to 30 measurements, these quantities may be regarded as valid estimates of $\bar{x}$ and σ for the population. Limits of 95% and 99.8% which are based on $\pm$ 1.96σ and 3.0σ are usually designated as warning and action limits and these limits are usually set, and are based on the sensitivity and importance of the measurements. Special attention should be paid to one-sided deviations from control limits, because systematic errors cause deviation in one direction and may indicate an abnormally wide scatter. Therefore laboratories, production, test methods, or operator can be checked for consistency of measured results.

Control charts are designed to incorporate the entire process of the analytical measurement from sampling variability, instrument stability, calibration standards and sample preparation. The chart presents data in a framework that clearly shows whether corrective actions are necessary to ensure that results reported are correct, and allows extrapolation from sample results to conclusion about the whole population with known risks of error or misinterpretation.

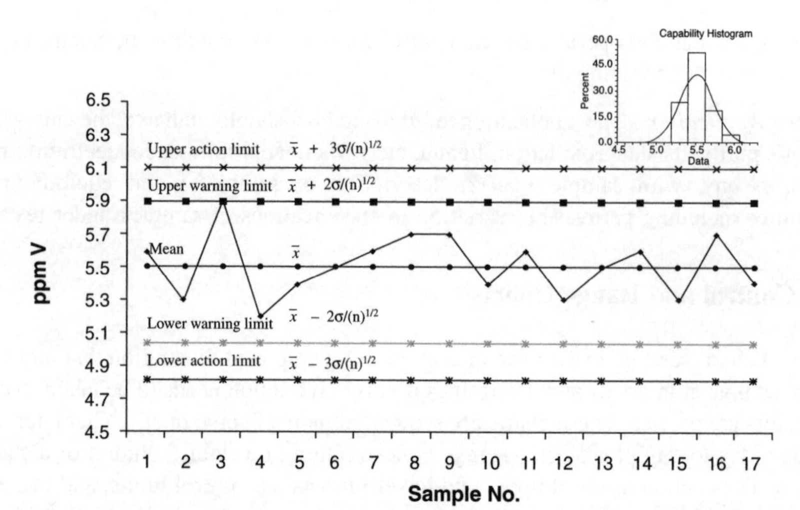

Figure 3.13 *Typical control chart showing upper and lower warning and action lines and capability histogram [15]*

A range control chart (see Figure 3.13) is a data analysis technique to test whether or not a measurement process is out of statistical control. These charts monitor the precision and the target value is the process capability. The range chart is sensitive to change in variation in the measurement process. It consists of a vertical axis for the range for each group and the horizontal axis is the sub-group designation. Horizontal lines are drawn at the mean value and at the upper and lower control limits. This chart is very similar to the chart in Figure 3.14 in which the range for each sample is plotted and compared with the predetermined limits. A serious fault can lead to Gaussian distribution, illustrated in Figure 3.15, where the process collapses from form A to form B, e.g. caused by change in viscosity of a solution, poor sample preparation, poor standard calibration curve, etc. The range of sample from B will have higher values than A. A range chart should be plotted in conjunction with the mean control chart.

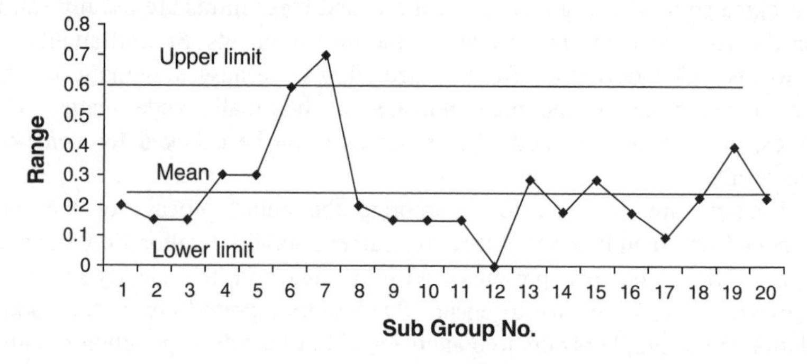

Figure 3.14 *Typical range chart showing the upper limits and lower limits*

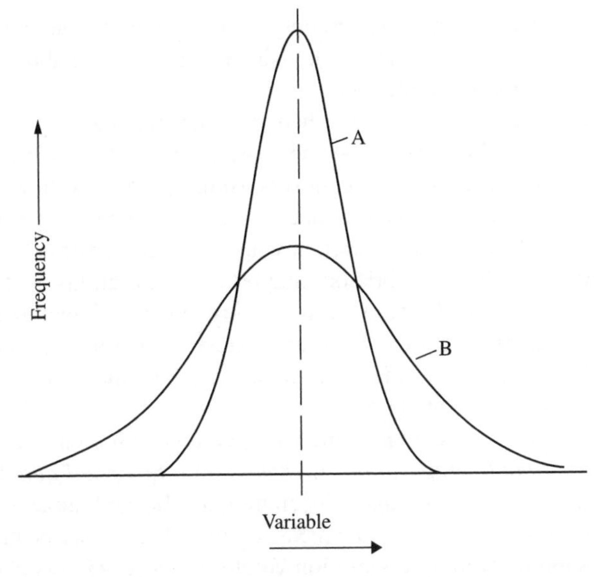

Figure 3.15 *Gaussian distribution of spread of analytical variables for higher and lower range values*

The idea of monitoring accuracy and precision was developed by Walter Shewhart [18] and the target value here was the known concentration of analyte in a control standard. The range graph monitors the precision, and the target value is the capability that it is necessary to establish in order to set up the control chart. Process capability will be limited by the random errors involved in measurements rather than error in preparing the standards.

3.10 Brief Outline of Measurement Uncertainty

> 'Measurement uncertainty is a parameter associated with the results of a measurement, that characterises the dispersion of the values that could reasonably be attributed to the measurand'. (Unknown source)

The parameter could be standard deviation or width of a confidence interval.

In analytical sciences, measurement uncertainty narrows down the differences between the actual measured value and the true value of a concentration of analyte. The actual measured value consists of two parts, the estimate of the true value and uncertainty associated with this estimation. Uncertainty of measurement is made of every component that is critical to the measured value, some of which may be evaluated from statistical distribution of the results of a series of measurements and can be characterised by standard deviations. The actual measured value does not coincide with the true value and may be considered as an estimate that may be larger or smaller than the true value. This is not an error but rather an inherent part of any measurement. Hence it is true to state that:

$$\text{Measured value}(M) \pm \text{Measurement uncertainty}(U) \qquad (27)$$

Determination of measurement uncertainty must be carried out with accepted and recognised methods that have already been validated and these methods are intrinsically tied to all possible errors in the method.

A measured value is complete only when it is accompanied by a statement of its uncertainty and is required in order to decide whether or not the result is adequate for its intended purpose. The uncertainty value must be suitably small to show that the reported results can be accepted with confidence and to ascertain whether or not it is consistent with similar results. There is an uncertainty in the concentration of the calibration samples used both in synthetic calibration samples and calibrations of standard addition. Weighing and volumes, which are a must in most analytical methods, must include weighing errors; volumes must include volume errors to take into account uncertainties associated with these steps of the analysis. These and others must also be included in the overall calculation of the analytical error.

An analytical method can be divided into two parts when it comes to determining the uncertainty value – *sample preparation* and *measuring the actual quantity.* The sample preparation consists of several stages depending on the technique used in order to achieve a solution suitable for the actual measurement. These may be broken down into homogeneity of sample, weight, dissolution volumes, matrix effects chemical interferences, volatility, ashing, digesting, extraction, complex, etc. The uncertainties associated with the actual measurement need to take into account wavelength stability, electronic noise, detector noise, gas flow variation, pump speed variation, computing records, etc.

It is important to note the differences between *error* and *measurement uncertainty.* Errors are *differences* in measurements while uncertainties are a *range of measurement.* The importance of measured uncertainty values quoted with the results improves the reliability of a result and adds confidence in the decision and reporting of the same. Knowledge of the uncertainty value also give credence to strive to reduce the uncertainty value associated with measurements which will facilitate better and more confident reporting.

In estimating the overall uncertainty, it may be necessary to take each source of variance and treat it separately to obtain the contribution from the source. Each of these is referred to as an uncertainty component. Expressing this as a standard deviation, an uncertainty component is known as a standard uncertainty. For a measured result y, the total uncertainty, called combined standard uncertainty and denoted as $u_c(y)$, is an estimated standard deviation equal to the positive square root of the total variance obtained by combining all the uncertainty components, using a propagation law of components.

Calculation of combined *standard uncertainty* from the individual components is carried out using $u_c(y)$ of a value y and the uncertainty of the independent parameters x_1, $x_2, x_3, \ldots, x_n$ on which it depends is:

$$u_c[y(x_1, x_2 \ldots)] = \left[\sum c^2 u(x_i)^2\right]^{1/2} = \left[\sum u(y, x_i)^2\right]^{1/2} \tag{28}$$

In combining the total variance for a sample preparation and analysis each has their own list of uncertainties. The overall uncertainty (U_o) at a specific confidence limit is selected and the value calculated using:

$$U_o = Z\delta/\sqrt{n} \tag{29}$$

where δ is the standard deviation of the measurement, Z is the percentile of standard normal distribution and n is the number of measurements.

If the variance due to sample preparation is negligible (i.e. $\delta_s^2 = 0$) and most of the uncertainty is due to the analytical stage only, then:

$$n_a = (Z\delta_a/U_a)^2 \tag{30}$$

If the uncertainty due to the analysis is negligible (i.e. $\delta_a^2 = 0$) then:

$$n_s = (Z\delta_s/U_s)^2 \tag{31}$$

If both sample preparation and analysis are contributing significantly to measurement uncertainty then the overall measurement uncertainty is:

$$U_o = Z(\delta_s^2/n_s + \delta_a^2/n_a)^{1/2} \tag{32}$$

Care must be taken that n_s and n_a must be chosen based on experimental experience and judgements.

The following simplified equation may also be to calculate the uncertainty:

$$U_o = 2[(\Sigma S_i)^2]^{1/2} \tag{33}$$
$$(n)^{1/2}$$

where ΣS_i is the summation of all significant estimated errors. The most common are weighing, dilution volume, measurement, electronic noise and quoted errors on controls, standards and solvents.

The expanded uncertainty is the final calculation and it involves multiplying the combined uncertainty by a chosen coverage factor k, and is required to provide an interval which may be intercepted to include a large fraction of the distribution of values which could be attributed to the measurand. The coverage factor can be considered by the level of confidence required and underlying distributions. For most purposes the k value is set to 2, particularly if the number of degrees of freedom is greater than 6. If fewer than 6, then the k value should be set equal to the two tailed value of the Student's test for the number of degrees of freedom associated with that contribution.

Figure 3.16 is a basic flow chart for the estimation of measurement uncertainty.

Estimation of a measurement uncertainty value is a simple process provided the following rules are applied:

1. *Select measurand.* Note what is being measured including measured quantities, constants, and calibration standards upon which it depends.
2. *Identify uncertainty components associated with sample preparation.* List the possible sources of uncertainty as described above.
3. *Quantify uncertainty associated with measuring method.* Calculate the value of the uncertainty component associated with each potential source ensuring that all sources are taken into consideration.
4. *Calculate combine uncertainty and expanded uncertainty.* All calculations carried out in step 3 must be expressed as standard deviations, and calculated as combined standard uncertainty. If necessary, a coverage factor should be applied to give the expanded uncertainty.

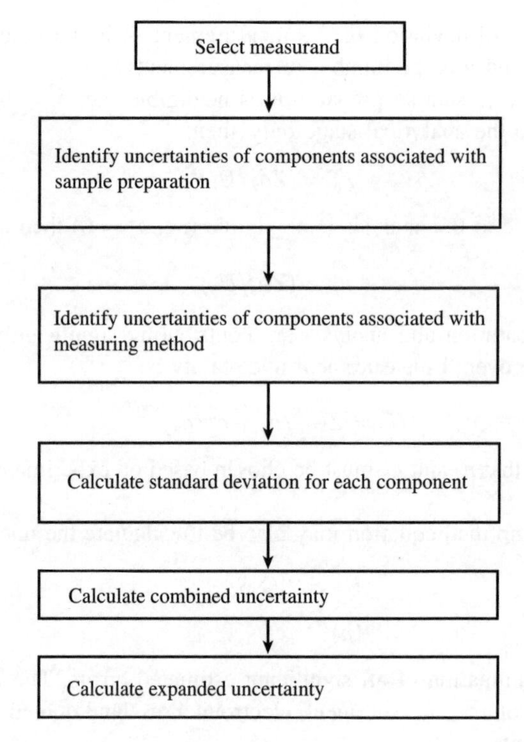

Figure 3.16 Schematic diagram of a flow chart for the estimation of measurement uncertainty in an analytical measurement

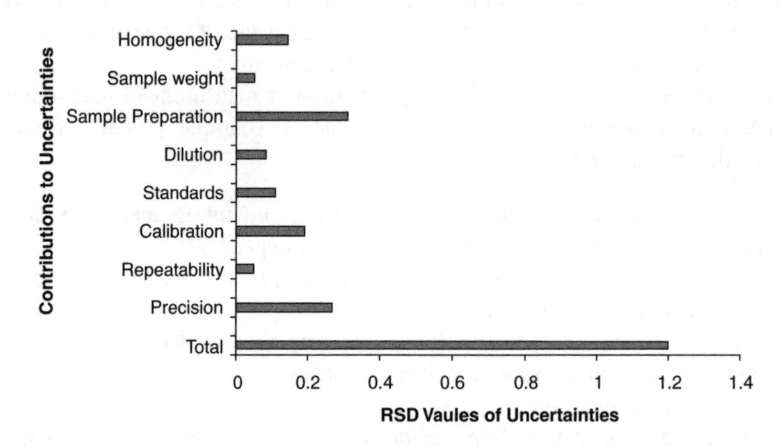

Figure 3.17 Graphical illustration of measurement uncertainty for individual sources with an analytical method for metal analysis using atomic spectroscopy

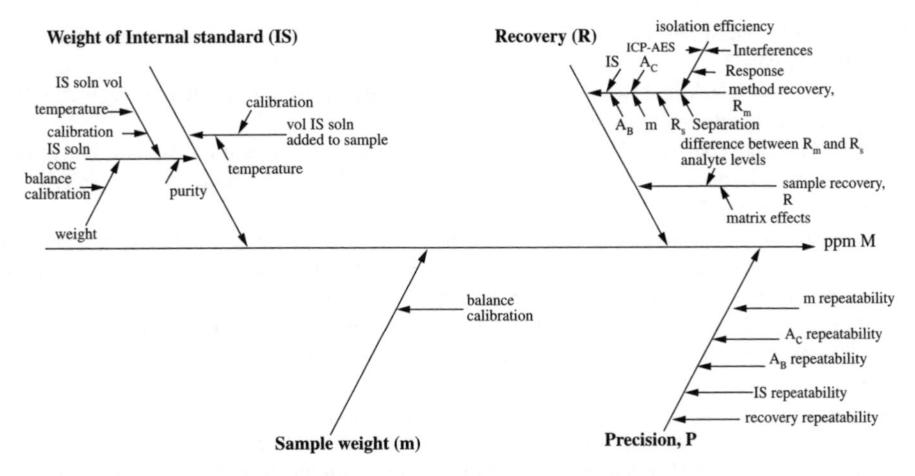

Figure 3.18 *Fish bone diagram showing the important stages for metal analysis using ICP-OES from sampling through to reporting. All steps are not applicable to every analysis but each method can have a similar design associated with that method*

Figure 3.17 is a graphical illustration of uncertainty components associated with metal analysis using atomic spectroscopy.

The 'fish bone' diagram shown in Figure 3.18 illustrates the main points that need to observed in sample analysis at each stage from sampling to reporting results.

References

[1] Meites, L. (Ed.) (1963) *Handbook of Analytical Chemistry*, 1st edition, London: McGraw-Hill.

[2] Keith, L.H., Crumpett, W., Deegan, J. Jr, *et al.* (1983) *Analytical Chemistry*, **55**, p2210. The interface between analytical chemistry and the law is discussed by: Harris, E. (1992) *Analytical Chemistry*, **64**, p665A; Gasselli, J.G. (1992) *Analytical Chemistry*, **64**, p667A; and Kaffner, C.A. Jr, Marchi, E., Morgado, J.M. and Rubio, C.R. (1996) *Analytical Chemistry* **68**, p241A.

[3] Malissa, H. and Schoffmann, E. (1955) Low pressure oxygen combustion, *Mikrochimica Acta*, **1**, p187.

[4] Kingston, H.M. and Jassie, L.B. (1988) *Theory and Practice: Introduction to Microwave Sample Preparation*, American Chemical Society, pp7–29.

[5] Berthelot, M. (1892) Combustion of samples using poor oxygen at elevated temperatures, *Ann. Chim. Phys.*, **26**(6), p555.

[6] Hempel, H. and Angew, Z. (1892) Combustion of samples using oxygen at elevated pressures, *Chemistry*, **13**, p393.

[7] Schöniger, H. (1955) Preparation of organic compounds for analysis using an enclosed combustion flask, *Mikrochimica Acta*, **1**, p123.

[8] Berthelot, M. (1892) Combustion of samples using pure oxygen at elevated temperatures, *Ann. Chim. Phys.*, **6**(26), p354.

[9] Gorsuch, T.T. (1959) Sample preparation using dry ashing method, *Analyst* (London), **84**, p135.

[10] Babington, R.S. (1962) 'Method of atomising Liquid in a Mono-Dispersed Spray', US Patent No. 3,421,692.

[11] Miller, J.C. and Miller, J.N. (1988) *Statistics for Analytical Chemistry*, 2nd edition, Chichester: Ellis Horwood.
[12] Barnett, Fassel, V.A. and Knisely, R.N. (1968) Theoretical principles of internal standardisation in analytical emission, *Spectrochimica Acta*, **23A**, p643.
[13] Ruzicka, J. and Hansen, E.H. (1988) *Flow Injection Analysis*, 2nd edition, Volume 62: A series of monographs on application of analytical chemistry and its applications, London: John Wiley & Sons, Ltd, pp15–19.
[14] Brennan, M.C. and Svehla, G. (1989) Flow injection determination of boron, copper, molybdenum, tungsten and zinc in organic matrices with direct current plasma optical emission spectrometry, *Fresenius Zeitschrift fur Analytische Chemie*, **335**, pp893–899.
[15] Brennan, M.C. and Svehla, G. (1992) *Novel Electroanalytical and Atomic Spectrometric Techniques in the Characterisation of Anaerobic Adhesives*, Cork: University of Cork, pp247–340.
[16] Thompson, M. (2000) Validation study, *Analyst* **125**, 2020–2025.
[17] Caulcutt, R. (1983) *Statistics in Research and Development*, London: Chapman & Hall.
[18] Shewhart, W.A. (1921) *Economic Control of Quality of Manufactured Product*, New York: Von Nostrand.

4

Analysis of Plastics, Fibres and Textiles for Metals Content Using ICP-OES

4.1 A Brief History of Natural and Synthetic Plastic Materials

Throughout history man has made use of natural polymers and fibres such as waxes, bitumen (tars) and horn material. Before the age of plastics, materials such as wood, metal, stone, ceramic and glass were commonly used to make items of general use. In the 18th century the properties of these materials were improved by purification and modification to make several advanced household and industrial items. By the 19th century, with the exploitation of chemical scientific knowledge and demands from industry for new materials not provided by nature, the stage was set for the development of a range of new materials; the discovery of plastics had a huge impact.

Charles Goodyear [1], an American scientist, worked with gutta-percha, a gum from natural tropical trees, and Thomas Hancock, a British scientist who simultaneously and independently developed a process for the vulcanisation of rubber (1839) by reacting it with sulphur and heat, are credited with the first deliberate attempt to chemically modify a natural polymer to produce a moulding material. Gutta-percha was used to protect and insulate the first submarine telegraph cables. The combined and independent efforts of these men helped to lay the foundation for the manufacture of synthetic materials using chemistry.

At the International Exhibition of 1862, Alexander Parkes [2] first introduced an organic derivative of cellulose that could be moulded when heated and retain that shape when cooled. He invented the treatment of cellulose fibres with nitric acid forming cellulose nitrate that was the first semi-synthetic plastic material. He used this to make decorative brooches, trinkets and knife handles. He claimed that it could do anything that rubber could do but could be produced at a lower price. He formed a company making a

A Practical Approach to Quantitative Metal Analysis of Organic Matrices Martin Brennan

range of products but, ironically, it failed due to high costs. In 1876, the Merriam family set up the British Xylonite Company who added the plasticiser camphor to cellulose nitrate that enabled the mixture to be moulded by rollers to form different shapes. These materials were considered to be the best substitute for a range of scarce natural products and the company went on to achieve success making items such as combs, collars, cuffs and a host of other popular and necessary items for everyday use.

In 1909 the first true plastic was developed by reacting phenolic-formaldehyde (Bakelite) in the presence of a (propriety) catalyst. It was done by L. Hendrik Baekeland [3] who coined the word 'plastic', a substance he created from coal tar. The material had excellent heat resistance and low electrical conductivity and when blended with mica, clays, asbestos, etc., had considerable strength and resistance. In recent years, these materials have been used as part of manufacture to make casings for clocks, toasters and radios, among other things.

The flammability and explosive nature of cellulose nitrate had prevented its use for mass production and moulding techniques. The development of cellulose acetate in ~1925 solved this problem and it found uses as safety film and was doped to stiffen and make waterproof fabric to be used on the wings and fuselage of early aeroplanes. Later development saw it being sold as a moulding powder in various degrees of hardness that could be quickly and economically moulded into shapes by special design injection mouldings – a key process in plastics technology.

The 1930s saw the introduction of the 'poly' generation and the first of many such thermoplastics was poly(vinyl chloride) or PVC which became commercial reality with the introduction of a plasticiser. At about the same time Du Pont Chemicals also launched the polyamide nylon 66 after studying the network structure of silk. A few years later German researchers developed nylon 6 from caprolactam. In the UK, ICI developed and produced polyethylene, a material vital to the success of radar technology during the Second World War. ICI also made a valuable wartime contribution with the development of poly(methyl methacrylate) or PMMA which was used to make shatterproof and protective screens.

Later (1935–45), new materials such as silicone were developed as water repellent and heat resistant paint. The development of epoxy resins offered a structural material for boat and car bodies. Poly(tetrafluoroethylene) (PTFE), polycarbonate, poly(ethylene terephthalate) (PET), polypropylene, polyurethane, ABS and acetals are the latest additions to find their way into plastics technology. Studies have also been carried out with the use of fillers and plasticisers as part of the next generation of materials.

Plastics are a large family of materials that can be softened and moulded by heat and pressure and they all consist of giant molecules built up from thousands of smaller molecules, combined to form repeating structures. They are composed of polymer/large molecules consisting of repeating units called monomers. In the case of polyethylene they are polymerised from the ethylene molecules and form long chains of carbon atoms in which each carbon is bonded to two hydrogen atoms. These polymers can be in the form of LDPE (low density polyethylene), LLDPE (linear low density polyethylene) and HDPE (high density polyethylene). The major difference is the degree of branching of the polymer chain. Both LLDPE and HDPE are linear, unbranched chains and the LDPE chains are branched. Some of the polymers are made using a transition metal catalyst, e.g. $TiCl_3$ (Zeigler-Natta metallocene catalyst). LDPE is made using a peroxide (benzoyl peroxide which also requires trace transition metals to promote the free radical peroxide formation) to initiate the polymerisation. Polymers can have 10 000 to 100 000 atoms or more in their molecules. Modern plastics are synthetic materials made from chemicals derived from coal or oil.

Advances in electronics and automotive engineering depend heavily on plastics, e.g. computers, vehicles, etc. The aerospace industry would grind to a halt without the advancement of plastic composites. The ability of plastics to be moulded into very complex shapes gives the design engineer the opportunity to reduce the cost of manufacture and assembly. Thermoplastic and thermosetting plastics reinforced with glass, carbon and fibres are used extensively in racing boats, aircraft and cars.

A modern series of new plastics are based on transition metals (e.g. Fe, Ti, Cr, Zn, V) to form polymers and possess unusual properties such as variable oxidation states, and ligand exchange on the metal atom. They have reduced UV absorption and visible radiation and exhibit electrical conductivity. Examples include cyclopentadienyl and arene metal π polymeric complexes that act as electron rich aromatic system and are very reactive to a range of monomers to form polymers.

Monomeric vinylic metal π complexes undergo polymerisation reactions to form polymers that have limited applications because of their poor thermo-mechanical properties but they absorb UV without degrading in the process. Such polymers can be used as a UV-resistant coating and act as catalysts in other monomer reactions. Metallocene methylene polymers are high temperature resistant materials and have uses as ablative materials for space capsule heat shields. These materials use Fe and Ru in their metal polymers.

Specific polymers containing trialkyltin esters (or trialkyl amino tin) can be used as cross linking agents, and can be used to prepare polymers by radical initiated co-polymerisation of monomers such as methylacrylate and methyl methacrylate to form special useful polymers. They are used as long-term anti-fouling agents in paints to prevent growth of fungi and barnacles on ship bottoms and shore installations. These tin polymer complexes hydrolyse slowly to release, at a controlled rate, minute amounts of trialkyltin (stannanols) which are toxic to marine organisms. The rate of release is slow enough to be effective in destroying organisms but is insufficient to harm marine life.

Organometallic condensation polymers behave as organic acid chlorides when prepared by copolymerising a difunctional metal halide with a difunctional Lewis base that contains a methylene. These polymers are white flaky powders and have various applications, including: (i) bactericides and fungicides; (ii) catalysts; (iii) semiconductors; (iv) uranium recovering agents from seawater; and (v) polymeric dyes. Metal phosphinates are powders which are used as thickening agents in silicone adhesives which can also function as a grease and have antistatic properties.

4.2 A Brief History of Chemistry of Plastics [4]

The definition of a polymer is a chemical containing many units of the same molecule joined together in a chain-like structure by chemical reaction, forming plastics. Each link of the chain is the ‘mer’ or basic unit usually consisting of carbon, hydrogen, oxygen and silicon, and is joined either directly and/or using catalysts (most of which are metal-based) under controlled pressure and temperature conditions. To form the chain many 'mers' are joined or polymerised together. Polymers have been in existence for a long time e.g. waxes, tar, horn, shellac, resins from trees, etc. In recent centuries, and with scientific understanding, these have been classified as polymers and chemical modification and treatment using heat and pressure have been used to make useful items like ornaments, jewellery, combs and basic household items.

Polymers have their own chemical and physical characteristics and the following is a brief list of their distinctive characteristic behaviour:

1. Most polymers are resistant to most chemicals and are used as packaging for shampoo, toothpaste, perfumes, cleaning agents, etc.
2. Polymers are light with a range of strengths and can be used to manufacture a range of modern household goods, clothes and shoes.
3. Polymers can be moulded into various shapes and sizes. They can be used to make parts for cars, and be added to adhesives and paints. Some plastics can be stretched, behave as elastomers and form thin threadlike fibres to be used for very intricate parts, e.g. medicine.
4. Polymers have a limitless range of colours that can be used to make attractive components and toys for children.
5. Polymers have many inherent properties that can be intensified by a range of additives to increase their use and applications.
6. Polymers are good thermal and electrical insulators. This is evident in handles of kitchen items and electrical cords, outlets, wiring, refrigerators and freezers.
7. Disposal of waste plastics is expensive. Plastics deteriorate but never decompose completely and they involve a high percentage ($\sim$42% by weight) of waste. Fortunately, in recent years recycling of plastics has become an important technology. Some plastics can be blended with unused virgin plastics to reduce the cost of waste disposal.
8. Some waste plastics/polymers are not suitable for recycling and can be burned by controlled combustion to produce heat energy. The controlled combustion will efficiently burn both the material and their by-products. Some plastics leave ash residues that contain metals from catalysts, stabilisers, colorants, etc.

4.3 Chemical Structure of Plastics [5]

The element carbon makes up the backbone of most polymers/plastics and hydrogen molecules are bonded into the link chains and are classified as hydrocarbons. Other elements such as oxygen, chlorine, fluorine, nitrogen, silicon, phosphorous and sulphur are found as part of the molecular make-up of polymers; examples include PVC which contains chlorine, nylon which contains nitrogen, polyesters and polycarbonates which contain oxygen, and Teflon which contains the element fluorine. Elements such as silicon and phosphorous can also be used as the backbone for a range of inorganic type polymers with hydrogen or elements completing the molecular chain.

Polymers are usually structured like spaghetti noodles piled together. There are two main types of polymers, amorphous or crystalline. Amorphous polymers are prepared by

```
H       H       H       H       H       H
|   H   |   H   |   H   |   H   |   H   |
--C   |   C   |   C   |   C   |   C   |   C----
|   C   |   C   |   C   |   C   |   C   |
H   |   H   |   H   |   H   |   H   |   H
    H       H       H       H       H
```

Figure 4.1 *Basic polymer structure of polyethylene*

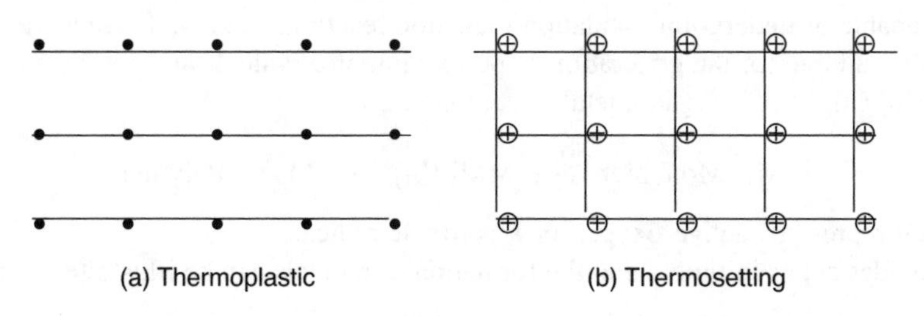

Figure 4.2 *Structures of (a) thermoplastic and (b) thermosetting plastic materials*

the 'control and quenching' polymerisation process that forms long chains which arrange themselves to give transparent materials. The non-transparent polymers are crystalline arrangements that have a distinct pattern. The basic structure of polyethylene polymer is shown in Figure 4.1. The degree of control in the manufacturing process can determine the degree of crystallinity and consequently the amount of light that can pass through the polymer. The higher the degree of crystallinity the less light will pass through it.

Polymers can be divided into two main groups – *thermoplastic* and *thermosetting*. The *thermoplastic group*, shown in Figure 4.2(a), can be melted by heat and reformed over and over again; this property allows for easy processing and recycling. The structure is a network of polymer chains joined to each other and the heating/melting of this type of plastic softens and breaks the intermolecular forces holding them together so that the chains can slide over each other. Hence, these types of plastics are not made to withstand high temperatures.

The *thermosetting group*, shown in Figure 4.2(b), cannot be reheated as it damages the plastic, forming a useless mass. These type of plastics form a network of cross-linked polymer chains and are moulded during their manufacture when the cross links are formed. They have higher thermal durability and resist acid or basic attacks when compared with thermoplastics.

Inorganic polymers exist, such as silicones, which contain alternating silicon and oxygen with inorganic and/or organic groups attached to the silicon atom. Poly(phosphazenes) are also inorganic polymers which contain nitrogen and phosphorous. Silicate minerals and glasses are true inorganic polymers.

4.4 Polymerization Process of Plastics

Polymerisation is a process of reacting the same monomer together chemically to form a three-dimensional network of polymer chains. There are many forms of polymerisation and the main two, particularly for plastic products, are (a) addition reactions and (b) condensation reactions.

4.4.1 Polymerisation by Addition Reactions

These are chain growth polymerisations and involve adding a monomer to a growing polymer chain one at a time using a peroxide radical generated by transition metals or metals capable of undergoing oxidation/reduction reactions readily. There are a range of metal salts suitable for the generation of peroxy initiator radicals and can be Ni, Co, Cu, Fe, Ti, Mn, Cd, Pb, Sn organometallic complexes.

$$n(\mathrm{M}), \text{ Monomer} \overset{[\mathrm{O}]}{\rightarrow} (\text{-MMM}^{\bullet}\text{-}) \rightarrow (\text{-M-})_n, \text{ Polymer} \quad (1)$$

where [O] represents active oxygen or a peroxide radical.
The peroxides act as initiators and the formation of radicals can be illustrated as follows:

$$\mathrm{ROOH} + \mathrm{M}^{2+} \rightarrow \mathrm{RO}^{\bullet} + \mathrm{OH}^{-} + \mathrm{M}^{3+} \ \text{(1) fast} \quad (2)$$

$$\mathrm{ROOH} + \mathrm{M}^{3+} \rightarrow \mathrm{ROO}^{\bullet} + \mathrm{H}^{-} + \mathrm{M}^{2+} \ \text{(2) slow} \quad (3)$$

The addition polymerisation reactions follow a three-stage standard reaction – *initiation, propagation* and *termination.*

Initiation reactions are usually started by an active free radical such as peroxide (-O-O-), e.g. benzoyl peroxide is a good inititator for the free radical addition polymerisation of styrene to produce polystyrene; $AlCl_3$ is an initiator for the cationic addition polymerisation of isobutylene to form isobutyl synthetic rubber or azobisiso-butyronitrile compounds (-N=N-) (abbreviated to AIBN). *Propagation reactions* are the continuing process and, eventually, lead to the *termination* stage that occurs by combination or disproportionation. This usually occurs when the free radicals combine with themselves and signals the end of the polymerisation process. All polymers formed by this process are thermoplastics. Table 4.1 is a list of common polymers prepared by the addition process.

Table 4.1 *Reaction monomers used to prepare polymers by the addition process*

Monomer	Chemical name	Plastic
Ethene	Poly(ethene)	Polythene
Propene	Poly(propene)	Polypropylene
Phenylethene	Poly(phenylethene)	Polystyrene
Chloroethene	Poly(chloroethene)	Poly(vinyl chloride)

4.4.2 Polymerisation by Condensation Reactions

This reaction involves reacting different monomers to form polymers and large molecules with the elimination of, e.g. water or HCl, in the process. The same rules apply as for the addition process and the polymers formed by this process can be used to prepare both thermoplastic and thermosetting plastics. Table 4.2 is a list of common polymers prepared by the condensation process.

Table 4.2 Reaction monomers used to prepare polymers by the condensation process

Monomer	Chemical name	Plastic
Amides	Polyamide	Nylon
Esters	Polyester	Terylene
Urethane	Urethane	Polyurethane

4.5 Additives in Plastics

Plastics contain numerous additives whose purpose is to alter the physical characteristics of the basic material to suit uses in a range of products. Some plastics use one or several metal binding additives along with a host of non-metal additives and all have an important role in the shelf life, colour, porosity, cost and durability of plastics. These additives can be fillers, colorants and stabilisers that include inorganic and organometallic compounds. Fillers are added to synthetic resins to improve the toughness and durability. Almost all thermosetting plastics contain fillers as reinforcement, e.g. polyester resins are reinforced with glass fibres, mica, etc. Thermoplastic materials are also supplemented with fillers to be used to manufacture products for floor coverings, mats, sheeting, etc., which are required to improve their endurance under stress, elasticity and flexural strength.

Fillers, stabilisers and colorants can be inorganic or organic, and are added as part of identification, stability and durability, they must be non-leaching and meet health requirements. Some plastics products are required to keep moisture and air out of sensitive products and these are usually high-density plastics with low permeation properties. Some products require plastics with high permeability to allow air (oxygen) as part of product stability, e.g. anaerobic adhesives. Healthcare products require that some containers do not allow air or moisture to permeate the surfaces (bacteria or other micro-organisms) and in some extreme cases may require that the inside be coated with an inner layer of approved lacquer or other very low permeable material.

Vinyl stabilisers can either be liquids or solids and formulated as single or mixtures into PVC products. They can be combinations of metal salts and organic additives added to improve shelf life, prevent degradation and control colour of finished products. Depending on the application, the stabiliser compositions are based on organometal salts of Sn, Ba, Cd, Pb, P, Ti, etc. Analysis of these salts is often required to provide certification that the metallic composition of the product agrees with established specifications of the products.

Inorganic fillers used in plastics include asbestos, talc, mica, diatomaceous earth, kaolin and others that can be monitored by metal analysis of one or more of the elements present in the filler. Organic fillers containing metals are also used to great extent in plastics. Thermosetting plastics are usually formulated with inorganic fillers to manufacture a range of items such as jewellery, brooches, household items, tools, and so on. The addition of metal stabilisers is important and they are added as part of maintaining product quality, improving shelf life, improving heat tolerance, as UV absorbers, antioxidants, pigments, colorants, etc. Heat and UV light can seriously damage some plastics, particularly the softer

PVC, polyolefin thermoplastics. Metal salts are added to plastics for one or more of the following reasons:

(a) rigid gelling and plastification
(b) higher melt viscosity
(c) improved tensile strengths and extensibility
(d) aids maintaining homogenous mixture
(e) optimises thermoforming properties
(f) reduce plating out
(g) flame retardants
(h) impact modification
(i) colorants
(j) stabilisers
(k) plasticisers
(l) UV absorbers.

Table 4.3 provides a list of a range of metal salts and complexes used in plastics (this list is not exhaustive).

Major and trace levels of metal initiators or catalysts can be present, such as Al, Ti, Fe, V, Cr, Fe, Cu, Ca, Zr, B, K, Na, Li and others which originate from organometallic salts. Catalysts may also include oxides and acetals that coordinate with such metals as Co, Zn, P, Pb, Co, Mn, Ge and Sb. Some of these metal salts can be used in the trans-etherification or polymerisation reactions to manufacture a range of polymers, e.g. metal acetals. Catalysts also play an important role in the selective PET polymerisation reactions.

Table 4.3 *Random selection of metal salts used as organometallic stabilisers, UV absorbers, catalyst, colorants, etc., in plastics*

Function of metal added to plastics	Type	Metal(s) detected
Plasticisers	Tributyl, tri-n-hexyl, or triphenyl phosphorous	P
Fillers	Asbestos, mica, TiO_2, CaSi (complexes), SiO_2	Al, Si, Ca
Organometallics used in plastics	Mn (Blue), Sr (chromates), Fe (pigment), Ni (azo complex), silanes	Mn, Sr, Fe, Ni
Organometallics as stabilisers	Dibutyl tin(IV) mercaptides, dibutyl tin(IV) mercaptoesters, dibutyl tin(IV) dilaurates	Sn
Salts of organic acids (surfactants)	Ba, Cd, Mg, Ca, Sr, Al, Zn and Pb laurates or maleates	Ba, Ca, Mg, Sr, Al, Zn, Pb
Salts of inorganic acids (stabilisers)	Pb (carbonates), Pb (sulphates), PbSi complexes, zinc borates, phosphate flame retardants	Pb, Si, Zn B, P
UV absorbers	Pb (phosphites)	Pb
Other metals detected in plastics	S, B, Na, Cu, Fe, Zn, Mn, Ca, Li, Al, Ni, Cr, Hg	S, B, Na, Cu, Fe, Zn, Mn, Ca, Li, Al, Ni, Cr, Hg

Plastics can be doped with metals to give materials that are conductive and are used in such applications as flat panel displays, antistatic packaging and rechargeable batteries. Their conducting power is not as good as the pure metal but does approach values close to metals. A range of dopants that are used successfully in plastics include AsF_5, I_2, SbF_5, $AlCl_3$, $ZrCl_4$, $MoCl_5$ and WCl_5. Conducting polymer complexes have also been prepared from natural rubber doped with iodine that has a conductivity several orders higher than that of the same rubber not doped with iodine. The conductivity appears to be essentially electronic with charge carriers hopping between sites of different polymer chains. This shows the importance of electron releasing groups in establishing the electric conductivity of these systems. However, most conducting polymers are unstable and their electrical conductivity degrades with time due to reaction with air (oxygen) and water.

4.6 Methods of Sample Preparation for Metal Content of Plastics, Fibres and Textiles

Plastics, fibres and textiles are solid organic polymeric compounds with molecular weights varying from 5000 to over 100000 with different degrees of rigidity and strengths. Analysis of plastics for metal content usually requires special care as they may contain major or trace levels. Metal content of plastics is required for one or several of the following reasons: quality assurance, contamination and leaching of toxic metals (especially where toys are concerned), waste disposal and health and safety requirements. Both natural/synthetic fibres and plastics may contain metals from residues of catalysts; stabilising agents, plasticisers and colorant, etc. and some metals may affect, favourably or unfavourably, their stability and other physical characteristics.

The first step in analysing plastics for metals content in polymers by ICP-AES technique is that they must be prepared in solutions that are suitable for nebulization. There are four general methods applicable for sample preparation for metal analysis by ICP-AES and they are: solvent dissolution of some plastics; dry ashing using a muffle furnace; acid digestion using a microwave oven; and oxygen bomb combustion.

4.6.1 Sample Preparation Using Dissolution Method [6]

Some natural and synthetic polymers are soluble in a range of solvents that are suitable for ICP-AES nebulisation for metal analysis. Other plastics are soluble in solvents that are not suitable for ICP-AES nebulisation can be made suitable by adding a slight excess of GAC and nebulised as normal against standards prepared in the same solvent mixtures. Effects of nebulisation efficiency due to the soluble plastics may need additional analytical methods, e.g. method of standard addition or internal standard to correct for signal response and quantification. Soluble polymers that require heat and stirring as part of sample preparation are usually more time consuming. Cellulose fibres can be hydrolysed with a little heat in an HCl solution and made up to mark with water. Some of the disadvantages of dissolving polymers in solvents are listed as follows:

(a) Certain elements blended in the plastic may not be soluble in the solvent, particularly if the added organometallics have undergone a chemical change during the processing, causing them to convert to insoluble inorganic salt. (Insolubles from plastics

Table 4.4 *List of polymers and solvents required to dissolve them. Some of these solutions are finally made up to mark with GAC*

Polymer	Solvent	Metal salt
Polyether	Methanol	Metal naphthanates and others
PVC	Dimethyl acetamide	Metal naphthanates and others
Polycarbonate	Dimethyl acetamide	Metal naphthanates and others
Polyacrylonitrile	Dimethyl formamide	Metal naphthanates and others
Cellulose acetate	MIBK	Metal naphthanates and others
Polystyrene	MIBK	Metal naphthanates and others
Cotton	Fuming H_2SO_4	Metal naphthanates and others
Cellulose	50% conc. HCl + H_2SO_4	Metal naphthanates and others
Wool	HCl or ammonia	Metal naphthanates and others

may be reproducibly quantified while stirring and nebulising and analysed using a slurry method as described in Section 3.5.10.)

(b) Some elements may volatilise, particularly where the solvent must be heated to assist solubilisation.
(c) Dissolved polymers may precipitate in the spray chamber or nebuliser causing erratic measurements.
(d) The nebuliser, spray chamber and torch must be thoroughly cleaned with solvent between measurements to avoid sample contamination or blockage.
(e) Some solvents can damage the nebuliser or spray chamber and must be used sparingly.

Table 4.4 gives a list of polymers and solvents suitable for dissolving some plastics. In some cases the metal is double precipitated into aqueous medium by dropping the organic solution containing the dissolved plastic in to a rapid stirring aqueous medium of ammonia for Pb, Ag, Sn and aqueous HCl for Cu, Fe, Cr, Ni, Sb, etc. where the metals are transferred quantitatively to the basic or acidic medium.

Method. Samples of polyether, PVC, cellulose acetate, and wool were analysed by carefully cutting the plastic or wool into sizes not greater than 4 × 4 mm using stainless steel scissors. An accurate weight of approximately 0.5 g samples of polyether, PVC, cellulose acetate and wool are analysed by dissolving in 20.0 ml of appropriate solvent listed in Table 4.5 and made up to 100.0 ml with glacial acetic acid (GAC). Each polymer solution is 'spiked' with elements of interest to determine their percentage recovery. Each solution is analysed with and without stirring (top clear liquid of settled particulates) to study the difference between the soluble sample/ metals and insoluble sample/metals of these products. Analysis with stirring is a 'slurry method' to determine the total metal content of each solution.

Results. The results in Table 4.6 show the detection and percentage recovery of each metal added. Analysis before and after stirring using the slurry technique was used to determine the insoluble fillers and plastisers, etc. Plastics that are surface treated with metals as colorants or protective coating may be contacted with 2.0 M HNO_3 or 2.0 M HCl for a period of time to dissolve the metal salt, and analysed against standard calibration curves prepared in the appropriate aqueous solution.

Table 4.5 *List of polymer samples that were analysed for each metal as listed. Each plastic sample was 'spiked' with metal standards*

Sample	Metal(s) detected	Added metal salt at 20 ppm each	Solvent	Preparation method
Polyether	Ca, Pb, Zn and Co	20 ppm Ca, Pb, P, Zn, Co, Cr, Fe, V, Sn and Cu	Methanol	0.5 g dissolved with heat at 60°C × 4 h
PVC	Cr, Fe, V, Co and Sn	20 ppm Ca, Pb, P, Zn, Co, Cr, Fe, V, Sn and Cu	Dimethyl acetamide	0.5 g dissolved with heat at 80°C × 4 h
Cellulose acetate	Ti, Fe, Cu and V	20 ppm Ca, Pb, P, Zn, Co, Cr, Fe, V, Sn and Cu	MIBK	0.5 g dissolved with heat at 60°C × 4 h
Natural wool	P, Sn + others	20 ppm Ca, Pb, P, Zn, Co, Cr, Fe, V, Sn and Cu	1.25 M NaOH	0.5 g dissolved with heat at 100°C × 6 h

Table 4.6 *Results of metal analysis using ICP-OES for polymers before and after 'spiking' using slury method and with and without stirring*

Sample (method used)	Metal	No addition (ppm)	+20 ppm addition	% Recovery before stirring	% Recovery after stirring
Polyether (Std. Add.)	Ca	16.0	34.9	120.0	94.5
	Pb	<1.0	19.5	1.6	98.0
	Cu	61.2	79.8	60.9	94.5
	Zn	22.0	43.2	23.7	106.0
	Sb	45.0	66.2	45.4	103.4
PVC ('Y' Int. Std.)	Cr	<1.0	20.6	<1.0	103.0
	Fe	6.8	26.8	8.9	103.0
	V	<1.0	20.9	<1.0	105.0
	Co	<1.0	19.6	<1.0	98.0
	Sn	33.6	54.2	35.2	103.0
Cellulose acetate ('Y' Int. Std.)	Ti	<1.0	19.4	<1.0	97.0
	Fe	17.6	38.4	24.8	104.0
	Cu	3.8	23.1	5.6	96.5
	V	<1.0	20.7	<1.0	103.5
	Ca	220	242.0	219.0	110.0
Natural wool (Cal. curve)	Pb	3.6	24.0	7.8	102.0
	P	<1.0	20.5	<1.0	102.5
	Zn	4.6	25.6	10.6	105.0
	Co	2.3	23.1	2.0	104.0
	Cr	1.6	22.1	2.9	102.5
	Fe	16.8	36.2	39.6	97.0
	V	<1.0	20.9	<1.0	104.5

4.6.2 Sample Preparation Using Dry Ashing Methods

Polymers that are insoluble in solvents may be dry ashed in the presence of metal retaining agents (e.g. *para*-toluene sulphonic acid, PTSA) and the ash residue dissolved in 0.1 M HCl acid solution made up to a known volume. Polymers containing non-volatile major or trace levels of metals can be prepared by dry ashing as a simple method for removing the organic matter while at the same time retaining the metals of interest. For trace analysis it is possible to ash the sample in the same vessel repeatedly for two or three times accumulating the metal concentration. The ashing method requires very little of the analyst's time and is suitable where metals can be retained in the vessel during the heating cycle.

Method. A known weight of a sample of 4 × 4 mm chopped plastics is added to a clean platinum dish with 0.15 g of PTSA and placed in a muffle furnace or microwave asher using a slow and low temperature ramping program (see Figure 4.3). The ramping program avoids loss of sample caused by 'spitting' and loss of elements of interest.

Note: Crucibles used for ashing are usually made of silica, porcelain or platinum. The latter is the best because trace metals may be lost in the pores of silica or porcelain dishes due to the irreversible absorption of the analyte into their walls. Platinum dishes do not absorb the metal atoms which are readily totally dissolved in appropriate solution. Blanks must also be prepared and analysed for each series of samples.

A major disadvantage of furnace/microwave ashing is that it is carried out in an opened vessel that could lead to loss of some elements by volatilisation even in the presence of retaining agents. The vessel needs to be opened as the sample needs oxygen from air to assist its burning/ashing. The opened vessel is also prone to atmospheric contamination by airborne particles and 'spitting' from other samples in the oven. The diagram shown in Figure 4.3 is an illustration of time versus temperature ramping control that is required to avoid 'spitting' and cross contamination by other samples being ashed in the same oven. It is also required to allow time to complete its charring, burning and ashing stages sequentially in order to totally remove the organic matter while at the same time retaining the metals in the vessel.

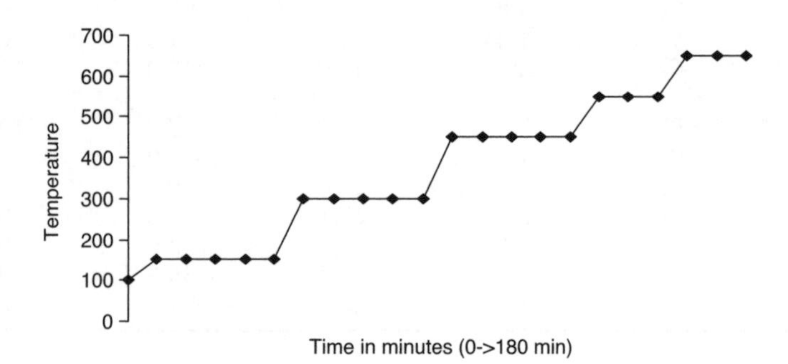

Figure 4.3 *Time and temperature ramping diagram for ashing using a furnace or microwave oven*

Controlling the temperature at which the ashing takes place can reduce losses due to 'spitting of sample' and volatilisation of metal(s). The determination of Pb, Sn, Cd, Zn and Se are some of the low volatile metals that must be ashed slowly using a ramping range of temperatures below 500°C in order to achieve total recoveries. The presence of a retaining additive, e.g. PTSA, may help to retain most metals as sulphates. The resulting ash is dissolved in 0.25 M HCl or HNO_3 as are the standards used to quantify the elements.

4.6.3 Sample Preparation Using Microwave Acid Digestion Method

Digestion of samples in a closed, heated and pressurised vessel has many advantages over the open vessel dissolution method. The samples are digested in microwave ovens using toughened fabricated high temperature polymeric vessels free of metal contamination (see Figure 3.3). The toughened vessel allows high pressure (maximum 800 psi) and high temperature (maximum 220°C) to be applied without danger or damage. The closed vessel reduces evaporation, so that minimum acid digestion solution is required, reducing blanks, and is safe to work with provided it is used according to the manufacturer's instructions. The closed vessel eliminates the loss of volatile elements, a serious problem when using open vessels, particularly with some dry ashing methods. The modern microwave digesters are usually fully automated and have multiple functions, such as ramping and holding of temperature and pressures, which allow sample preparation to be safe and achieve successful digestions. The precise parameters associated with each method may be recorded for future analysis of similar samples. The ovens are fitted with exhaust scrubbers so as to avoid toxic fumes damaging other laboratory equipment and also to avoid harming laboratory personnel.

Modern microwave digestion systems monitor both pressure and temperature in the containers with automatic shut-down safety valves. The oven temperature and pressure can be set and when this is achieved, the power is switched off. This continues for several stages of the set programs until completion of run and it will turn off automatically. The microwave vessels are available commercially and toughened to handle most samples.

A micro-evaporator can also be attached to modern microwave digesters and are designed to reduce the acid solutions to a low or even a dry state. The principle is that samples are exposed to microwave energy under vacuum to accelerate volume reduction. An 'auto-detect' feature of the CEM MAR 5 system is that a detector is installed to inform the operator when the evaporation is complete and automatically stops operation. Acid fumes are neutralised by an integrated vapour scrubber system.

Method. Analysis of nylon for metals content is carried out by adding 0.75 g of finely cut samples to Teflon vessels suitable for microwave digestion. Then 10.0 ml of conc. HNO_3 is added to the vessel containing the sample and allowed to predigest for 30.0 min prior to microwave heating and pressure digestion according to parameters in Table 4.7.

After microwave digestion the mixture is allowed to cool for 1 h and the contents are reduced to ~10.0% of the original volume using a micro-vap attachment. The reduced liquid is transferred to a 50 ml volumetric flask and the vessel is washed

Table 4.7 *Heating and pressure parameters for microwave acid digestion of nylon in HNO_3*

Stage	Power (W)	% power	Ramp time (min)	Pressure (psi)	Temp. (°C)	Hold (min)
1	600	100	10	140	130	5
2	600	100	10	150	150	5
3	600	100	10	350	160	5
4	600	100	10	500	180	5
5	600	100	10	600	210	5

with deionised water up to mark. Standards containing 0.0, 0.5, 1.0 and 2.0 ppm Zn, Fe, Cu, Ca, Na, P, and Ba are prepared in 0.5 M HNO_3. [If a higher concentration is present, the standards concentration must be changed to suit the level of metal(s) in the final solution.]

Method for Microwave Acid Digesting of Polyurethane, Polyphenylene Sulphite, Polysulphone and HDPE. Sample preparation for analysis of polyurethane, polypropylene terephthalate, polyphenylene sulphite, polysulphone, HDPE, PVC, polyethers, cellulose acetate, and natural wools for metal content is as for nylon with an additional step involving charring the sample prior to digestion.

The sample is prepared in a suitably clean vessel by weighing ~0.75 g accurately of finely cut pieces ($<4 \times 4$ mm) of plastic/polymer and adding 5.0 ml of conc. H_2SO_4. The mixture is allowed to predigest for 30.0 min prior to charring as illustrated in Table 4.8.

Digestion of Sample. To the charred samples add 10.0 ml of conc. HNO_3 and 2.0 ml H_2O_2. The conditions in Table 4.9 are applied to the digested samples.

After completion of digestion the mixture is evaporated to ~10.0% of its original volume using the micro-vap attachment to the microwave oven. The final mixture is allowed to cool and transferred to 50.0 ml volumetric flasks with washing using deionised water up to mark. The metals content is determined against 0.0, 0.5, 1.0 and 2.0 ppm of Ca, Pb, Cu, Zn and Sb of a multi-element standard prepared in 0.5 M HNO_3. [If a higher concentration is present, the standards concentration must be changed to suit the level of metal(s) in the final solution.]

Table 4.8 *Parameters used for charring in microwave oven*

Stage	Power (W)	% power	Ramp time (min)	Pressure (psi)	Temp. (°C)	Hold (min)
1	600	100	2	0	0	0
2	600	0	1	0	0	0

Table 4.9 Parameters for acid digestion using microwave oven

Stage	Power (W)	% power	Ramp time (min)	Pressure (psi)	Temp. (°C)	Hold (min)
1	600	100	10	150	140	5
2	600	100	10	175	150	5
3	600	100	10	200	160	5
4	600	100	10	300	180	5
5	600	100	10	500	210	5

4.6.4 Sample Preparation Using Oxygen Bomb Combustion Method

Digestion of samples using high pressure oxygen bomb combustion is an excellent technique for sample preparation, particularly trace metal analysis. This technique can be applied to most plastics provided that small sample (~0.25 g) of fine grain sizes of plastics are used. The solutions obtained are clean and easily analysed for metal content against standards prepared in the same solution added to bomb.

Procedure for Bomb Combustion. A sample size of 0.25 g of finely cut plastic pieces or beads is placed in a shallow platinum dish along with 0.25 g of paraffin oil (to aid combustion) and the dish is placed into a wire sample holder attached to the lid of the bomb. A 10 cm length of nichrome or platinum wire is connected between the electrodes and 5.0 ml of water or 0.05 M NaOH is added to the bottom of the vessel. The bomb is assembled according to the manufacturer's instructions and filled with oxygen to 30 atm. The pressurised oxygen vessel is completely submerged in a water tank and checked for leaks. Assuming no leaks are detected, the bomb is fired to combust the sample. After combustion the bomb is allowed to cool and the excess burned gas is released as CO_2 and H_2O slowly, until the cap can be opened with ease by hand. The inside wall is washed with deionised water into a volumetric flask and made up to mark with deionised water. The metal analysis is carried out using 0.0, 0.5, 1.0 and 2.0 ppm multi-element standards prepared in the sample solution added to the bomb. [If a higher concentration is present, the standards concentration must be changed to suit the level of metal(s) in the final solution.)

4.7 Comparative Study of Methods of Analysis of Plastic Samples for Metals Content

The choice of which method to use may be decided by the type of polymer/plastic material and the metal required for analysis. With the analysis of a completely unknown sample it may be necessary to carry out trial and error tests of different methods with and without 'spiking' before accepting the final sample preparation method. Once confidence is achieved in the method of analysis, the procedure is noted for future reference.

Plastic composition consists mainly of a combination of polymer, stabiliser(s), plasticisers, colorants and fillers. Metals in plastics usually occur in such substances as

Table 4.10 *Results of analysis of selected plastic materials for listed metals by ICP-AES after microwave acid digestion (A), dry ashing to 650°C with PTSA (B) or high pressure oxygen bomb combustion (C)*

Plastic	Metal	Preparation method			Preparation method +250 ppm of each metal added			% Recovery		
		A	B	C	A	B	C	A	B	C
PVC	Cd	660	185	669	898	199	915	95	—	98
	Pb	390	78	396	644	215	623	102	—	91
	Ca	550	539	565	811	793	806	104	102	96
	Si	110	118	113	356	365	365	98	98	99
	Cr	55	61	58	316	310	313	104	100	102
	Fe	95	102	92	351	358	365	102	105	109
HDPE	Cr	155	149	159	409	412	421	102	105	105
	Ti	285	291	288	544	512	534	104	101	98
	Pb	120	14	127	390	68	384	108	—	103
	V	44	43	41	293	288	291	99	99	100
PET	Sb	89	67	91	346	102	339	103	—	99
	Co	53	49	55	306	299	311	101	100	102
	P	196	61	203	442	313	461	98	101	103
	Pb	40	6	43	297	88	293	103	—	100
PMMA	Zn	78	72	74	326	330	336	98	104	105
	Fe	35	37	33	291	283	289	102	98	102
	Ca	290	299	287	531	539	529	96	96	97
	Al	158	149	161	401	402	392	97	101	93

fillers TiO_2, china clay, $CaCO_3$ and asbestos; in stabilisers containing Pb, Cd, Ba, K, stearates, $PbSO_4$, $PbCO_3$; organometallic compounds; Sn stabilisers of dibutyl tin dilaurate, dibutyl tin maleate and dioctyl tin dilaurate. Other metals salts used in plastics in major, minor and trace quantities are S, Ca, P, Zn, Fe, Co, B, Al, Si, Mn, Cr, etc. The identification and quantification of metals can assist in identification of additives in an unknown plastic formulation. Table 4.10 records the results of a study of the three combustion methods of sample preparation for the analyses of plastics for metals content. The following plastics were studied: PVC, PET, HDPE and PMMA.

The metal salts are added to plastic products as organometallic, metallocenes or inorganic salts. The ashing method of sample preparation is only suitable for elements that do not volatilise during the heating cycle. The results obtained and recorded in Table 4.10 are reproducible for most elements with the exception of Pb and Sb in which some loss had taken place using the opened dry ashing method even in the presence of PTSA.

Sample preparation using the bombing combustion method has an advantage over the microwave digestion method in that it does not require very strong multi-acidic solutions and may be combusted in water only. If trace metal analysis is required, the bomb method is the best because the resulting solution is easy to use for sensitive reproducible analyses.

In developing a method for metal analysis of plastic samples it may be necessary to perform sample preparation by dissolution and three combustion methods to decide which one is the most suitable. The metal, level and the sample determine the most suitable method.

4.8 Study of Leaching of Metals from Plastics

Plastic materials are moulded to make all kinds of objects, e.g. bottles, boxes, flexible air containers, parts of items for handles of knifes, spoons, toys, etc. and they must comply with government health and safety regulations with regard to potential leaching of chemicals and metals. These regulations are rigorously applied where the plastics are to be used in contact with foods, pharmaceutical/medicinal products, adhesives, bone repairs, toiletries and household detergents.

Plastic plays a major role in the cost-saving packaging and storing of a range of manufactured products and their ability to leach chemicals and metals is considered part of a strict health and environment requirement. They are rigorously tested using product stimulants to determine whether additives in plastic can migrate to the product. As a general rule the best plastics are those that do not contain dangerous metals, however this may be impossible as they are used in the manufacturing process of some plastics as catalysts, fillers, stabilisers or colorants. Most high quality plastics contain some metals but the healthy tendency is to use metals that are not going to (i) leach from the plastic or (ii) if they do leach, that they are regarded as harmless.

The health and safety requirements for plastics used with sensitive products require that chemical analysis including metal be carried out and a Certificate of Analysis supplied with each batch. The temptation to use toxic elements is great because the best available catalysts are those comprising toxic elements, such as Cd, Se, Cr, Hg and Pb salts, and they are cheap. Table 4.11 is a list of metals and their maximum concentration permissible in a range of products and as part of waste disposal requirements.

Table 4.11 *Random list of permissible concentrations of metals that may be acceptable to a range of products*

Metal	Maximum conc. (ppm)	Metal	Maximum conc. (ppm)	Metal	Maximum conc. (ppm)
Al	100	Co	25	P	10
As	0.20	Cs	0.25	Pb	0.5
Ba	50	Cr	0.25	Sb	0.5
Be	20	Fe	25	Se	0.25
Cd	0.25	Hg	0.25	Zn	100

4.8.1 Study of Leaching of Metals from Children's Toys

The following is a laboratory stimulation of potential metal leaching from plastics, carried out by contacting the plastics with the product it is used to store, or using acid solutions as close as possible to a common acidity of most products. According to BSI 5665: Part 3: 1989 [EN 71-3], the upper limits of ingestion of toy material by mouth shall not exceed 0.1 μg for As to 25 μg for Ba per day depending on the metal from plastics to which children are exposed. Similar plastics used to make children's toys were contacted with 1.25 M HCl for 48 h at 40°C. The acid solution was analysed for metal content using ICP-AES.

Method. Portions of red and yellow coloured HDPE, polypropylene, PET and clear PMMA plastic materials are comminuted into sizes approximately 4 × 4 mm using a diamond edge cutter.

Weigh accurately approximately 2.5 g of the comminuted HDPE, PET and PMMA plastic into separate 100 ml screw capped bottles and contact each with 50 ml of 1.25 M HCl with stirring for 48 h at 40°C. To a second set of bottles weigh accurately approximately 2.5 g of the same plastics and 'spike' with 50 ppm of Al, Cd, Fe, Pb and Zn. Contact the samples and 'spiked' metals with 50 ml of 1.25 M HCl using the same conditions. The pH of all solutions is maintained at less than 1.0 by drop-wise addition of 2.0 M HCl over the period of contact. The solutions are filtered through a membrane filter of pore size 0.45 μm and the resulting clear liquids are analysed for metals content against standard calibration curves of multi-element standards 0.0, 0.5, 1.0 and 2.0 ppm of metals of interest in the same acid solutions to give the results shown in Table 4.12.

Table 4.12 *Results of analysis of 'metal spiked' HDPE, PET and PMMA plastics showing the results and percentage recoveries*

Plastic	Metal found (ppm)	Added metal (ppm)	Recovered	% Recovery
HDPE				
Al	43	50	97	108
Cd	<0.2	50	49	98
Fe	65	50	118	106
Pb	<0.2	50	53	106
P	<0.2	50	47	94
Zn	77	50	126	98
PET				
Al	12	50	61	98
Cd	<0.2	50	51	102
Fe	21	50	74	106
Pb	16	50	69	106
P	<0.2	50	44	88
Zn	27	50	81	108

Table 4.12 *(Continued)*

Plastic	Metal found (ppm)	Added metal (ppm)	Recovered	% Recovery
PMMA				
Al	128	50	181	106
Cd	<0.2	50	49	98
Fe	65	50	118	106
Pb	58	50	104	92
P	<0.2	50	46	92
Zn	54	50	101	94

The results in Table 4.12 show that metals can be leached from the plastics. This simulated experiment shows that all plastics need to be monitored carefully to ensure that they are safe. It is highly probable that if metals are leached other chemicals are also leached from plastics.

4.9 Analysis for Toxic Metals in Plastics and Non-Electrical Additives Used in Electrical and Electronic Components as Required by RoHS [7]

The RoHS ('Restriction of use of certain hazardous substances in electrical and electronic products') is a European Union (EU) directive applied to all electrical and electronic equipment and parts that restricts the use of elements Pb, Cd, Hg and Cr(VI) among another 47 dangerous substances. The directive is applied for environmental regulations concerning all disposals and requires removal of toxic chemicals including metals from electrical and electronic equipment. These restrictions are part of waste disposal requirements when the products are no longer used (End of Life Vehicle, ELV). RoHS legally only affects European countries but because of the global nature of the electronics market, it is becoming a worldwide standard. These restrictions are also applied to plastics used in electrical and electronic components.

The directives apply to all materials used in the make up of electrical and electronic components and include paint, plastics, adhesives and rubbers, plastic parts, protective coating materials, epoxy adhesives, cyanoacrylate adhesives, polyurethane adhesives as well as complete computer motherboards. Computer and peripheral equipment are made up of different materials consisting of plastic, ferrous and non-ferrous metals, trace heavy metals, glass, foams, rubber, carbon powder and additives. Other non-metal materials posing environmental and health problems present in electrical and electronic products are beyond the scope of this book and will not be discussed here. The supportive components used in these products fall within the following ten product categories in the directive which are:

1. Large household appliances
2. Small household appliances
3. IT and telecommunications equipment
4. Consumer equipment
5. Lighting equipment
6. Electrical and electronic tools

7. Toys, leisure and sports equipment
8. Medical devices
9. Monitoring and control instruments
10. Automatic dispensers.

Other items such as medical devices and meters that use plastic parts are exempt from the RoHS requirement until such time as the EU sees fit to come up with specifications before including them. Waste Electrical and Electronic Equipment (WEEE) are defined as those requiring electricity or electromagnetic fields to operate them and most of the modern equipment contains plastics to some extent.

The RoHS directive that came into effect on 1 July 2006 states that all electrical products including plastics used in these products must be compliant by that date. The maximum concentration values (MCVs) are limits set by the EU commission for each restricted hazard substance as stated in the waste disposal commission and amended directive report 2005/618/EC. THE MCV of selected toxic metal limits for each homogenous material including plastics is set at 0.1% ($1000\,\mu g\,g^{-1}$) for Pb, Hg, Cr(VI) and 0.01% for Cd ($100\,\mu g\,g^{-1}$). The homogenous materials according to the RoHS directive are aimed at including all parts that make these components, metals plastics, inks, solders and adhesives. Therefore, an individual of a major part such as plastics in the electric/electronic component cannot be mechanically separated from the total part and must be included. Countries outside the EU that are selling similar products into the EU market must comply with the EU RoHS directive and provide Certificates of Analysis to show that their products satisfactorily meet these requirements. Most countries outside the EU are now tightening their specifications in order to be accepted by reducing or removing the hazard substances and supplying their products according to this directive. The challenge for modern manufacturing companies today is to redesign their products using cleaner, safer and environmentally friendly materials.

Typical plastics used in electronic and electrical appliances are polyethylene, polypropylene or polyethene terphthalate, and these are studied here as part of the RoHS requirement for the presence of toxic metals. This method is to show that analysis of these plastics used in electrical and electronic equipment is essential, especially if the origin of the plastic is unknown and the supplier is unable to state whether or not they are free of these metals. The metals are measured against calibration standards curves for each metal and may also include additional attachments for improving limits of detection such as ultrasonic nebulisers for Cd, Pb and Cr and the cold trap method for Hg.

The appropriate method is chosen for sample preparation for toxic metal content of plastic products by using the microwave acid digestion technique. The method includes a recovery study to show that metals present are not lost or contaminated during the sample preparation. The samples are prepared by cutting the plastics into $4 \times 4\,mm$ sizes using high grade stainless steel scissors. They are digested according to preconditioning method 2 of EPA 3050B (1996) suitable for organic based matrices with references to BS EN 1122:2001 method B, using a heated and pressurised acid decomposition in a sealed container (suitable for microwave acid digestion). An ultrasonic nebuliser is included for Cd, Pb and Cr, and a continuous cold vapour trap method for Hg so as to improve the sensitivity of measurements.

These regulations apply to electrical and electronic manufacturers, retailers and distributors, local authorities, waste management industry, exporters and re-processors and businesses and other non-household users of WEEE.

Disposal of waste electrical products has been a concern of manufacturing countries worldwide, and to effectively commence a process of safe disposal of such materials containing toxic metals required by the Montreal Protocol, EPA US and Japanese manufacturing industries have suggested the general reduction of use of all types of hazardous compounds from all household and plastic goods. The EU also passed the directive 2002/95/EC that restricts the use of certain hazardous substances in electrical and electronic products. The directive 2002/96/EC works to regulate the disposal of these products and is progressive in their banning. In today's global economy these regulations affect us all, as the regulations of one country may be different to regulations in another country and an international uniform agreement has yet to be accepted. Detection and measurements of these toxic compounds and metals particularly at low level is becoming increasingly important on a worldwide basis.

The banned metals in electrical and electronic products are analysed using AES methods. Other banned compounds are determined by other techniques such as FTIR, GC/MS, GC/ECD and UV/VIS spectrophotometry.

Toxic metals are those that harmfully affect the biological function and disrupt the essential physiological process. The toxic effects can be traced to their ability to disrupt the function of essential biological molecules such as proteins, enzymes and DNA. In some cases they displace chemically-related metal ions that are required for important biological functions such as cell growth, division and repairs. Enzymes, such as cysteine amino acids, contain a reactive sulphur atom necessary for their function. Certain toxic metals have a high affinity for sulphur and bind tightly to the cysteine molecule inhibiting the enzyme's functionality. Toxic metals can replace essential elements from proteins e.g. Cd can displace the essential element Zn a zinc protein salt which is required as part of a health supplement. Similarly Pb can replace Ca in specific sites in bones if it is required. Toxic metals such as Ni and Ar whose toxicity is well known have a long-term cumulative effect which can build to high concentrations and can cause health problems later in life.

The distinction between metals having harmful effects and beneficial effects needs to be further defined as insufficient study has been carried out, e.g. the use of the metal Hg in dentistry, selenium as a heath supplement; Cr^{3+} is essential in maintaining blood sugar levels while Cr^{6+} is a harmful lung carcinogen; Be^{2+} is a light but toxic metal sometimes used in electrical and electronic components.

4.9.1 Method for Metal Analysis of Plastics and Non-Electrical Additives Used in Electrical and Electronic Products

Samples of adhesives, protective paints and carefully ground motherboard components (after cutting into fine pieces and milled to fine particle sizes) are usually prepared for metal analysis according to preconditioning method 2 of EPA 3050B (1996) with reference to BS EN 1122:2001 method B, using a pressurised decomposition in a sealed container (microwave acid decomposition method). An extension of the method included the use of ultrasonic nebuliser, hydride generator or cold vapour trap method (for Hg content) as a means of improving sensitivity of measurements for respective metals.

Method. Weigh accurately approximately 0.75 g of plastic pieces of sizes 4×4 mm into a microwave vessel followed by 5.0 ml of conc. H_2SO_4 and charred as in Table 4.8. The charred sample is then predigested for 1 h in 10.0 ml of conc. HNO_3 and 2.0 ml H_2O_2. The mixture is digested at elevated temperature and pressure using the microwave oven. Finally, the acid volume is reduced using a micro-vap attachment. The digested samples are then allowed to cool for 1 h, the contents transferred to a 25.0 ml volumetric flask and the digestion vessel washed with diluted deionised water into the 25.0 ml volumetric flask and diluted to mark with the water.

A second vessel containing approximately the same known weight of plastic sample is 'spiked' with 100 ppm Cd, Pb, Cr, Sn and 10 ppm Hg and the mixture prepared as above. (The concentration of 'spiked' metals is added at one-tenth of the maximum concentration that RoHS allows.)

A third solution is also prepared without sample using the same concentration of acid/H_2O_2 and digested and prepared as blank.

All samples are analysed for Cd, Pb, Cr against standard calibration curves prepared from 0.0, 0.5, 2.5, 5.0 and 10.0 ppm of each metal in 0.25 M HNO_3. The ultrasonic nebuliser is used for the determination of Cd, Pb and Cr while the continuous cold vapour trap method is used for the determination of Hg. The recovery of each metal is determined for each metal. The Hg forms the vapour ion of the metal in solution after reduction with $SnCl_2(Sn^{2+} + Hg^{2+} \rightarrow Sn^{4+} + Hg^0)$ and the metallic mercury is swept to the plasma torch by the argon gas. This method is sensitive for Hg and has the advantage that it removes the analyte from the main solution and has very low limits of detection.

Results. Results are shown in Table 4.13 and 4.14. Figure 4.4 is a flow chart for analysis of plastics for the presence of toxic metals.

Table 4.13 *Results of analysis of PE, poly(phenylethene) (PPE) and PET for RoHS listed metals*

Sample	Cd (ppm)	Pb (ppm)	Cr (ppm)	Hg (ppm)
PE	<1.0	16	2.3	<1.0
PPE	<1.0	<1.0	4.3	<1.0
PET	<1.0	6.6	<1.0	<1.0

Table 4.14 *Results of recovery study of 'spiked' metals*

Sample	Cd (ppm)	Pb (ppm)	Cr (ppm)	Hg (ppm)
PE	e.g. 97.6(98%)	121(107%)	101.5(99%)	42(84%)
PPE	95.4(95%)	95(95%)	109(105%)	44(88%)
PET	95(95%)	111(104%)	93(93%)	41(82%)

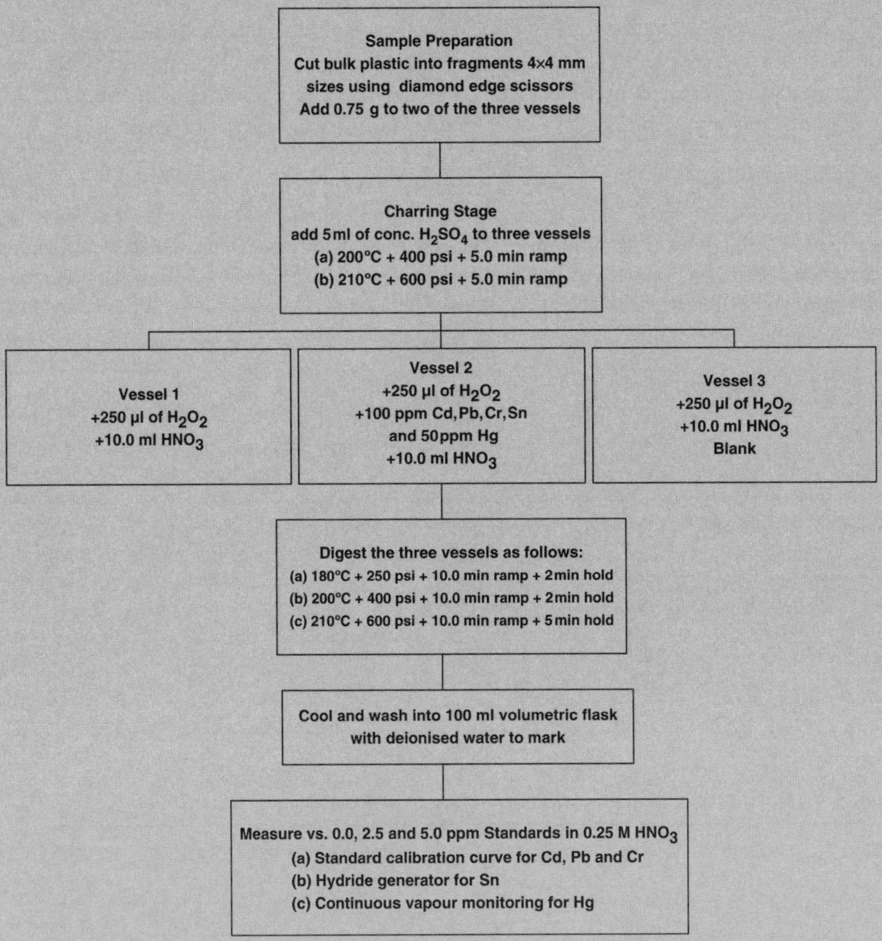

Sample Preparation
Cut bulk plastic into fragments 4×4 mm sizes using diamond edge scissors
Add 0.75 g to two of the three vessels

Charring Stage
add 5 ml of conc. H_2SO_4 to three vessels
(a) 200°C + 400 psi + 5.0 min ramp
(b) 210°C + 600 psi + 5.0 min ramp

Vessel 1
+250 µl of H_2O_2
+10.0 ml HNO_3

Vessel 2
+250 µl of H_2O_2
+100 ppm Cd, Pb, Cr, Sn and 50 ppm Hg
+10.0 ml HNO_3

Vessel 3
+250 µl of H_2O_2
+10.0 ml HNO_3
Blank

Digest the three vessels as follows:
(a) 180°C + 250 psi + 10.0 min ramp + 2 min hold
(b) 200°C + 400 psi + 10.0 min ramp + 2 min hold
(c) 210°C + 600 psi + 10.0 min ramp + 5 min hold

Cool and wash into 100 ml volumetric flask with deionised water to mark

Measure vs. 0.0, 2.5 and 5.0 ppm Standards in 0.25 M HNO_3
(a) Standard calibration curve for Cd, Pb and Cr
(b) Hydride generator for Sn
(c) Continuous vapour monitoring for Hg

Figure 4.4 *Flow chart for sample preparation for analysis for the presence of toxic metals as required by the RoHS, WEEE and ELV directives of plastics used in electrical and electronic equipment*

Method. Weigh accurately approximately* 0.5 g of each sample [PVC, protective coating material (PCM), epoxy, cyanoacrylate (CA), polyurethane adhesives (PU) and grounded computer motherboard (CPU)] into a digestion vessel. Samples are pre-digested in 10.0 ml of conc. HNO_3, 5.0 ml conc. HCl and 2.0 ml of H_2O_2 for 2 h at room temperature. The samples are digested at elevated temperature and pressure using the microwave acid digester. The digested samples are diluted to 50.0 ml with deionised water in plastic volumetric grade 'B' flasks and analysed immediately against 0.0, 0.5, 2.5, 5.0, and 10.0 $\mu g\ ml^{-1}$ of multi-element standards listed in Table 4.13 in the same acid solution.

The samples are also prepared with 100.0 $\mu g\ ml^{-1}$ spikes of Cd, Pb, Cr and 10.0 $\mu g\ ml^{-1}$ Hg in the same way as above. The 'spiked' and unspiked samples are analysed for mercury (Hg) content using the cold vapour trap. The metals Pb, Sn, and Sb are analysed using a hydride generation method and Cd, Cr and Ni are analysed using an ultrasonic nebuliser against standard calibration curves of each metal.

Note: The following is the suggested reaction of Hg^{2+} with stannous chloride ($SnCl_2$) at room temperature: $SnCl_2 + Hg^{2+} \rightarrow Sn^{4+} + Hg^{o} \uparrow$. The conditions in Tables 4.15 and 4.16 are applied to the microwave acid digester. The mercury ions in solution are reduced to metallic mercury and are swept out of the solution by argon gas to the plasma torch for measurement. The advantage of this technique is that it removes the analyte from the sample solution resulting in lower limits of detection.

Table 4.15 *The samples were digested using the microwave digestion conditions in Table 4.15 using 10.0 ml of conc. HNO_3 and 2.0 ml of 30% H_2O_2 initially. The digestion was completed with conc. HF using the microwave acid digestion conditions in Table 4.16*

Stage	Power (W)	% Power	Time (min)	Pressure (psi)	Temp. (°C)	Hold (min)
1	1200	100	10.0	250	150	5.0
2	1200	100	10.0	400	170	5.0
3	1200	100	10.0	500	180	5.0
4	1200	100	10.0	600	190	5.0
5	1200	100	20.0	600	210	20.0

Table 4.16 *Microwave acid condition for rapid digestion after HF addition*

Stage	Power (W)	% Power	Time (min)	Pressure (psi)	Temp. (°C)	Hold (min)
1	1200	100	10.0	200	150	5.0
2	1200	100	10.0	300	160	5.0

Results of analysis. To confirm that this method is suitable for measuring these elements, the test was repeated using the same weights for sample preparation but 'spiked' with

*The expression 'weigh accurately approximately' means to allow the sample weight to be close to the required weight (approximately), e.g. a weight of 10.0 g is required but any weight between 9.5 and 10.5 g is acceptable providing the weight is recorded accurately.

Table 4.17 *Analysis of non-electrical additives in electrical and electronic equipment analysed for toxic metals as required by RoHS and WEEE directives*

Sample	Cd ($\mu g\,g^{-1}$)	Pb ($\mu g\,g^{-1}$)	Hg ($\mu g\,g^{-1}$)	Cr ($\mu g\,g^{-1}$)
PVC	<1.0	<1.0	<1.0	<1.0
PCM	<1.0	<1.0	<1.0	<1.0
Epoxy	<1.0	<1.0	<1.0	<1.0
CA	<1.0	<1.0	<1.0	<1.0
PU	<1.0	<1.0	<1.0	<1.0
CPU	<1.0	467	<1.0	36.6

100.0 $\mu g\,ml^{-1}$ of each metal listed in Table. The sample and 'spiked' were pre-digested and digested using the same conditions above. This test was designed to show that all metals are retained during the sample preparation. The results and percentage recoveries are shown in Table 4.18.

Table 4.18 *Table of results and recoveries of 'spiked' metals after sample preparation using microwave acid digestion*

Sample	Cd ($\mu g\,g^{-1}$)	Pb ($\mu g\,g^{-1}$)	Hg ($\mu g\,g^{-1}$)	Cr ($\mu g\,g^{-1}$)
PVC	106.0(106%)	98.9(98%)	11.6(112%)	107.6(108%)
PMC	98.0(98%)	105.7(106%)	9.8(98%)	110.2(110%)
Epoxy	91.2(91%)	98.0(98%)	9.35(94%)	10.6(106%)
CA	86.2(86%)	90.3(90%)	11.0(110%)	11.2(112%)
PU	103.7(104%)	100.3(100%)	9.55(96%)	9.76(98%)
CPU	109.5(110%)	564.0(97%)	11.16(112%)	11.2(112%)

The recovery of 'spiked' metals was close to 100% for most elements. It shows that this method is acceptable and that the results are reasonably accurate. The percentage recovery for Pb, Cr and Ni is close to 100%. This test shows that none of the listed toxic elements are present or lost during the sample preparation or analysis. The level of toxic metals in PVC, PCM, epoxy, CA and PU are negligible compared with levels in the CPU boards. These are a cause for concern and care should be taken when disposing of all obsolete CPU boards containing these metals, particularly those made for electronic components in the early days of computer manufacturing. These boards have the highest levels of toxic metals present, as shown in Table 4.17.

4.10 Conclusion

Devising the best procedure for analysis of common and complex plastics materials for metals content may need to be studied by trial and error using the range of methods available. The following criteria are important when preparing samples for metal analysis:

(i) precision and accuracy required;
(ii) detection limits;
(iii) availability of suitable sample preparation equipment; and
(iv) availability of sensitive detection/measuring devices.

The methods available in most literature are not necessarily suitable for all samples and often do not include the most modern sample preparation techniques. Most methods often require proof of their reliability by including recovery studies to support confidence in the analysis. Analysis of an unknown plastic material with unknown levels of metals often requires extreme care in all stages of analysis when compared with a known plastic and using a well tried and developed method that has background information. In some cases where measurements of extremely low levels of metals are required, entirely different and more sensitive methods are often used taking special precautions in avoiding contamination and interference.

A study of the leaching ability of metals from plastics is important particularly where children's toys, foods and pharmaceuticals are concerned. A potential plastic to be used for any of these products must be analysed for total and leaching metal content by contacting the plastics with a stimulant similar to the product. Such plastics are strictly monitored and are rejected if a trace concentration of one or more listed toxic elements is detected.

The analysis of plastics for hazardous metals e.g. Cd, Pb, Cr(total) and Hg is now essential on all products including plastics used in electric and electronic equipment sold within the European states. Directives for these products have been issued under 'Packaging Directive', 'End of Life Vehicle' (ELV), 'Reduction of Hazardous Substances' (RoHS) and 'Waste Electrical and Electronic Equipment' (WEEE), and have been introduced to control the risk to health and waste disposal in the environment. The listed metals have been used extensively in the past as pigments, stabilisers and catalysts, especially PVC, and these toxic elements can be released over time into the environment. Therefore, polymer producers are advised that future products be lower than EU values or free of these metals.

The digestion procedures of most plastics demand a choice of acids for charring, i.e. $H_2SO_4 + H_2O_2$ and digestion, e.g. conc. HNO_3. Nylon is an example of digestion without the 'pre-digestion or charring stage' that can be digested with ease in conc. HNO_3 only under microwave conditions, while most others require more rigorous treatments. Samples are only considered metal compliant when their levels are less than the EU directives maximum allowable values.

References

[1] Hosler, D., Burkett, S.C. and Tahkanian, M.J. (1999) Prehistoric polymers: rubber processing in Mesoamerica, *Sciences*, **284**(5422), pp1988–1991.

[2] Plastics Historic Society and Encyclopaedia Britannica (2005) History of plastic, *Bulletin for the History of Chemistry*, **30**(1).

[3] Kettering, C.F. (1945) *Biographical Memoirs of Leo Hendrik Baekeland 1863–1944*, National Academy of Sciences of USA.

[4] Nicholson, J.W. (2006) *The Chemistry of Polymers*, London: Royal Society of Chemistry.

[5] Young, R.J. (1987) *Introduction to Polymers*, London: Chapman & Hall.

[6] Price, W.J. (1979) *Spectrochemical Analysis by AAS*, London: Heyden & Sons Ltd, pp243–246.

[7] EU Directives 2002/95/EC, 2000/53/EC, 91/338/EEC, 2002/96/EC, 'Restriction of Use of Certain Hazardous Substances in Electrical and Electronic Products.' BS EN 1122:2001 method B.

5

Metal Analysis of Virgin and Crude Petroleum Products

5.1 Introduction

Crude oil or petroleum product is a substance on which modern life depends. One-third of the world's power and most organic chemicals come from it. It is used as fuel in automobiles, aircraft and ships, and provides heat and light to homes, schools and businesses. It can be distilled to form lubricants to keep machinery running smoothly and it can give bitumen residues that are used as tar to surface roads and runways for aircraft. The derivatives from crude oils can also be used to make plastics, fibres, drugs, detergents, paints, polishes, ointments, solvents, insecticides and weed killers.

The origin of crude petroleum is uncertain, but it is believed to have been formed millions of years ago by the decay of marine plants and animals that had become covered with layers of rock, silt and mud. Pressure of layers of rocks and heat from the earth's core caused decomposition of the organic matter and resulted in the formation of crude petroleum; this theory is supported by the evidence of methane gas which comes mainly from the decaying of animal and vegetable matter aided by some aerobic bacteria. Crude petroleum is a naturally occurring complex mixture made predominantly of hydrocarbons and is a term for unprocessed oil that originates in the earth. It can also be described as a fossil fuel because it is made naturally from decaying plants and animals living in old seas millions of years ago. It can be clear to black in colour and exist as a liquid of low viscosity to semi-solid states. The process cycle is believed to involve petroleum generation and migration from the source rock to reservoirs, maturation and alteration, which changes the composition of the petroleum after it has accumulated in the natural reservoir.

The relative abundance of members of homologous series is often similar to living systems e.g. isoprenoids, porphyrins, steranes, hopanes, etc., and the preference for odd numbered long chain normal alkanes is well documented [1]. The isotopic composition

A Practical Approach to Quantitative Metal Analysis of Organic Matrices Martin Brennan

of oils, the elemental composition and the presence of petroleum-like material in recent sediments are consistent with low temperature origin of the crude oils in the ground and there is evidence of a biological source of these materials. It has been shown by reactions studied under laboratory conditions that the formation of crude petroleum is induced by temperature and time i.e. a slight increase in temperature increases the rate of reaction, which suggests that it follows a first order reaction. The occurrence of the crude oils can also be created by use of marine organisms. As these organisms died they sank to the bottom of the seabed and were buried in the sand and mud forming an organic rich layer that turned into sedimentary rock, and the process repeated itself layer upon layer. The seas withdrew over millions of years and the sedimentary rock contained sufficient oxygen to completely decompose the organic material. The bacteria may have broken the trapped material into organic substances rich in carbon, hydrogen and sulphur. The weight of layers above it caused partial distillation of the organic remnants, transforming them into crude oils and natural gases, all differing in chemical composition.

The crude oils lodged in the earth's rock basins are now such a commercial commodity that the financial world operates around them. The spin-offs have been such a commercial success that advancements into the future are certainly a real possibility.

The earliest use of oils goes back to 4000 BC when they were used for setting jewellery, mosaics, adhesives, embalming and for cementing (e.g. walls of Babylon, the Pyramids). Although these oil materials were discovered a long time ago it is only since the 1920s that they have made an impact on the world market. In the 18th century early exploratory equipment was used to search for water and brine water, a source of salt; the workers were disappointed to have struck oil. An early use for these oils was to run kerosene lamps (1850) and they became part of an initial worldwide cheap source of light. Large wealthy mechanised industries e.g. Shell, Mobil Oils, etc., have been set up worldwide in order to keep pace with modern demands for oils and their products produced from crude oils.

5.2 Brief Introduction to Refining Process in the Petroleum Industry [2]

The complexity of a refining process varies from one refinery to another. In general, the more sophisticated the refinery the better its ability to upgrade crude oil into high value products and the cheaper they are to produce. Oil refineries perform three basic steps: fractionation, conversion and treatment.

The first stage in the refining fractionation process involves separating the hydrocarbon mixtures into fractions of different boiling points using tall fractionation columns that are 80 m high and contain intervals of horizontal trays with holes in them. The temperature decreases from bottom to top allowing separation of each fraction through their different boiling points, drawing off using pipes from each tray. Inside the columns the liquid and vapours separate into components. The lightest fractions are made up of petrol and light petroleum gas (LPG) which rise to the top where they are condensed back to liquids. Medium density liquids e.g. kerosene and diesel oil, stay in the middle. Heavy liquid separates lower down and the heaviest fractions fall to the bottom and are drawn off and used as bitumen (tar). The separated fractions are then transferred into streams for conversion aided by metal catalytic cracking using heat exchange and pressure to crack the heavy hydrocarbon molecules into lighter ones. Fluid cracking is an alternative method that uses intense heat and a metal catalyst to crack heavy fractions into petrol molecules. Hydro-cracking is also used by some oil refineries and operates by using lower temperatures and high pressures aided with a metal catalyst to generate the lighter molecules. The crude is added at the base of the column so that the process is continuous.

The gasoline and naphtha fractions are the fractions from which the petrol (or gasoline) is made using a 'cracking' process and is required in great quantities to meet worldwide demands. The methods of cracking, polymerisation and reforming are carried out in the refineries using metal catalysts which involve breaking down a long chain of hydrocarbon molecules with high boiling points into smaller molecules with lower boiling points, also in the presence of metal catalysts. They are reformed by turning straight chain hydrocarbons into branched chain isomers that are more suitable as a fuel for automobiles. A molecule of $C_{12}H_{26}$ could yield but-1-ene, pentanes, propenes and a host of side products such as alkanes and methanes.

The large-scale manufacture of chemicals from petroleum and natural gases has given rise to a large number of chemical manufacturing companies throughout the world. Chemicals other than petroleum products are also obtained from these products e.g. manufacture of plastics, pharmaceuticals and speciality chemical compounds.

The petroleum industry often characterises crude oils according to their geological origin, each having different and unique properties. They can vary in consistency from a light volatile fluid to a semi-solid and are classified according to the US environmental protection agency as crude oil Types A, B, C and D as follows:

Type A crude oils. These are light, volatile, highly mobile, and clear and spread rapidly on a solid plane or water surface. They have strong odours, high evaporation rate, can penetrate porous surfaces, do not adhere to surfaces, can be flushed with water and are flammable. These oils are toxic to humans, fish and other biota. Most refined products, and many of the high quality light crude, can be included in this class.

Type B crude oils. These are non-sticky oils which can be waxy or have a heavy oily feeling and are less toxic than oils in Type A. They adhere weakly to surfaces and can be removed by strong forceful flushing.

Type C crude oils. These are viscous and sticky, tarry brown products and flushing will not remove them from surfaces. Their density is close to water and some will sink to the bottom. Their toxicity is low and includes some residual fuel oils blended as medium to heavy crude.

Type D crude oils. These are non-toxic and do not penetrate porous surfaces and are black, heavy, semi-solid, tarry bitumen. They also contain traces of residual oils, heavy crude oils and lighter paraffin oils. Analysis of this type of crude oil for metals content, particularly toxic metals, is important because it is used for road surfacing, roofing, children's playgrounds and other uses that could have environmental concerns.

5.3 Metals in Crude Oils and Petroleum Products

The list of metals present in major or trace level in petroleum products can vary from alkali, alkaline, transition and refractory types depending on the source and refining process. The most common metals detected at varying concentrations are vanadium, iron, nickel, copper, sodium, potassium and others in trace quantities, and they vary from the location of source and how long the crude was in the earth. The chemical forms in which these metals occur is largely unknown but as oil deposits occur in the presence of sea water/oil emulsion it is expected that alkali and alkaline metals are in abundance. Other metals may be present as suspended inorganic compounds, as finely divided clays and other mineral matters produced from the reservoir rocks in contact with the crude oils. Co-ordination compounds of vanadyl and nickel porphyrins, for which the metals are obtained from the earth's source have been detected and quantified. Other metals present as organometallic complexes are usually found at lower concentrations. Metal porphyrins with low molecular weights tend to be volatile and those with high molecular weights tend not to be volatile. Most of the metallic species in petroleum products are present as volatile organometallic compounds, inorganic salts, fine metallic particles and colloidal suspensions. Knowledge of total concentration of all metals including toxic metals present in oil products is necessary for production, safety and environmental purposes.

The analysis of metals content in crude oils and feed-stocks is important and requested by most oil companies as some metals can poison expensive, added catalysts used as part of the cracking process or can form volatile organometallic compounds that can transfer from one stage of a process to another. Metals such as sodium, potassium, cobalt, nickel, copper, iron and vanadium are a few that can reduce the catalytic efficiency of the cracking distillation plant. These metals are often present mainly as organometallic compounds that are part of the hydrocarbons during the multi-fractionation process. Some metals can cause corrosion problems, reducing the efficiency of the refining plant while the presence of copper in the refined products promotes formation of gums and lacquers during heating or storage. It may also accelerate the deterioration of these products and reduce their storage stability.

The composition of crude oils differs in different parts of the world as does the type of metal present and concentration. The different metal contents can be useful for identification of the source of oil and also help in determining those responsible as part of forensic and environment pollution investigations. Wear metals in lubricating oil are also monitored to provide early detection and correction of major problems in the various machine parts.

5.4 Requirements for the Determination of Metal Content in Virgin and Crude Oils

Crude petroleum is a naturally occurring complex material composed mainly of the elements hydrogen, carbon, sulphur, phosphorous and nitrogen, as well as a host of organometallic and free inorganic metal salts or complexes in the crude composition as contamination from the earth's surrounding minerals. The crude oil is present as solid (asphalt), liquid (crude oil) and gas (natural gas).

The presence of major and trace metals in crude oils and their products is of considerable interest to the oil industry. This interest derives from the association of metals with the geological origin of the crude and the influence of contamination by metal in the refining or processing of petroleum products, or the effects of metals on the utilisation or performance of the finished products. Varying concentrations of the mineral constituents remain with the oil as it flows on its course through the refinery and, in some cases, this flowing may reduce the heavier mineral content by virtue of 'salting out' en route. Other fractions may see a considerable increase in concentration through pickup from storage vessels or by reactions of side mineral products as a consequence of the formation of metal soaps of sulphonic acids or naphthalenic acids in the treatment process.

The influence of metals in fluid catalytic cracking such as Ni, V, Fe, and Na from the feed-stock may result in expensive catalyst replacements and affect the quality and distribution of the cracked products. The build-up of deposition metals from the continuous process of fuels can cause refinery problems in the heat exchangers, equipment deterioration and sometimes complete failure. On the other hand, controlled quantities added during the processing to generate petroleum products may increase the yields. Some products such as heating oils, detergents, lubricating oils and inhibitors owe their improved characteristics to the presence of some metal compounds.

The list of metals detected in petroleum products is considerable and the most common detected are Fe, Na, Ni, V, Cr, Mg, and Si; other metals at a lower concentration are Sb, Ar, Be, Cd, Co, Pb, Mn, Hg, Mo and Se. The elements Fe, Ni and V are the most important as they are present in all crudes at a higher concentration than other metals and, as expected, can be extracted from the earth's mineral source. The important requirement of knowledge of metals in crude oils and their products has resulted in a considerable effort to develop analytical methods for accurate measurements of the true concentration of all metals present.

To evaluate the influence of given metals in oil products it is fortunate that analytical methods are available that are sensitive, accurate and informative. Analytical methods associated with oils involve step-by-step procedures from sample preparation to suitable solutions for measurements against certified standard calibration curves. The sample preparation step is very important in the analysis, and the method used is decided by the concentration of metal present and whether or not it is soluble in the oil or present as particulates. Preparation by simple dilution may not be sufficient for very low concentrations of elements as the dilution may inhibit the ability to measure low levels due to non-detection, precipitation or settling out of the metals of interest.

Dry ashing of crude oils can cause serious loss of ash or elements through volatility of some metals, even in the presence of metal-retaining compounds. The methods using microwave acid digestion or bomb combustion are suitable for sample preparation for most trace metal analysis because they are retained in solution. This includes those that are volatile. Unfortunately, these methods are time-consuming and can be erroneous, and require experience skilled operators, but are necessary because they are precise, accurate and quantitative.

Analysis of crude and lubricating oils for metal content can be carried out using sequential or simultaneous ICP-OES. If only Fe, Ni and V are required and present in a suitable concentration, it may be possible to prepare the samples using the solvent dilution method, providing the concentration is quantifiable and does not settle or precipitate out of solution. These metals (Fe, Ni and V) are measured against certified standards prepared in the same solvent. If, however, the true concentration at trace level and results are required for toxicological and environmental reasons it may be necessary to use other tedious sample destructive methods, such as oxygen bomb combustion, microwave acid digestion or, in a few cases, aqueous acid extractions.

5.5 Wear Metals and Metal Contaminants in Lubricating Oils [3]

Lubricating oils are used in light and heavy machinery and the analysis for metal content before and after use is very important. Information gained from the analysis is invaluable in establishing the extent of wear of an engine and whether an oil change is required, particularly for large expensive engines. Large engines are used in large vehicles such as buses, lorries, cranes etc., and for people and cargo transport systems, and it is important to monitor carefully the concentration of wear metals in order to extend their life. Additives such as colour improvers, antifoaming agents, oxidation inhibitors, corrosion inhibitors, detergents, pour-point depressants and viscosity index improvers containing metallic salts can also contribute to metals in engine oils.

The increase in metals content can also be indicative of foreign contamination from water coolant leaks, antifreeze, roadside clays, sand, etc. The metal analysis of used oil is usually checked after a period of use or mileage covered and analysis can reveal trends relative to time or distance travelled and warn whether corrective actions need to be carried out before serious catastrophic failure occurs. The metal contaminant can also chemically degrade oil and fluids. In deserts or sandy areas where machines and engines operate, local prevailing dust can give rise to contamination when fine sand dust or clays gets into the engines. These can cause abrasion within the gaps of the cylinder piston and walls of the engine leading to serious damage. Therefore, a metal scan for these metals is essential in order to determine the extent of wear metal or contamination and aid in decision making about whether or not replacement is necessary, to ensure a prolonged life-span of the engine.

The source of metals in engine oil is due to wear from cylinder liners, piston rings, valve train, crank shaft, rocker arms, spring gear, washers, nuts, pins connecting rods blocks, oil pump, gear bearings, brake bands, clutch and shift spools. Hydraulic fluids can also be contaminated by wear metals from pump housing, vanes, cylinder boxes, rods, valves and pistons. Finally, the oils used in turbines can give rise to wear metal contamination that derives from reduction gear, shaft, bearing, piping and the housing case of the turbine motor. Table 5.1 shows metals commonly detected in wear oils and possible reasons for their presence.

Table 5.1 *List of metals present in used engine oils and possible source of contamination from other metal parts associated with engine design*

Metal	Reason for presence
Cu	Ductile, good thermal and electrical conductance
Sn	Alloy with Cu + Pb for sacrificial bearing linings
Al	Light metal, resistance to corrosion and temperature
Cr	Added to increase hardness and corrosion resistance
Pb, Sb	Used for sacrificial wear surfaces, babbits
Si	Silicon presence as contaminant, abrasive properties on engine
Ag	Reduces friction, good thermal and bearing plate, susceptible to Zn based additives
Fe,Ti,Ni, Cd	Wear metals from engine block
Mo	Wear metal, lube friction modifier

Table 5.2 *List of metals present in used engine oils associated with contaminants and additives used in functioning of the engine*

Metal	Possible source
Na, Al, Si	Corrosion inhibitor may indicate coolant leakage, ingested road salt, sea salt, antifoaming agents
B	Corrosion inhibitor, antifreeze, antiwear/antioxidant, coolant leakage, grease
Mg	Detergent/dispersive additive, alloy element in steel
Ca	Detergent/dispersive additive, alkaline, high sulphur fuelled engine
Mo	Anti-wear additive, alloy in bearing and piston rings
Ba	Corrosion and rust inhibitor, dispersant, detergent, antiwear, alloy bearing, etc.
Zn	Galvanise casing, corrosion inhibitor, thrust washers
P	Antiwear, corrosion inhibitor, antioxidant, EP additive
Ti	Wear metal for aircraft engines, bearings, paints
V	Fuel contaminant, alloy metal for steel
Cr	Coolant treatment

Metal additives other than those detected in oils and hydraulic fluids from engines are associated stabilisers, colorants or results of manufacturing process, may also be present. Table 5.2 is a list of metals detected and possible reasons for their presence.

5.6 Brief Outline of the Determination of Metals in Organic Materials Using Atomic Spectroscopy Methods [4]

Major and trace levels of metals in organic materials are relatively easy to measure directly using AAS, ICP-OES and ICP-MS. The AAS can tolerate a wider range of solvents than ICP-OES. In fact, most organic solvents may enhance the sensitivity in AAS when compared with measuring the same concentration of metal in aqueous solution. The increase in sensitivity of organic solutions with the AAS technique is due to the reducing nature of the resulting AAS flame and the ease of vaporisation of the solvent that improves the nebulisation efficiency and a slight increase in the temperature of the flame. Samples not soluble in aqueous media but which form solutions with organic liquids can be used as a simple method of sample preparation. In the case of ICP-OES it does not have this luxury as it can only tolerate a limited number of solvents compared with AAS and does not contribute to analytical signal sensitivity. The many types of oil products and chemicals obtained from them that require metal analysis are raw materials used in paints, oil and colour, plastics, adhesives, pharmaceutical formulations, cosmetics, etc. The main advantage of ICP-OES over AAS is that most metals are more sensitive and have a wider linear working range.

The solution obtained from dissolving an oil in a solvent is the simplest sample technique and involves dissolving a known weight of oil or fluid in a suitable solvent that is compatible, stable and noise free, and can be used for nebulisation with an ICP-OES plasma torch. Crude, lubricating oils and hydraulic fluids are soluble in a few solvents that are compatible with ICP-AES, e.g. kerosene, propylene carbonate, tetralin and decalin. The excitation of elements in solutions can be viewed either with radial or axial torches with the latter giving higher intensity readings and lower limits of detection.

Solvents that have low oxygen content require an auxiliary oxygen gas to assist the combustion of oxygen-starved organic solution. An oxygen module accessory is required to supply a controlled volume of oxygen to the argon line and is also required to reduce the molecular band emissions, and prevents carbon build-up in the tip of the auxiliary tube of the plasma torch when using organic solvents containing little or no oxygen. However, if a solvent such as glacial acetic acid (GAC) that contains approximately 54% oxygen is used, this oxygen module may not be required. The GAC has a further advantage over other solvents in that it behaves reasonably similarly to water; the only difference is that it forms a slightly elongated blue/green bullet in the centre of the plasma. The GAC solvent can be blended with other oil soluble solvents up to 25% maintaining the solubility of the oil and adding oxygen to the plasma to assist in the decomposition of the organic sample/solvent.

The plasma employs direct serial coupling of the radio frequency generator to plasma for a highly efficient transfer of energy to the plasma torch. The energy required for organic solvent is higher than that for aqueous samples and hence requires a higher wattage power to maintain the plasma. For a 40.0 MHz free running radio frequency the generator is set between 1.2 and 1.5 kW in order to maintain the plasma torch compared with 0.9–1.1 kW required for aqueous samples during the analysis; this is coupled with the direct power control system which rapidly responds to sample-induced changes in the plasma impedance. This produces stable, robust and sustaining plasma that is ideal for analysing difficult samples.

Greases and waxes which are semi-solid extracts from residues after cracking of crude oils which are the heaviest contain an abundance of metals that are usually removed by chemical methods. The metal clean 'virgin' grease is 'spiked' with molybdenum to be used as lubricant where liquid oils cannot be used. The original grease contains several metals, particularly Li, Na, Ca, Mg, K, B, P, etc., which are detrimental to the greasing properties. Analysis of this grease is essential to ensure that the grease is clean of these metals prior to 'spiking' with molybdenum. Grease samples can be dissolved in kerosene, decalin and tetralin and analysed for metals against multi-elemental certified standards prepared in the same solvent. An internal standard 'Yttrium' is used to correct for viscosity effects.

5.7 Application of Atomic Spectroscopic Techniques in the Analysis of Virgin and Wear Oils for Metals Content

The application of atomic spectroscopy methods to the analysis of petroleum products is important to the oil industry. All oil samples must be prepared in solution form and be at a concentration so as to be detected to quantify all metals of interest with accuracy and precision. Solutions containing petroleum products in organic solvents may be measured directly or with the use of internal standards to correct for viscosity effects. It is important that the selected solvent dissolves the oil and products and does not cause erratic flickering of the plasma, or quenches it. It is also important that the same solvent can be used to prepare calibration standards. The following methods are common sample preparation methods for metal analysis of crude and lubricating oils.

5.7.1 Choice of Solvents Suitable for Metal Analysis of Crude and Lubricating Oils Using ICP-OES [5]

The difference between AAS and ICP-AES for metal analysis of crude oils and lubricating oils after dilution is the compatibility of solvents. Solvents that are compatible with AAS may not be compatible with ICP-AES. Selection of the most suitable solvent for metal analysis using ICP-OES is important in terms of stability and reproducible measurements. Table 5.3 is a short comparative list of common solvents used in atomic spectroscopy and their behaviour using AAS and ICP-OES and the respective solvent.

Table 5.3 *Comparative list of suitable solvents for AAS and ICP-AES*

Solvent	AAS	ICP-AES
n-Heptane	Yes	(quenches plasma)
p-Xylene	Yes	Poor (wears tubing rapidly)
Dioxane	Yes	Poor (wears tubing rapidly)
White spirit	Yes	OK (wears tubing rapidly)
Kerosene	Yes	OK (wears tubing rapidly)
Tetralin	OK (noisy)	OK
Propyl alcohol/white spirit (20:80)	Yes	OK (noisy)
MIBK	Good	(quenches plasma)
Decalin	Poor (weak signals)	Good
Toluene	Poor (noisy)	Noisy + wears tubing

5.7.2 Selection of Representative Samples in the Study of Metal Analysis of High Viscosity and Low Viscosity Oil Blends

Due to variations of properties of crude oils from around the world it is difficult to obtain an ideal standard crude oil sample that could be used as a control standard. Fortunately, base oil blends with high viscosity and low viscosity free of metals are available from MBH (Conostan, London). These oils, which are representative of a wide range of crude and lubricating oils, are used to study analytical methods in terms of solubility, precision and accuracy using ICP-OES. Table 5.4 lists the properties of two such oils that are used as part of this study.

Table 5.4 *Properties of high viscosity and low viscosity oils that are close to crude and lubricating oils*

Base oil	Specific gravity	Viscosity (cSt) at 40°C	Pour point (°C)	Flash point (°C)	Total metal content $\mu g\,ml^{-1}$
Conostan 254 C 20	0.81–0.86	16	−7	175	<0.1
Conostan 254 C 75	0.86–0.89	69	−15	215	<0.1

5.7.3 Physical Properties of Selected Solvents for Dissolving High Viscosity and Low Viscosity Oils for Metal Analysis

Classification of solubility of crude and lubricating oils and compatibility with ICP-OES is limited to a narrow range of solvents. The most popular solvents employed for the metal analysis of virgin and used lubrication oils are kerosene, xylene, toluene, decalin, tetralin and a special ICP blend (BDH Conostan) premisolv ICP solvent. These solvents will dissolve Type A and Type B crude oils, listed in Section 5.2, up to 20% while Type C, to a maximum concentration of 5% before de-asphalting occurs. Type D crude oil is unsuitable for dissolving and samples for analysis must be prepared by destructive methods, e.g. ashing, microwave acid digestion or bomb combustion. Standards used to determine the concentration of metals in crude oils must also be prepared in the same solvent to dissolve the oil. Table 5.5 is a list of properties of three common solvents available for ICP-OES used as part of this study.

Table 5.5 *Physical properties of solvents used for dissolving Conostan high viscosity oil and low viscosity oil blends for analysis of oils using ICP-AES*

Solvent/property	Kersosene	Decalin	Tetralin
Formula	Petroleum	$C_{10}H_{18}$	$C_{10}H_{12}$
Molecular weight	—	138.3	132.2
Density (g/mL)	0.8	0.9	0.97
pH	—	—	—
Odour	strong	mild aromatic	mild aromatic
Appearance	pale yellow	clear	clear
Boiling point (°C)	150	190	207
Melting point (°C)	−20	−40	−45
Vapour density (g/m^3)	4.5	4.8	4.6
Flammability	very	moderate	moderate
Contact	non	mild	mild
Flash point (°C)	37–65	57	87
Cost	cheap	expensive	cheap

5.7.4 Methods of Sample Preparation for Metal Analysis of High Viscosity and Low Viscosity Oil Blends

Sample preparation of crude and lubricating oils for metal analysis serves many purposes that can vary from sample to sample and the demands of the efficient operation of refining plants. The following sample preparation methods are appropriate to most lubricating oils and are considered optimum in terms of the determination of the concentration of metals present with a high degree of accuracy and precision.

The viscosity range of crude and lubricating oils can vary significantly from different location sources and the distilled version. The determination of metals is necessary prior to refining and 'cracking/fractionation'. The metals are present as inorganic particulates and organometallic compounds in variable concentrations. Metal analysis can be useful for identifying the source of oil by comparing with analytical results of previous known crude oil samples. This information is also useful in assisting in forensic and environmental investigations.

Cost associated with the maintenance of engine and machine wear is significant, hence diagnostic methods for determining the condition of engines and machinery is important. Careful analysis of such lubricating oils for the extent of wear metals is carried out for the purpose of 'engine life' monitoring and can assist in diagnosing if replacement of oil or engine is required or not.

The concentration of metals that are detrimental to catalysts added can vary between $\sim$20.0 ppm for Fe to $\sim$100 ppm for Ni and 1000 ppm for V. The presence of these metals necessitates the need for analysis of these metals to determine their concentrations prior to the cracking process. The 'best' method to analyse these oil samples needs to be rapid and accurate. Careful selection of the method either from experience or by trial and error may be applied depending on the metal and the concentration. Sample dissolution in a solvent or solvent mixture is considered the easiest but may not be suitable for low limits of detection. Destructive sample preparation methods, i.e. oxygen bomb combustion, microwave acid digestion followed by pre-concentrating may be required for trace analysis and/or with the aid of a hyphenated system, e.g. ultrasonic nebuliser. Samples prepared by destructive methods are dissolved in aqueous solutions that have very low matrix and spectral interferences.

Unfortunately, it is not feasible with the design of the modern torch to pump the liquid oils directly as the lower volatiles and the dense fraction of the oil would 'quench' the plasma.

Acid extraction of crude and lubrication oils may also be considered as a method of preparation but it is tedious and requires a 'total hands on approach' that may involve repeated extraction to ensure total removal of metals of interest. This method of sample preparation is prone to errors, gives poor recovery and is not very popular.

Kerosene is a good solvent for use with ICP-AES but is prone to noisy plasma. The solvent tetralin (1,2,3,4-tetrahydronaphthalene) has been used by workers involved in metal analysis of crude and lubricating oils with success. The solvent decalin (decahydronaphthalene) was also found to be a good solvent for metal analysis of crude oils but it is very expensive and not used extensively. The analytical performance of these solvents was studied for stability over an extended period of time to determine the effect of varying viscosities. The solvents toluene and xylene are also good solvents for dissolution but have high background to noise ratio and will not be discussed further.

5.7.5 Long-Term Study of Metal Analysis Using Kerosene, Teralin and Decalin Solvents Using ICP-OES

The three solvents selected as part of this study are found to be suitable for dissolving and analysing crude and lubricating oils for metal content using ICP-AES. In this study the solvents kerosene, tetralin and decalin were used as part of study of metal analysis for metal content of 'spiked' high and low viscosity Conostan 20 and 75 blend oils.

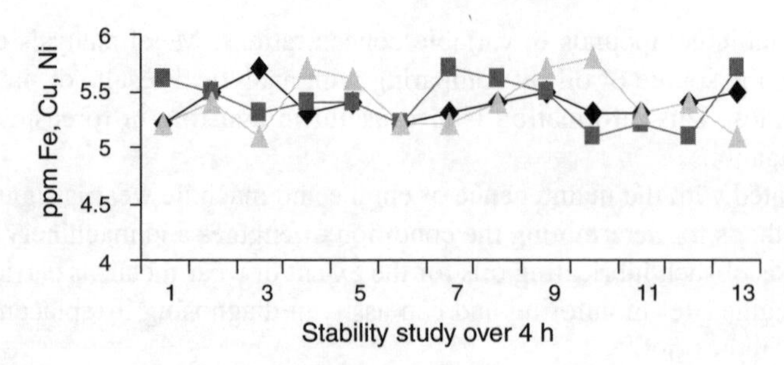

Figure 5.1 *Long-term stability diagram for Cu, Fe and Ni in kerosene, tetralin and decalin solvent. The relative standard deviation over 4 h for each metal is less than 2.5% for tetralin and decalin. The solvent kerosene gave a standard deviation of less than 4.5% for each metal*

The long-term stability study is a useful test to evaluate the effects of analysis over a period of time particularly where several samples have to be analysed at the same time. The test is carried out by preparing copper, iron and nickel metals in kerosene, tetralin and decalin and analysing each element against standards prepared in the same solvent and monitoring over a period of 4 h taking readings every 30 min (Figure 5.1) using the instrument parameters in Table 5.6.

Table 5.6 *Parameters for ICP-AES for long-term study of the performance of metal analysis using kerosene, tetralin and decalin solvents*

Parameter	Setting
Power	1.25 W
Plasma gas flow	15.0 L min^{-1}
Auxiliary gas flow	1.5 L min^{-1}
Spray chamber	HF resistant cyclonic
Radial torch	Fixed assembly
Nebuliser	Cross flow
Integration time	3.0 s
Delay time	60.0 s
Stabilisation time	30.0 s
Sample uptake rate	0.5 ml min^{-1}
Sample read rate	0.25 ml min^{-1}
PMT	850 V

5.7.6 Comparative Study of Non-Destructive Methods of Analysis of Metals 'Spiked' in High Viscosity and Low Viscosity Oil Blends Using ICP-OES

5.7.6.1 Standard Calibration Curve Method. Samples of metal free blank oils of viscosities 20 and 75 (Cat. No. 254 C 20 and 254 C 75), respectively, were 'spiked' with 5.0 $\mu g\,g^{-1}$ (ppm) of Fe, Ni, V, Cu, Na, Ti, and Co using the multi-elemental standards in oil available from MBH Conostan (Cat. No. 252 C 12-500). These metal free blank oils are grit/particle free and dissolved in kerosene, tetralin and decalin as follows:

Method for Preparation of Samples. To three 100.0 ml volumetric flasks accurately weigh approximately 10.0 g of metal free conostan oil 20 blend and dissolve one in ~50 ml kerosene, the second in ~50 ml tetralin and the third in ~50 ml decalin. Repeat this for the metal free conostan 75 blend. To each of the six flasks add 1.0 ml of 500 ppm certified oil based multi-element standard containing the elements Cu, Fe, V, Ni, Na, Ti and Co and dilute to mark with the respective solvent. Each solution is shaken to dissolve the sample and 'spiked' with standards to give 5.0 ppm of each metal for each solution.

Method for Preparation of Standards. Three multi-element stock standard (500 ppm Cu, Fe, Na, V, Ti, Co and Ni metal Cat. No. 252 C–S 12-500 available from Conostan MBH) is used to prepare 0.0, 2.5, 5.0 and 10.0 ppm of each metal by dissolving 0.0, 0.5, 1.0 and 2.0 ml of the stock standard in kerosene, decalin and tetralin, respectively.

Table 5.7 summarises the concentration of sample and 'spiked' metal prepared in the three solvents.

Table 5.7 *Samples of Conostan oil 20 and 75 blends containing listed metal 'spiked' and solvents studied*

Sample	Oil sample	Wt of oil (g)	Solvent	'Spiked' 5.0 $\mu g\,ml^{-1}$ (ppm)
A	Blank oil 20	10.0	Tetralin	Fe, Ni, V, Cu, Na, Ti and Co
B	Blank oil 75	10.0	Tetralin	Fe, Ni, V, Cu, Na, Ti and Co
D	Blank oil 20	10.0	Decalin	Fe, Ni, V, Cu, Na, Ti and Co
E	Blank oil 75	10.0	Decalin	Fe, Ni, V, Cu, Na, Ti and Co
G	Blank oil 20	10.0	Kerosene	Fe, Ni, V, Cu, Na, Ti and Co
H	Blank oil 75	10.0	Kerosene	Fe, Ni, V, Cu, Na, Ti and Co

Low viscosity and high viscosity oil blends are prepared in kerosene and analysed against standards prepared in kerosene. Oil blends prepared in decalin are analysed against standards prepared in decalin. Finally oil blends prepared in tetralin are analysed against standards prepared in tetralin. The instrument is calibrated and equilibrated prior to analysis and the radial torch adjusted to achieve maximum signal response. Standard pump speed, gas flow rate and rinse out times are set as per manufacturer's instruction selecting the most sensitive line for each metal. Each sample is read six times allowing a 2 min wash out time between each measurement to remove memory effects. Extraction unit is switched on at full power to remove any toxic metal ions and vapours.

Table 5.8 *Results of analysis of Conostan 20 oil blend 'spiked' with the listed metals. The results in brackets are scatter for six measurements and symbol σ is used to denote the spread of results over six measurements*

Solvent	Cu	Fe	V	Ni	Na	Ti	Co
Kerosene	5.6	5.6	6.4	6.5	5.8	6.1	7.3
($n = 6$)	*(4.9–6.2)*	*(4.9–6.3)*	*(5.5–7.2)*	*(5.3–7.6)*	*(4.7–6.9)*	*(5.6–6.6)*	*(6.7–7.9)*
σ	*1.3*	*1.4*	*1.7*	*2.3*	*2.2*	*1.0*	*1.2*
Tetralin	4.7	5.2	5.5	5.7	5.4	4.8	5.4
($n = 6$)	*(4.3–5.1)*	*(4.7–5.6)*	*(5.0–5.9)*	*(5.3–6.1)*	*(4.7–6.0)*	*(4.4–5.2)*	*(4.9–5.9)*
σ	*0.8*	*0.9*	*0.9*	*0.8*	*1.3*	*0.8*	*1.0*
Decalin	4.7	4.9	5.0	4.8	5.1	5.2	5.6
($n = 6$)	*(4.4–5.0)*	*(4.5–5.3)*	*(4.7–5.2)*	*(4.4–5.2)*	*(4.6–5.6)*	*(5.0–5.4)*	*(5.2–6.0)*
σ	*0.6*	*0.8*	*0.5*	*0.8*	*1.0*	*0.4*	*0.8*

Results. Table 5.8 shows the results of analysis of the low viscosity 20 oil 'spiked' with 5.0 μg ml^{-1} (ppm) for metals (see Table 5.4) against standard calibration curves generated for each metal.

Table 5.9 shows the results of analysis of the low viscosity 75 oil 'spiked' with 5.0 μg ml^{-1} (ppm) for metals against standard calibration curves generated for each metal.

5.7.6.2 *Standard Calibration Curve with Internal Standard Method.* The above experiments were repeated using samples prepared in Table 5.7 and include yttrium (Y) as an internal standard. The Y internal standard (obtained dissolved in oil available from MBH Conostan) was added to each sample and standard at a concentration equal to the highest concentration of metal, i.e. 10.0 ppm. The internal standard can be used to quantify each of the metals in each solution.

Table 5.9 *Results of analysis of Conostan 70 oil blend 'spiked' with the listed metals. The results in brackets are scatter for six measurements and symbol σ is used to denote the spread of results over six measurements*

Solvent	Cu	Fe	V	Ni	Na	Ti	Co
Kerosene	3.2	4.4	3.4	2.6	3.5	3.1	2.6
($n = 6$)	*(2.9–3.5)*	*(3.2–5.6)*	*(3.0–3.8)*	*(1.9–3.3)*	*(2.2–4.7)*	*(2.4–3.8)*	*(2.1–3.2)*
σ	*0.6*	*1.4*	*0.8*	*1.4*	*1.5*	*1.4*	*1.1*
Tetralin	2.7	2.8	2.9	3.4	3.2	3.3	3.2
($n = 6$)	*(2.1–3.3)*	*(2.3–3.3)*	*(2.1–3.6)*	*(2.3–4.4)*	*(2.8–3.6)*	*(2.7–3.9)*	*(2.3–4.1)*
σ	*1.2*	*1.0*	*1.5*	*2.1*	*0.8*	*1.2*	*1.8*
Decalin	3.1	2.5	2.4	3.4	2.8	3.2	2.6
($n = 6$)	*(2.0–4.2)*	*(1.9–3.1)*	*(1.8–2.9)*	*(2.1–4.7)*	*(1.9–3.7)*	*(2.4–4.0)*	*(2.0–3.2)*
σ	*2.2*	*1.2*	*1.1*	*2.1*	*1.8*	*1.6*	*1.2*

Table 5.10 Results of analysis of metal-free blank oil of viscosity 20 'spiked' with metals listed in Table 5.7 using Y as internal standard. The results in brackets are the scatter of six measurements of the same sample. The symbol σ is used to denote the spread of results

Solvent	Cu	Fe	V	Ni	Na	Ti	Co
Kerosene	5.3	4.7	5.2	5.4	5.8	4.8	5.3
($n = 6$)	(4.9–5.5)	(4.5–4.9)	(4.8–5.5)	(5.2–5.7)	(5.5–6.1)	(4.6–5.0)	(5.0–5.6)
σ	0.6	0.4	0.7	0.5	0.6	0.4	0.6
Tetralin	5.2	5.3	4.5	4.2	3.8	4.8	4.4
($n = 6$)	(4.6–5.8)	(4.7–5.9)	(4.0–5.0)	(3.3–4.9)	(3.1–4.7)	(4.0–5.6)	(3.9–4.9)
σ	1.2	1.2	1.0	1.6	1.6	1.6	1.0
Decalin	4.7	4.9	5.2	4.8	5.1	4.9	4.6
($n = 6$)	(4.4–5.0)	(4.7–5.1)	(4.9–5.4)	(4.6–5.0)	(4.7–5.5)	(4.7–5.1)	(4.2–5.0)
σ	0.6	0.4	0.5	0.4	1.0	0.4	0.8

Table 5.10 shows the results of analysis of the low viscosity 20 oil 'spiked' with 5.0 $\mu g\,ml^{-1}$ (ppm) for metals listed in Table 5.7 using Y as internal standard against standard calibration curves generated for each metal.

Table 5.11 shows the results of analysis of the low viscosity 75 'spiked' with 5.0 $\mu g\,ml^{-1}$ (ppm) for metals listed in Table 5.7 using Y as internal standard against standard calibration curves generated for each metal.

5.7.6.3 Analysis Using Method of Standard Addition. Sometimes it is not possible to overcome interference effects using standard calibration curves or internal standard methods of analysis. However, a third method involving the standard addition may be used to achieve reproducible results. Under these conditions all solutions will have the same matrix composition, so influence of the matrix will be the same. It must be emphasised that this method only corrects for the slope of the calibration curve i.e. element measurements and not for effects of sample plasma noise, shifts, etc. The oil

Table 5.11 Results of analysis of metal free oil of viscosity 75 'spiked' with metals listed in Table 5.7 using Y as internal standard. The results in brackets are the scatter of six measurements of the same sample. The symbol σ is used to denote the spread of results

Solvent	Cu	Fe	V	Ni	Na	Ti	Co
Kerosene	5.2	5.4	5.4	5.1	5.9	5.1	5.6
($n = 6$)	(4.9–5.5)	(5.2–5.6)	(5.0–5.8)	(4.8–5.6)	(5.4–6.4)	(4.8–5.3)	(5.2–6.0)
σ	0.6	0.4	0.8	0.8	1.0	0.5	0.8
Tetralin	5.5	5.2	5.6	5.4	5.2	5.3	5.2
($n = 6$)	(5.2–5.8)	(4.9–5.5)	(5.2–6.0)	(5.0–5.8)	(4.9–5.5)	(5.1–5.5)	(5.0–5.4)
σ	0.6	0.6	0.8	0.8	0.6	0.4	0.4
Decalin	5.1	5.2	4.8	4.9	5.2	5.1	5.5
($n = 6$)	(4.9–5.1)	(4.9–5.5)	(4.6–5.2)	(4.7–5.1)	(4.9–5.5)	(5.0–5.2)	(4.8–5.8)
σ	0.2	0.6	0.6	0.4	0.6	0.2	1.0

samples must not have any positive or negative response other than due to elements present in the sample. This method is tedious as it involves preparing a minimum of three samples and must be used to generate a calibration for each metal. The principle is that a standard solution of known concentration of metal is added to the sample requiring analysis so that matrix effects are eliminated, and the sample response is increased linearly by increasing 'spiking' of a known concentration of metal. A calibration curve is generated for each metal and is extrapolated to the negative concentration line. Modern advanced computer programs can carry out calculations associated with these methods with ease.

Method. The sample (in this case the metal-free blank Conostan 75 viscosity oil is 'spiked' with known concentration of metals listed in Table 5.7 and is divided into four aliquots. To the four flasks add known increasing concentrations of the standard control stock solution (500 ppm of each metal) to 10.0 g of sample of to give 0.0, 2.5, 5.0 and 10.0 $\mu g\, g^{-1}$ of multi-elemental standard when diluted to 100 ml in each solvent. The preparation is carried out using the solvents kerosene, tetralin and decalin made up to 100 ml. The samples are analysed and the linear curve is extrapolated to the negative concentration line to determine the concentration of each metal in the original 'spiked' sample.

Results. Table 5.12 gives the results of the standard addition study.

Table 5.12 *Results of standard addition study (results in brackets are percentage recovery)*

Metal	Kerosene	Tetralin	Decalin
Cu	4.4 (88)	4.3 (86)	4.7 (94)
Fe	4.8 (96)	4.6 (92)	4.4 (88)
V	4.9 (99)	4.9 (98)	5.1 (102)
Ni	5.3 (106)	5.4 (108)	5.1 (102)
Na	3.9 (78)	3.8 (76)	3.3 (66)
Ti	4.9 (98)	5.2 (104)	5.3 (106)
Co	3.9 (78)	4.2 (84)	4.6 (92)

5.7.6.4 Conclusion to Study of Non-Destructive Methods of Metal Analysis of Oil Products. The results in Table 5.9 show accurate results for analysis of 'metal spiked' low viscosity (Conostan 20 blend) oil when analysed against a standard calibration curve in solvents kerosene, decalin and tetralin, respectively. The scatter of results for the six measurements of each sample is acceptable. The results for higher viscosity (Conostan 75 blend) oil gave consistently lower values, which illustrates the effect of viscosity on the nebulisation efficiency.

The method using Y as internal standard gave good results for both viscosities after analysis in each solvent with recoveries ranging from 88 to 106% which are acceptable.

The method of standard addition also gave good results returning between 78 and 108% recovery after analysis in each solvent. The method was tedious and time-consuming.

The element yttrium was selected as internal standard because of its rare presence in crude oils and it behaves similarly in terms of atomisation efficiency to most elements present in crude oils.

5.8 Analysis of Type C and D Fractions for Metal Content Using Dry Ashing Method [6]

The dilution methods of sample preparation of petroleum fractions for metal analysis are limited to the concentration of metals in each fraction. The dilution method is applicable for routine monitoring crude and lubricating oils, providing the concentration of metals is quantifiable using standard calibration curve, internal standard or standard addition method. The method is unsuitable for low concentration of metals especially as those can build up accumulatively, causing poisoning in all catalytic fractionation plants. The low concentration of toxic metals may be undetected by dilution methods and may escape monitoring if a more sensitive method is not used.

The ashing method is applicable only for heavier oils i.e. Type C and Type D samples for metals that do not volatilise or sample 'spit' during the ashing program. This method can only be used where the metal(s) concentrations are very low and will be retained in the presence of a retaining compound (PTSA) during the dry ashing to 550–600°C. The resulting ash is dissolved in 1.0 M HCl or HNO_3 and conc. HF where necessary. A large sample size may be ashed and dissolved in a small volume of mild acid solution allowing low levels of metals to be detected.

Method. Into two platinum clean dishes weigh accurately approximately 2–5.0 g of Type D fraction (tar fraction) and into one add 50.0 ppm of Fe, Cu, Si, P and Co and the second contains no metal standard. To both dishes add 0.1 g of PTSA and ash using a ramping program to a maximum temperature of 550–600°C. After completion of the ashing the dishes are allowed to cool to room temperature. The resulting ash is dissolved in 10.0 ml of 1.0 M HCl and 1.0 ml of conc. HF with a little heat. The acid solution containing the dissolved ash is transferred to a 25.0 ml volumetric flask and made up to mark with deionised water. Standards containing 0.0, 1.0, 2.0 and 5.0 $\mu g\,ml^{-1}$ are prepared in the same acid mixture(s) and diluted to mark with deionised water.

Results. The results obtained by this experiment were between 95 and 109% which is indicative of a good recovery for the listed elements using the ashing method. The metals Pb and Sn were included to test for their volatility in this type of sample (Table 5.13). The method works well for some metals but not for volatile metals. These samples can also be prepared using microwave acid digestion or bomb combustion methods.

Table 5.13 *Results of analysis with and without metal 'spike' of Type D tar sample*

Metal	Added (ppm)	Total metal found (ppm)	Results found (ppm)	% recovery
Fe	50.0	108	54	108
Cu	50.0	53.7	1.2	105
Si	50.0	169	127	84
P	50.0	93	46	94
Co	50.0	55	0.6	109
Pb	50.0	2.1	−47.9	loss
Sn	50.0	1.4	−58.6	loss

5.9 Analysis of 'Metal Spiked' Oil Blends Using Microwave Acid Digestion for Metals Content

Microwave acid digestion method is suitable for all fractions of petroleum products. The method involves complete breakdown of samples in a specially designed microwave digestion vessel under enclosed conditions and a mixture of appropriate strong acids. The highfrequency microwave oven can achieve a temperature of up to ~220 °C and pressure to 800 psi, which are necessary to decompose the wide variety of oil fractions. Some distillation fractions may contain insoluble grit and particulates that are in suspension and must be dissolved in strong mineral acids including conc. HF if necessary. Sample preparation using the acid digestion method are the most informative, precise and accurate as it involves total destruction of the sample while retaining the metals. The following 'spike' and 'non-spike' metal samples of 20 and 75 viscosity oil blends were prepared for metal analysis using the microwave acid digestion technique as follows:

Method

Vessels 1 and 2

Weigh accurately approximately 1.0 g each of 20 and 75 viscosity of metal-free blank oil blend into plastic vessels suitable for microwave acid digestion followed by 10 ml of conc. HNO_3.

Vessels 3 and 4

Weigh accurately approximately 1.0 g each of 20 and 75 viscosity oil blend into a plastic container suitable for microwave acid digestion and add 0.5 ml of 500 ppm multi-element standard containing the metals of interest followed by 10.0 ml of conc. HNO_3.

Acid blank vessel

A third vessel is also prepared containing the same acid but no sample or metal 'spike' standard.

The mixtures in the five vessels are allowed to predigest at room temperature for 1.0 h. The microwave temperature and pressure conditions selected are shown in Table 5.14.

Table 5.14 *Parameters and conditions for microwave acid digestion of low and high viscosity oil blends*

Stage	Power (W)	% power	Time (min)	Pressure (psi)	temperature (°C)	Hold (min)
1	600	100	10	150	140	5.0
2	600	100	10	200	150	5.0
3	600	100	10	350	160	5.0
4	600	100	10	500	180	5.0
5	600	100	10	600	200	10.0

On completion of the digestion cycle in Table 5.14, the vessels are allowed to cool to room temperature and the digested samples transferred to 50 ml volumetric flasks and washed with deionised water up to mark. The final dilution of the 'spiked' samples should contain 5.0 $\mu g\ ml^{-1}$ (ppm) of each metal. Multi-elemental standards of 0.0, 2.5, 5.0 and 10.0 $\mu g\ g^{-1}$ (ppm) of each metal are prepared in 0.5 M HNO_3. The analysis is carried out after selecting the most sensitive line and instrument parameters. The sensitivity is enhanced using a CETAC ultrasonic nebuliser. The results are calculated by subtracting the blank oil blends results from the 'spiked' oil blends readings (Table 5.15).

Note: In the case of real crude oils, samples may contain clays, SiO_2, etc., that may be insoluble using this acid. To dissolve these salts concentrated H_2SO_4 and/or HF would be sufficient. The acid concentrations may be reduced using a CEM micro-vap attachment.

Table 5.15 *Results of analysis of oils of viscosities 20 and 75 after acid digestion in a microwave oven. Recoveries are in brackets. All results are in $\mu g\ ml^{-1}$ (ppm). The blank oil blends contained no detectable metals*

Metal	Blank 20 oil	'Spiked' 20 oil	Blank 75 oil	'Spiked' 75 oil
Cu	<0.001	4.8 (96%)	<0.001	5.2 (104%)
Fe	<0.001	5.2 (104%)	<0.001	5.1 (102%)
V	<0.001	4.9 (98%)	<0.001	4.9 (98%)
Ni	<0.001	4.9 (98%)	<0.001	4.8 (96%)
Na	<0.001	5.1 (102%)	<0.001	4.7 (94%)
Ti	<0.001	5.1 (102%)	<0.001	5.2 (104%)
Co	<0.001	5.1 (102%)	<0.001	5.3 (106%)

5.10 Analysis of 'Metal Spiked' Oil Blends Using High Pressure Oxygen Combustion for Metals Content

Digestion of samples using high pressure oxygen bomb combustion is an excellent technique for sample preparation, particularly for the presence of trace volatile metals. The method breaks down the sample under violent combustion conditions using high pressurised oxygen in an enclosed vessel. This technique can be applied to almost all crude oil fractions resulting in solutions that are clean and easy to analyse. They do not contain a high concentration of acids or bases and metals content is determined against standards prepared in the same solvent added to the bomb. The disadvantage associated with this technique is that it requires skilled operators to achieve precise and accurate analysis. The high viscosity 75 oil blend is combusted with and without the addition of a control standard (0.5 ml of the 500 ppm multi-element standard) using the parameters described for the microwave digester (Table 5.14). Volatile metals Sn and Pb were included at the same concentration as the multi-elemental control standard.

Method. Weigh accurately approximately 1.0 g of metal-free 75 viscosity oil blend into a shallow platinum dish and place it on a wire sample holder attached to the lid of the bomb. A 10 cm length of nichrome or platinum wire is connected between the electrodes and 5.0 ml of water or 0.05 M NaOH is added to the bottom of the bomb. To the water add 0.5 ml of 500 $\mu g\, ml^{-1}$ (ppm) multi-element standard (252C-S21-500) which gives 5.0 $\mu g\, ml^{-1}$ (ppm) of each metal including the extra two metals Sn and Pb. The bomb is assembled according to the manufacturer's instructions and filled with oxygen to 30 atm. The pressurised oxygen vessel is completely submerged in a water tank and checked for leaks. Assuming no leaks are detected, the bomb is fired to combust the sample. After combustion the bomb is allowed to cool and the excess resulting gases (mainly CO_2 and H_2O) are released slowly through a control vent for 2.0 min. The vessel is opened and contents are transferred to a 50 ml volumetric flask, and the vessel is washed with deionised water and made up to mark. The metal analysis is carried out against standards of 0.0, 2.5, 5.0 and 10.0 ppm of each metal prepared in deionised water.

The sample preparation is also repeated using the same oil but no 'spiked' metal standards are added and used as sample blank.

Results. Table 5.16 gives the results of metal analysis.

Table 5.16 *Results of metal analysis of high viscosity 75 oil blend 'spiked' with metals using high pressure oxygen combustion*

Metal	Blank 75 oil	'Spiked' 75 oil	% recovery
Cu	<0.001	4.9	98
Fe	<0.001	4.9	98
V	<0.001	5.1	102

Table 5.16 (Continued)

Metal	Blank 75 oil	'Spiked' 75 oil	% recovery
Ni	<0.001	5.1	102
Ti	<0.001	5.2	104
Co	<0.001	4.9	98
Pb	<0.001	4.5	90
Hg	<0.001	4.6	92

5.11 Comparative Study of Analysis of Trace Levels of Toxic Metals Using Microwave Acid Digestion and Oxygen Bomb Combustion

The benefit of sample preparation techniques using microwave acid digestion and bomb combustion is that the sample is totally enclosed during the decomposition. These methods remove matrix interference and generate aqueous solutions, which can be analysed using ICP-OES. Sub-trace concentrations can be detected when hyphenated attachments are used, e.g. ultrasonic nebuliser, hydride generation or continuous cold vapour method. These methods are essential where trace levels of toxic elements are present that need to be identified and quantified.

The following comparative study was carried out to test whether or not these sample preparation methods are capable of detecting low levels of $0.25\,\mu g\,ml^{-1}$ or $250\,\mu g\,L^{-1}$ (ppb) of these metals. The sample of high viscosity Conostan 75 oil blend was 'spiked' with $0.25\,\mu g\,ml^{-1}$ of As, Cd, Hg, Pb, Cr, Se and Sn and digested as follows:

Method

1. Sample preparation by microwave acid digestion

Weigh accurately approximately 1.0 g of Conostan 75 high viscosity oil blend into a plastic container suitable for microwave acid digestion. To this add 0.125 ml of 100 ppm multi-element standard containing As, Hg and Se, respectively (MBH Cat. No. 255C 100) and 0.025 ml (25 µl) of 500 ppm multi-element standard containing Cd, Pb, Cr and Sn, respectively (MBH 252C 12-500). The oil blend and metals are pre-digested in 10.0 ml of conc. HNO_3 for 1 h.

They are digested using the microwave conditions in Table 5.17.

Table 5.17 *Microwave conditions for digestion of Conostan 75 high viscosity oil blend 'spiked' with toxic metals*

Stage	Power (W)	% power	Time (min)	Pressure (psi)	Temperature (°C)	Hold (min)
1	600	100	10	150	140	5.0
2	600	100	10	200	150	5.0
3	600	100	10	350	160	5.0
4	600	100	10	400	180	10

After digestion the vessels are allowed to cool to room temperature and the solution transferred to a 50 ml glass volumetric flask and washed with deionised water up to the mark.

The metals As, Cd, Cr, Pb and Se are analysed against multi-element standards of 0.0, 0.1, 0.25, 0.5 and 1.0 ppm prepared in 0.1 M HNO_3 using ICP-OES with the CETAC 5000AT ultrasonic nebuliser attachment. The element Hg is measured using the continuous cold vapour method.

2. Sample preparation by bomb combustion

Weigh accurately approximately 0.75 g of Conostan 75 high viscosity oil blend into a shallow platinum dish which is placed on a wire sample holder attached to the lid of the bomb. A 10 cm length of nichrome or platinum wire is connected between the electrodes. Then 5.0 ml of water or 0.05 M NaOH is added to the bottom of the vessel. To the aqueous solution add 0.125 ml of 100 ppm multi-element standard containing As, Hg and Se, respectively (MBH Cat. No. 255C 100) and 0.025 ml (25 μl) of 500 ppm multi-element standard containing Cd, Pb, Cr, and Sn, respectively (MBH 252C 12-500).

The bomb is assembled according to the manufacturer's instructions and filled with oxygen to 30 atm. The pressurised oxygen vessel is completely submerged in a water tank and checked for leaks. Assuming no leaks are detected, the bomb is fired to combust the sample. After combustion the bomb is allowed to cool and the excess gases (mainly CO_2 and H_2O) are released slowly through a vent for approximately 2.0 min. The vessel is opened and contents are transferred to a 50 ml glass volumetric flask and the vessel washed and made up to mark with deionised water. The metals are analysed against multi-element standards of 0.0, 0.1, 0.25, 0.5 and 1.0 ppm prepared in 0.1 M HNO_3 using ICP-OES with the CETAC 5000AT ultrasonic nebuliser attachment. The element Hg is measured using the continuous cold vapour method.

Results. Table 5.18 gives the results of the comparative study.

Table 5.18 *Results of comparative study of analysis of 250 μg kg^{-1} of 'spiked' toxic metals after sample preparation using microwave acid digestion and oxygen bomb combustion. Results in brackets are percentage recovery*

Metal	Microwave + micro-vap	Bomb combustion	Nebuliser aids
As	119.0 (47.6%)	249.0 (99.6%)	USN
Cd	243.0 (97.2%)	246.0 (98.4%)	USN
Cr	247.0 (98.8%)	252.0 (100.8%)	USN
Hg	183.0 (74.4%)	246.0 (98.4%)	CCV
Pb	252.0 (100.8%)	255.0 (102.0%)	USN
Se	206.0 (82.4%)	253.0 (101.2%)	USN
Sn	244.0 (97.6%)	252.0 (100.8%)	USN

USN, ultrasonic nebulisation; CCV, continuous cold vapour.

5.11.1 Conclusion to Trace Analysis of Toxic Metals in Oil Products

The analysis of petroleum products for the presence of toxic metals is an environmental requirement and results of analysis must be made known to all concerned bodies including the public. In order to establish the true concentrations present, a reliable method must be employed in order to be certain of the results. A sensitive method must be available and the use of an ICP coupled with an ultrasonic nebuliser, hydride generation or cold vapour method for Hg offer the most sensitive and reliable methods available. If analysing for lower levels of these metals, a mass spectrograph with hyphenated attachments may be required.

In order to analyse crude petroleum products for these metals the sample must be prepared in such a way that the matrix interference is reduced or removed while at the same time retaining the metals of interest. The two best methods available are microwave acid digestion and high pressure oxygen combustion. The results tabulated in Table 5.18 as carried out in this study show that the microwave acid digestion method is not suitable for As, Hg and Se, which only gave 34–82% recovery, while the remaining metals were closer to 100% recovery for the 'spiked' metals. The high pressure oxygen bomb combustion gave recoveries of 95–106% for all the 'spiked' metals which is acceptable. The 'spiked' metals from the standards may not be representative of these metals present in a real crude fraction.

5.12 Extraction Method for the Determination of Metals of High Viscosity and Low Viscosity Oil Blends

The extraction method is probably the oldest procedure for separating analytes from organic phases and is based on selective partition between two phases, particularly where two insoluble liquids are involved. It can be used as a preconcentration step by taking advantage of large sample size to low extracting volumes. The solution used for extracting the metals can contain a complexing agent to stabilise the metal in the solution, and should contain all the elements of interest in a reduced volume. The acid solution or complexing agent used to extract the metals must be pure and contamination-free.

A high percentage of recovery is possible and recovery checks can be carried out by 'spiking' the sample with metals of interest. However, if all the metals present in the organic liquid are in a reasonable concentration it may be possible to dilute it in a solvent that is compatible with ICP-OES.

Extraction or separation of dissolved chemical component $[X]_A$ from liquid phase A is accomplished by bringing liquid solution of $[X]_B$ into contact with a second phase B that is totally immiscible. A distribution of the component between the immiscible phases occurs. After the analyte is distributed between the two phases, the extracting analyte is released and/or recovered from phase A for analysis. The theory of chemical equilibrium leads us to a reversible distribution coefficient as follows:

$$X_A \Leftrightarrow X_B \qquad (1)$$

This equilibrium constant (K_D) is called the Nernst distribution law:

$$K_D = \frac{[X]_A}{[X]_B} \quad (2)$$

The term in brackets denotes the concentration of X in each phase at room temperature and pressure. A successful extraction is the optimum condition for distribution of solute between phases and lies to the far right in Equation (1) with a large K_D.

The extraction method is labour intensive and time-consuming. Although the method can be used for extracting and preconcentrating, its use is decreasing because solvents of the required purity are expensive and the capabilities and sensitivity of modern analytical techniques have improved significantly in recent years.

5.13 Analysis of Old Lubricating Oil for Total Metal Content Using a Slurry Method with Internal Standard [7]

A useful method of analysis of crude and wear metals is the sample slurry method. It is well known that crude and wear oils contain soluble, suspended and insoluble metal particulates and analysis of oils containing the soluble and suspended metals can be successfully carried out using a slurry method, provided that they are less than 4.0 μm in size and at a concentration that can be detected. An internal standard can be added to the oil sample slurry and nebulised along with the sample for matrix correction.

It may be possible to break up the larger particles into finer particles by rolling a known weight of crude or wear oil in a Teflon bottle filled with zirconium beads. The bottle is sealed with a tight fitting cap and rolled using a mechanical roller for 12–24 h. The rolling will break the larger particles down to fine particles suitable for slurry analysis. The grounded oil is dissolved in a suitable solvent and any particles present are suspended in solution by continuous stirring using a magnetic stirrer and Teflon-coated magnetic bar during the analysis for metal content. The most suitable nebuliser for this type of analysis is the V-groove cross flow type as the sample supply orifice is large enough to accept slurry solutions. The advantages and disadvantages of using the slurry method of analysis are:

Advantages

1. Samples can be analysed against standards prepared in the same solvent used to dissolve the slurry/sample.
2. An internal standard can be added to the stirring slurry solution to correct for variation in matrix effect.
3. It avoids the use of hazardous acids required for digestion and solution.
4. It reduces loss of volatile elements (must be high enough for detection).
5. Several samples can be analysed.
6. It is contamination-free.
7. It is easy to use.

Disadvantages

1. Suspensions and particles must be <4 μm to avoid blocking the nebuliser.
2. Dispersions may take place during peristaltic pumping.

3. Extra washout times are required between samples to ensure total removal of previous sample.

The following experiment was carried out on a sample of mixed wear oils obtained from several used engine blocks, and the oil was stored in a large drum:

Method

Sampling

Approximately 1.0 kg of stirred oil in the drum is sampled and is further stirred under laboratory conditions using a large Teflon coated magnetic stirring bar and stirrer. Two 50 g samples (labelled A and B) are taken for metal analysis.

Sample A

Sample A is centrifuged at 2500 rpm for 20 min and approximately 10.0 g of the top clear layer is accurately weighed into a 100 ml plastic volumetric flask. The centrifuged oil is dissolved in tetralin solvent with 10.0 ppm yttrium internal standard added and made up to mark with tetralin solvent.

Sample B

Sample B (not centrifuged) is stirred thoroughly with Teflon stirring bar and stirrer and approximately 10.0 g of the oil is accurately weighed into a 100 ml plastic volumetric flask along with 10.0 ppm yttrium internal standard and made up to mark with the tetralin solvent.

Analysis. Multi-element calibration standards of 0.0, 5.0 and 10.0 ppm Fe, Ca, B, Cu and Al are prepared by dissolving 0.5 and 1.0 ml of 1000 ppm multi-element standard control stock in tetralin solvent to100 ml with the same concentration yttrium internal standard as for samples A and B. Both samples are stirred and nebulised to determine the metal content against a standard calibration curve and corrected with an internal standard using ICP-OES.

Both centrifuged and uncentrifuged samples were analysed by the bomb combustion method described in Section 5.10, as a comparison with results by the slurry method.

Results. The slurry method of analysis of wear oils can give reproducible results when compared with the bomb combustion method (Table 5.19). The random selected metals analysed for this sample were used to show that this method could be used as an alternative provided that the particle sizes of insoluble suspensions in the oil are suitably small. This method may be an alternative to sample preparation by tedious destructive methods.

Table 5.19 *Results of analysis of wear oil obtained from a local automobile factory using a slurry method and yttrium as internal standard. The results before and after centrifugations are indicative of suspended and dissolved metal content*

Metal 1	Sample A Slurry method 2	Sample A Bomb comb. 3	Sample B Slurry method 4	Sample B Bomb comb. 5	Undissolved metal 6 (Col. 4 – Col. 2)
Fe	18	23	306	317	288
Ca	73	69	1080	1102	1007
B	13	10.6	96	94	83
Cu	71	74	76	73	5
Al	6	9.5	453	449	447

5.14 Conclusion

The most suitable method for metal analysis of crude oils must be carefully considered when precise results are required. The metal content of different fractions of oil from the refinery industries must reflect the true concentration and be informative to manufacturing and environmental agents.

The concentration ranges covered by these methods are determined by the sensitivity of the instruments, the amount of sample taken for preparation and dilution volumes.

The following is a brief summary of good practice in all oil analysis:

1. Representative sample (may need to be heated and mixed while warm prior to sampling).
2. Accurate standards to generate calibration curves.
3. Linear calibration curve.
4. Contamination free solvent.
5. Sufficient rinse times between standards and samples.
6. Selection of wavelengths that give good detection limits.
7. Correction for spectral interferences (use alternative line).
8. Elements must not precipitate from standard or sample solutions.
9. Dilute samples sufficiently to minimise nebulisation transport effects caused by matrix effects.
10. Internal standards must be added at precisely the same concentration to standards, blank and samples.
11. Standards must be close to sample.
12. Avoid ashing of sample as sample preparation method especially where volatile elements need to be determined.
13. Removal or reduction of strong acids used to prepare samples by acid digestion techniques.
14. All 'spiking' of crude oils must be carried out with elements that are of interest.
15. Analysis may be carried out in duplicate giving near similar results before reporting.

The disadvantage of solution sample preparation of oil samples for metal analysis is the loss of sensitivity due to the dilution factor and the effect the sample/solvent may have on the signal response. It can also give rise to poor precision accuracy which is required for sub-trace analysis due to the matrix effect of the oil, especially where toxic metal analysis is required. Excellent results are obtained for all metals in all oil samples when prepared using microwave acid digestion or bomb combustion destructive methods.

The advantage of analysis using an internal standard for precise measurements allows several oil samples with a range of different viscosities to be analysed against the same standard calibration curve. The disadvantage is that it is unsuitable for trace levels of metals present in oil samples which are necessary to quantify for efficient manufacturing process and health and environmental purposes. The most suitable methods of sample preparation for trace analysis are microwave acid digestion and bomb combustion which results in aqueous solutions that are more stable, sensitive and reproducible in the plasma torch. The aqueous solutions can also be used by hyphenated techniques e.g. ultrasonic nebuliser, hydride generation or the continuous cold vapour mercury method with ease.

Sample preparation by ashing is time-consuming and may be erroneous especially where sub-trace analysis is required. The ashing can be carried out in the presence of sulphuric or sulphonic acids, e.g. H_2SO_4 and PTSA, to form metal sulphates to prevent loss in the heating cycle. Some metals may not be totally retained even in the presence of retaining agents or may enter the pores of the vitreosil or porcelain dishes which are difficult to remove, and some volatile elements may not form sulphates and may be lost. Platinum dishes may be attacked by acidic or basic solutions and must be thoroughly washed with water or mild acid or base solutions immediately after use.

The slurry method of analysis has the advantage of detecting all elements including most volatile elements present at reasonable concentrations. However, it can suffer poor reproducibility from oversize insoluble particles.

Trace elemental analysis can also be used to indicate the level of contamination of middle distillate fuels, e.g. turbine fuels. Metal contamination can cause corrosion and deposition on turbine components at elevated temperatures. Some diesel fuels have specification limits to guard against engine deposits, however they sometimes employ Mo or Ni as a catalyst for the refining process which eventually ends up in the finished products. There are several sources of multi-elemental contamination in naval distillate fuels. Sea water is pumped into the diesel tanks as ballast to immerse ships and submarines. Some oil transport ships have dirty tanks and contamination and corrosive products can also come from piping, linings and heat exchangers.

The presence of metals, e.g. Pb, Cu, Zn and Al, can accelerate oxidative deterioration of refined products in oils destined for firing in boilers which can lead to corrosive products and may end up as toxins in waste ashes. Analysis of petrol and diesel products for metal content cannot be determined by direct aspiration/nebulisation of the undiluted sample due to the plasma quenching and explosive nature of petrol. A simple method for determining metal content of these products is to evaporate the petrol to low volume under vacuum using a low temperature rotary evaporator. The sample can be reduced to as low as 70–90% of the original volume and retain the elements in the reduced petrol volume, including volatile elements. The resulting concentrate can be re-dissolved in ICP-OES compatible solvents, e.g. tetralin, isopropanol, etc., for metal analysis against standards prepared in the same solvent.

References

[1] Bray, E.E. and Evans, E. (1961) Study of sedimentary organic matter, *Geochimica et Cosmochimica Acta,* **22**(1), 2–15.

[2] Magnusson, F.S. (1978) Petrochemical Feedstocks, Chemicals and Rubber Program, Office of Basic Industrial Materials Division, Industry and Trade Administration, US Department of Commerce, Washington, DC, November.

[3] Fassel, V.A., Peterson, C.A., Kniseley, R.N. and Abercombie, F.N. (1976) Simultaneous determination of wear metals in lubricating oils by ICP-OES, *Analytical Chemistry,* **48**, pp516–519.

[4] Price, W.J. (1979) *Spectrochemical Analysis by Atomic Absorption*, Heyden, p234.

[5] Boorn, A.W., Cresser, M.S. and Browner, R.F. (1980) Evaporation characteristics of organic solvents aerosol used in analytical atomic spectrometry, *Spectrochimica Acta,* **75B**, pp823–832.

[6] Gorsuch, T.T. (1974) Dry ashing in oxygen atmosphere, *Analyst (London)*, **84**, p135.

[7] Brennan, M.C. (1992) *Novel Electrochemical and Atomic Spectrometric Techniques in the Characterization of Anaerobic Adhesives*, PhD Thesis, Cork: University College Cork.

6

Metal Analysis of Structural Adhesives

6.1 Introduction

Natural resins, gums and asphaltic pitches have been used as hot melt adhesives since ancient times to join many substrates, and some are still popular today. The disadvantage of these adhesives is that they have poor resistance to heat, moisture and biological changes. The modern adhesives made from synthetic organic chemicals are designed to be more durable. The phenolic resins are the earliest type of synthetic adhesives and have good resistance to moisture, heat and biological attacks, but they are brittle causing their bonds to shatter under stress or vibration.

In the modern day, adhesives are described as glues or pastes that are widely used on substrates as part of bonding applications. In the 1950s, high strength phenolics containing synthetic rubbers and flexible resins were used successfully on a large scale in the bonding of brake linings in the automotive industry. The disadvantage of these adhesives is that they react by liberating water and if this water becomes trapped it slowly weakens the bond and eventually fails and the substrates fall apart over time.

During the Second World War and later in the 1960s, synthetic monomer resins such as epoxy, methacrylate, urethane, cyanoacrylates and silicones were developed and entered the market making a significant impact in the adhesive industries. These polymeric-based monomers are used in specific applications with outstanding success and have led to substantial financial rewards and jobs for over 250 000 people worldwide. The industry's accepted definition of a successful adhesive is a substance capable of holding substrates together for the lifetime of the component being bonded. The bonding area is usually thin layers called bond lines which have the ability to transfer loads from one substrate to another. The physical state of the adhesive must be liquid or semi-solid initially at point of contact so that it can be applied readily, and then form a strong bond, with each substrate capable of transmitting stresses from one substrate to another.

A Practical Approach to Quantitative Metal Analysis of Organic Matrices Martin Brennan

For an adhesive to be successful it must compete with non-adhesive joining techniques such as welding, nuts and bolts, rivets and staples. Modern adhesive joining can be used to replace or supplement traditional joining methods only when the quality, performance and economics are acceptable.

6.2 Setting and Curing of Adhesives

Successful bonding requirements for adhesives are that they must be instant, permanent, and non-permeable. Some adhesives are designed with slower fixture times so as to allow repositioning of parts prior to permanent bonding. High viscosity adhesives are used to fill gaps and resist sag in vertical applications while low viscosity adhesives are required to fill small cracks and to bond at the same time. The terms thick and thin adhesives carry no precise definition, as adhesives must have a fast grasp and must wet the bonding area, and should not bleed or contaminate surrounding areas (blooming). Heat cure adhesives should cure at as low a temperature as possible but remain liquid during transport and storage. They should remain flexible at cryogenic temperatures but not creep at elevated temperatures and be easy to apply and safe to use. They should not require special complex equipment to apply and bond through oily surfaces or other contaminated substrates and, finally, be available at low cost.

The polymerisation of adhesives follows the basic rule of chemical reactions – initiation, propagation and termination steps – and because they are reactive they should truly cure. The following is a brief summary of typical adhesives used by consumers or for industrial applications.

6.3 Introduction to Modern Synthetic Adhesives [1]

6.3.1 Cyanoacrylate Adhesives

Eastman Kodak was the first to discover cyanoacrylate monomer during wartime research as part of a study to improve the quality of cyanoacrylate esters. A decade later, H. W. Coover and J.M. McIntire [2] and their team of scientists spent time following this up with more research focusing on acrylate polymers. They discovered the monomer ethyl cyanoacrylate after accidently bonding the prism of an Abbe Refractometer. In the 1970s studies to improve the stabilisers in these monomers led to extended shelf life of these products, thus offering this product to the public as an attractive and cheap method of assembling or repairing broken parts.

The chemistry of cyanoacrylate adhesives contains no co-reactants but can polymerise at room temperature on any substrate that is exposed to atmospheric moisture or alkaline surfaces. Synthesised cyanoacrylate esters can be methyl, ethyl, n-propyl, n-butyl, allyl, ethoxyethyl and methoxyethyl. The basic structure of the cyanoacrylate monomer is:

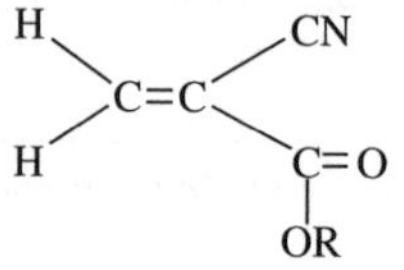

$$H_2C{=}C(CN){-}C({=}O){-}OR$$

Where R = CH_3
= CH_3CH_2
= $CH_3(CH_2)_3$
= OCH_3 etc.

The initiation reaction is the nucleophilic attack at the β-carbon of the monomer to generate a carbanion which during the propagation stage reacts with the monomer and continues to chain transfer to termination:

$$\mathrm{Nu} \rightarrow \mathrm{{>}C{=}C(CN)(COOR)} \rightarrow \mathrm{NuCH_2C(CN)(COOR){-}} \rightarrow \mathrm{NuCH_2\,C(CN)(COOR){-}CH_2\,C(CN)(COOR){-}}$$

Initiation Propagation Termination

Unreactive modifiers can be added provided they are very low in moisture or base interferences and in some cases are added as stabilisers. Other additives include inhibitors, thickeners, plasticisers, dyes or colorants and adhesion promoters.

The most popular route in synthesising cyanoacrylate monomers is the initial reaction between cyanacetate and an aldehyde (usually formaldehyde) in a solvent in the presence of an aromatic amine catalyst to form a prepolymer, which is cracked under azeotropic conditions to yield the monomer. The azeotropic removal of water or condensation is used to monitor the performance of the reaction. The addition of P_2O_5 at the end of reaction acts as a drying agent and neutralises the amine catalyst prior to cracking of the prepolymer. Stabilisers used in cyanoacrylate adhesives are latent acids or metal-based Lewis acids with the latter being superior. The concentration added is sensitive to shelf life and to performance of the adhesive on application. Too much stabiliser can cause the product to be over-stabilised; too little can cause it to be unstable in storage. The 'happy' level is pitched between the performances required for specific applications. The acid level in consumer adhesives tends to be fixed as most of these products require instant bonding. Therefore, controlling and monitoring the stabiliser levels must be precise with a high degree of accuracy.

To monitor these concentrations of stabilisers for precise, accurate measurements a method must be available that must detect levels as low as $600\,\mathrm{ng\,g^{-1}}$ and as high as $5000\,\mathrm{ng\,g^{-1}}$. The allowed tolerance can vary between $\pm 20\%$ for low levels to $\pm 10\%$ for higher levels without affecting the performance of the product. The lower concentration would be expected to cure faster than those containing higher concentrations of stabiliser. The different level of stabiliser is dictated by the application of these adhesives and must be formulated under very controlled conditions. ICP-OES is an ideal tool to monitor the metal type stabilisers using sample dilution, bomb combustion or microwave acid digestion methods.

6.3.2 Anaerobic and Acrylic Adhesives [3]

Acrylics, unsaturated polyesters and other monomer-based adhesives containing ethylenic unsaturation, cure by formation of free peroxide radicals formed with transition metal ion donators such as cobalt, iron, copper and nickel. In engineering applications the presence of transition metals on surfaces sees them act as initiators by forming the free peroxide radicals from the added peroxide compound(s). The base monomers usually contain inhibitors such as phenols or other materials classed as peroxide stabilisers to prevent premature curing caused by trace peroxides present in the monomer or adhesive while stored in containers. Therefore, before polymerisation and bonding can occur the peroxide radicals must exceed the concentration of these inhibitors. Anaerobic adhesives are reactive products which cure by redox using free radical polymerisation derived from the oxygen-free environment as found in tightly bonded substrates. These products are oxygen sensitive and the oxygen maintains the liquid state by inhibiting free radical polymerisation as follows:

$$R^{\bullet} + O_2 \rightarrow ROO^{\bullet}$$

The active peroxy radical reacts with molecular oxygen dissolved in the adhesive to form an inactive hydroperoxy radical before initiation and chain propagation can occur. Therefore, before polymerisation and bonding can occur the peroxide radicals must exceed the concentration of the inhibitors and oxygen. The concentration of oxygen is low but sufficient to maintain the liquid state of the product. In one part adhesive, it relies on reactions with metal surfaces to provide the redox initiation. Generally, these products require oxidising or reducing metals, such as transition metals.

Anaerobic adhesives contain reactive monomer(s), accelerators, stabilisers or inhibitors, fillers, colorants, modifiers, tougheners (rubbers, butadiene graft, etc.). The chemistry of anaerobic adhesive is complex but the schematic brief of the reaction may be shown as follows:

In the absence of oxygen:

$$\text{Initiator} \rightarrow R^{\bullet} \rightarrow \text{R-Mon}^{\bullet} \rightarrow\rightarrow \text{Polymer}$$

(monomer radical)

In the presence of oxygen:

$$\text{Initiator} \rightarrow O_2 \rightarrow \text{Inactive}$$

The characteristic feature of anaerobic polymerisation is the initiation process. An active metal surface promotes the redox decomposition of the peroxide molecule to form an initiator leading to rapid polymerisation. The free radical $RO^{\bullet}$ generated can add to the monomer to form a propagating radical species or be halted by oxygen or other inhibitor. The absence/presence of oxygen in anaerobic adhesives is limited by bond-lines and quickly used up by reacting with initiating or propagating radicals. Krieble and colleagues described the use of hydroperoxide, saccharin and amine combination on a variety of substrates. Traditional use of anaerobic adhesives is as thread-lockers, sealing and general metal bonding. The hydroperoxides react with transition metals in two ways:

$$\text{Fast:} \quad ROOH + M^{2+} \rightarrow RO^{\bullet} + OH^{-} + M^{3+}$$

$$\text{Slow:} \quad ROOH + M^{3+} \rightarrow ROO^{\bullet} + H^{+} + M^{2+}$$

Anaerobic adhesives require confinement on an active metal surface. Substrates such as plastic, cadmium and zinc are slow but can be improved with the use of accelerators or primers.

Acrylic adhesives (or structural adhesives) are essentially a two-part product: radical polymerisation is initiated by use of a second component which is supplied as a primer. The curative of these products is divided between the two parts with one half containing curative and the second half the initiator which when combined initiates the polymerisation and bonding. These types of adhesives can be designed to function with or without metals and can be used on most substrates including non-active substrates such as plastics and glass.

The chemistry of structural acrylic adhesives is based on the reaction of the hydroperoxide/amine aldehyde condensate to generate alkoxy radicals. An alternative catalyst system is based on the reaction between benzoyl peroxide and amine, redox couple forming the benzoyl radicals. The curing of acrylics and anaerobics are both inhibited by oxygen but less in the former because of the higher catalyst concentration, which generates radicals at a greater rate. The acrylics are more stable and can be packed in high density bottles whereas anaerobics require low density, oxygen permeable plastic containers. Some acrylics require metals as initiators but these depend on the formulation design. Metals play a big part in the successful use of these products, particularly the anaerobic grades, and monitoring the raw materials and finished products for their trace contamination is extremely important to ensure that these products meet their shelf life specification and product quality.

6.3.3 Epoxy Structural Adhesives

Epoxy adhesives are established adhesives with applications on a broad range of substrates and have been available for many years. They are made up of several components of which the most important is the resin containing the epoxy oxirane ring which is a three-member ring containing oxygen. Epoxy adhesives can be manufactured as a liquid or a high melting point solid and can be aromatic or aliphatic, cyclic, acrylic, mono-functional or poly-functional. Epoxy adhesives contain a variety of metal salts which are added at sensitive concentrations as curing agents, catalysts, tougheners and fillers. These metal salts need to be monitored in order to maintain product quality.

Additives (including metal-based) can improve the epoxy adhesive properties e.g. shelf life, cure speed, modulus, impact shock, thixotropy and, in certain cases, electrical conductivity. They can be formulated to meet low viscosity requirements suitable for free flowing for potting application, to solid adhesives useful for bonding large size machinery such as aircraft, shipping, large road haulers components, etc., and can cure from temperature ranging from sub-zero to 150°C. Most epoxies do not emit volatiles during the curing process which makes them an environmentally friendly product. These adhesives will bond metals, plastics, glass and wood and have very high cohesive strengths once bonding takes place. They display very little creep, sag or shrinkage because the molecules are cross-linked.

The reactivity of epoxy groups towards nucleophilic and electrophilic species can be explained through the release of ring strain in the three member oxirane group. Nucleophilic curatives such as amines or mercaptans attack the secondary ring carbon while electrophilic curatives behave as Lewis or Bronsted acids. The epoxy ring can be opened by hydroxyl or other epoxy group aided by tertiary amines, Lewis acids or co-reactants such as primary amines, mercaptans and dicarboxylic acids:

$$\text{NuH:} + \underset{\text{(O bridging } CH_2\text{–}CH)}{CH_2\text{—}CHCH_2O\sim} \longrightarrow \text{Nu—}CH_2\overset{OH}{\overset{|}{C}}HCH_2O\sim$$

The structures for iodonium salts and Bronsted acid precursors are:

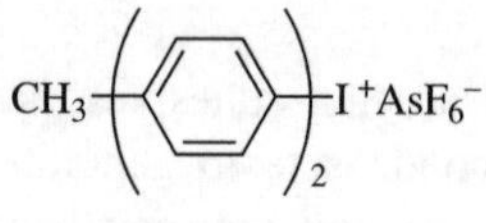

Iodonium salts

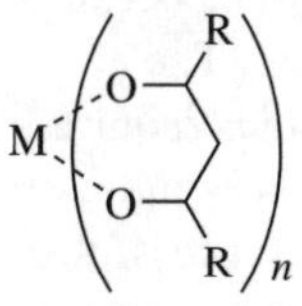

Bronsted acid precursors

Room temperature acid catalysis involving cationic acetylacetone salts of Si, B, Ge and P have been used successfully in modern epoxy chemistry.

Curing of these adhesives can also be activated by trace levels of water absorbed by the resin at room temperature.

Salts of tetra-amine phthalogyanine such as Cu, Co, and Ni have also been used as curatives with success to give cured resins with considerable improvements in heat resistance compared with resins prepared with conventional curatives. Evidence for this is the considerable reduction in cure temperature by use of the catalytic BF_3(MeNH) complex.

Lewis acid cationic type catalysts used to cure epoxies by heat were the earliest studied and found to display reasonable good latency with boric acid or organic salts of Al, Ti, Si, Zr, Si and Ge. The counter anions perfluorinatedphosphate, arsenate and antiminate salts gave gelation times using DGEBA resin ranging from 1 to 5 s at 150°C. Calcium glycerolphosphate salts have been used successfully to catalyse anhydride-based cure of epoxy resins at room temperature.

Reinforced fillers are added to improve the tensile and flexural strength of epoxies and fillers used with success are silica, asbestos, and alumina. In recent years, encapsulated fillers containing salts of Ni, Cu, Co and Fe have been found to give excellent reinforcing properties. Anti-corrosion fillers such as Al_2O_3 are also added to some epoxy formulations.

Conductive epoxies can be prepared by adding Ag^+ powder (very expensive) to the epoxy formulations. Recent conductive epoxy adhesives were prepared containing powder solder consisting of a blend of Bi-Pb-Sn alloy. Conductive epoxy adhesives must contain a high loading to function as a conductor and the blend of some metals that contain toxic metals must be used in enclosed spaces only. Flow-control fillers are also added to some epoxy formulations as colloidal Si, TiO_2, or Cr_2O_3 to reduce sag and improve modulus within the bonds. Metals such as tin can also cause serious damage to sensitive computer components due to volatility of certain tin salts caused by the heat generated during the operating of computers. These additives can be monitored with ease using ICP-OES.

6.3.4 Phenolic Adhesives

Phenolic adhesives are structural adhesives with specific applications, e.g. where wide gap bonding is required and where large structures need to be bonded. Phenolic resins are the product of a special reaction ratio of a phenol and formaldehyde in the presence of an organic catalyst. There are two main types of phenolic resins: phenol or methylol terminated. The phenol terminated are called novalacs, while the methylol terminated are called resoles (one step resins). Modern phenolic resins are prepared in the presence of metal carboxylates and these resins contain a large number of benzylic ether linkages and have open para positions which have good temperature stability and are usually of low viscosity.

The novalac phenolic resins react under weakly acidic and anhydrous conditions using metal catalysts of the divalent state, e.g. Ca, Mg, Zn, Pb, Cd, Co, Ni and Cu acetals, halides or sulphonates. The mechanism of this ortho-hydroxymethylation reaction has been attributed to the formation of chelate complexes as intermediates as proposed occurs at the initial and subsequent condensation reaction of the phenolic alcoholics. It has been shown that electropositive bivalent metals work best when pH is between 4 and 7. This may be described as follows:

$$M^{2+} + CH_2(OH)_2 \Leftrightarrow [M\text{-}O\text{-}CH_2\text{-}OH]^+ + H$$

Metal analysis associated with these adhesives involves monitoring raw materials for metal contamination and metal content of activators or initiators used as the second part of these adhesives.

Certain phenolic-formaldehyde resins can be made to cure in the absence of metals and they are ortho-ortho resoles, etc., and these adhesives will not be discussed further.

6.3.5 Polyurethane Adhesives

Polyurethane adhesives are the result of a reaction between an aliphatic diisocyanate with an aliphatic diamine or polyols/polydiols as follows:

$$OH\text{-}R\text{-}OH + OCN\text{-}R^1\text{-}NCO \rightarrow (\text{-}CO\text{-}NH\text{-}R^1\text{-}NH\text{-}CO\text{-}ROO)_n$$

$$\text{Polyol diisocyanate} \rightarrow \text{polyurethane}$$

The rate of reaction depends on the structure of both the polyol and the isocyanate. Aliphatic polyols are more reactive than aromatic and more stable. Polyurethanes are described as any polymer chain that has been extended by reaction with di- or polyisocyanate. The isocyanate group (-NCO-) can react with most compounds containing an active hydrogen atom.

Isocyanates also react with water yielding CO_2 and urea and the gas generated is the principal source of blowing in the manufacturing of low density foams. Amine reactions with isocyanates can act as chain extenders and curing agents in the polyurethane. The selection of the correct diamine is important in determining the rate of the reactivity with the alcohol to give a resulting polyurea segment which increases the potential of both primary and secondary cross-linking reactions. These reactions can give rise to the formation of soft and hard segments. The soft segments are long chain polyols, e.g.

polyether or polyester diol, and the hard segments are short chain polyols or diamine chain extenders, as shown by the following structure:

$$[\text{-CONHC}_6\text{H}_5\text{CH}_2\text{C}_6\text{H}_5\text{NHCOO(C}_4\text{H}_8\text{O)}_m\text{-}]_n\text{-}[\text{CONHC}_6\text{H}_5\text{CH}_2\text{C}_6\text{H}_5\text{NHCOOC}_4\text{H}_8\text{O}]_p$$

Soft segment Hard segment

Controlling the concentration of hard and soft segments can have an effect on the hardness, modulus, elasticity and elongation ability of the adhesive. Increasing the concentration of the hard segments of a polyurethane will increase its hardness and modulus but decrease its elasticity and elongation ability and vice versa by increasing the concentration of the soft segments.

These adhesives are structural adhesives and have found widespread use in both industrial and household environments. They are designed to bond metals, plastics, rubber, wood, glass and ceramics with thin or wide gaps. They display good water resistance and have good room temperature performance but are poor at elevated temperatures. The following is the simple urethane linkage structure:

-NHCOO-

Polymerisation occurs by addition of a hydroxyl-containing compound to an isocyanate group as follows:

$$\text{R-N=C=O} + \text{HA} \rightarrow \text{R-NH-CO-A}$$

This reaction shows the addition of an active hydrogen compound, HA, across the N=C double bond of an isocyanate. Isocyanates also react with water and the initial product is carbamic acid which is unstable, losing carbon dioxide, and generates an amine that reacts with the excess isocyanate to produce urea as follows:

$$\text{RNCO} + \text{H}_2\text{O} \rightarrow \text{RNHCOOH} \rightarrow \text{RNH}_2 + \text{CO}_2$$

$$\text{RNH}_2 + \text{RNCO} \rightarrow \text{RNHCONHR}$$

A list of metal catalysts associated with polyurethane adhesives will be discussed under polyurethane analysis. Some epoxy adhesives may contain metal salts that can act both as filler and catalyst. Some of these catalysts contain toxic metals that must be controlled and monitored for health and environmental requirements.

6.4 Metal Salts and Concomitant Metals in Adhesives

Metals in adhesives are associated with the addition of inorganic salts, organometallics, catalysts or contaminants. Inorganic salts are added to adhesives as fillers, thixotropic agents, toughening or thickening agents. Fillers are usually added as a non-adhesive substance to improve permanent bonds and strength and as a binding agent holding substrates together through adhesive forces. They are added to improve tensile and flexural strength, as corrosion inhibitors, for moisture resistance, flow control or electrical conduction. Toughening agents are added where ultimate strength in a bond is required; it can reduce shrinkage, sag and costs. Thickeners play an important role where adhesives need to be applied in non-horizontal planes and where large gaps need to be filled and, at the same time, secure the bond. Organometallics, as initiators are usually encapsulated and broken by force or heat. Once they are released they behave as cationic catalysts and initiate cure of the adhesive. In the majority of cases metal salts are prepared as a second part of a dual adhesive system and brought together at the time of bonding.

Raw materials used to manufacture adhesives must be free of trace metals contamination. Adhesives such as the anaerobic or acrylic type are extremely sensitive to trace transition metals [$<25.0\,\mu g\,kg^{-1}$ (ppb)] and all materials used to make these products must be totally free of these metals.

6.5 Metals Associated with Cyanoacrylate Adhesives

The industrial manufacturing process for cyanoacrylate monomers is designed to generate pure organic compounds free of metals. The prepared monomer is functionally very reactive and is polymerised by several mechanisms of which the most common is by anionic methods. In most applications the initiation is usually carried out by the nucleophilic contaminant (water or moisture) found on most surfaces. These adhesives differ from other adhesives in that they are monofunctional and can homopolymerise rapidly at room temperature. A number of modifiers have been added to impart a range of desired properties and these include stabilisers, inhibitors, thickeners, plasticisers, tracers, colorants and preservatives.

The metals in cyanoacrylate adhesives are usually added as stabilisers, conductors or tracers and need to be monitored carefully to ensure that they are at the correct concentration so that the adhesive remains a liquid during transport and storage. The formulation must also function effectively as an adhesive when applied to bond wood, metal, glass, paper, rubber, and leather and cloth substrates. The metal type stabilisers suitable for cyanoacrylate adhesives are generally Lewis acids and these function by accepting a lone pair of electrons forming a coordinate covalent bond. The Lewis Acid and Base theory is one of several acid-based reactions, and the term 'acid' is ambiguous and should be clarified as to whether it is a Lewis or Bronsted-Lowry acid. An electrophile (electron acceptor) is a Lewis acid while a Bronsted-Lowry acid is always a proton (H^+) donor, and any electrophile (including H^+) is reserved for those Lewis acids which are not Bronsted-Lowry acids.

Christe and Dixon [4] predicted that the strongest Lewis acid is SbF_5^- as it had the strongest fluoride affinity. They have also shown that fluoride is a 'hard' base while 'chloride' is a soft base. The following Lewis acids have been applied to cyanoacrylate adhesives as shelf stabilisers: $AlCl_3$, Fe(III)Cl_3, PCl_5, NbF_5 and $(CF_3SO_3)3Yb.xH_2O$ (ytterbium trifate). These salts can be used as single or multi-additive as stabilisers only, or as stabilisers and tracers. Metal tracers are formulated into cyanoacrylate products to prove the origin of manufacturing, particularly where information as part of criminal, forensic or environmental knowledge may be required. They must not affect the performance of the product.

Most cyanoacrylate adhesives are maintained in a liquid state at room temperature by the addition of free radical and anionic stabilisers at suitable concentrations so as not to interfere with the functionality of the adhesive. These stabilisers/inhibitors are added at controlled and critical concentrations and need to be monitored for every batch manufactured to ensure that they meet the product specification. A product containing as much as 10% difference in added concentration of stabiliser can affect the performance of the product. This is important where these adhesives are applied in a rapid production line requiring instant bonding of sensitive components. The most common anionic inhibitors used are sulphur-based, such as SO_2, H_2SO_4, MSA, HPSA, PTSA, and sultones.

Sultones were the earliest anionic stabilisers used in cyanoacrylates but fell from favour because of their potential carcinogenicity. Chelates of boric acid derivatives with polyhydroxy compounds also were considered as anionic inhibitors. Anionic inhibitors are normally added at concentrations between 0.001 and 0.01% depending on the application.

Typical free radical inhibitors added as hydroquinone monomer acrylether (HQ), or other hindered phenolics, etc., are employed at concentrations from 0.2 to 0.5% depending on the type of cyanoacrylate ester and its additives.

In recent years new anionic stabilisers with Lewis acid (e.g $AlCl_3$, BCl_3, Fe[III]Cl_3, PCl_5, NbF_5, $(CF_3SO_3)3Yb.xH_2O$) properties have appeared on the market and they have the added advantage of behaving as a gas to keep the neck and tip of bottles clear of polymerised cyanoacrylate adhesive, and retaining the liquid state of the product. These acids can be used singly or in combination at suitable concentrations to maintain stable products. Some stabilisers are gases at room temperature and are possibly retained by reaction with water present in cyanoacrylate adhesives. These stabilisers can be readily monitored using ICP-OES. The maximum concentration that can be added to cyanoacrylate adhesives is 0.0005 to 0.01% and the required level depends on the type of adhesive.

6.6 Non-Destructive Methods of Analysis for Metals Content in Cyanoacrylate Adhesives

The following is a comparative study of three dilution methods which can be used to determine the concentration of metal type stabiliser and/or tracer in cyanoacrylate adhesives using ICP-OES. The concentration of Nb in cyanoacrylate is normally in the range 0.08 to 2.0 $\mu g\,ml^{-1}$ (ppm), depending on the product.

6.6.1 General Method

Standard Calibration Curve. Dissolve known weights of cyanoacrylate adhesives in 50:50 glacial acetic acid and propylene carbonate to a known volume. (The weight of adhesive is dictated by the level of metal present.) A calibration curve for analysing cyanoacrylate adhesives using ICP-OES is prepared from standard solutions containing 0.0, 0.5, 1.0 and 4.0 $\mu g\,ml^{-1}$ (ppm) of the element of interest (Al, Fe, P, Nb or Yb) in 50:50 glacial acetic acid and propylene carbonate. A calibration curve is plotted as concentration versus intensity. The intensity of the analyte of interest in the sample is measured against this calibration curve and the concentration of metal is determined by multiplying the measured value by the dilution factor and compound molecular weight. A control sample of cyanoacrylate adhesive* containing a known concentration of metal is

*The control cyanoacrylate adhesive is a standard containing known concentration of the metal. A control chart may be generated to check the long-term accuracy of the method.

also weighed accurately and dissolved in the same solvent mixture. The control sample is used to check the accuracy of the method.

Complex solutions of cyanoacrylate adhesives containing variable levels of fillers need to be measured for stabiliser content. It is difficult to prepare standard solutions with similar composition to the sample. The method of standard addition may address some of these problems with a high degree of accuracy.

6.6.2 Standard Addition Method

Method. Add known and similar weights of cyanoacrylate sample to four grade 'B' plastic volumetric flasks followed by increasing concentrations of the analyte Nb (0.0, 0.5, 1.0 and 2.0 ppm) and dilute to volume with 50:50 acetic acid and propylene carbonate. Measure the intensity of each concentration and plot against the analyte concentration. The straight line is extrapolated to the negative concentration axis and the point where the calibration line cuts the concentration line is the concentration of analyte in the sample. Similarly, prepare the same concentration of standards without the sample and measure intensity for each concentration. Plot a calibration curve of intensity versus concentration and if this line is parallel to the sample curve then the standard addition method can be used to quantify the level of metal present. The control standard is also prepared in the same manner as for the sample.

The method of internal standard can also be used as it corrects for variable sample viscosities for both standards and sample.

6.6.3 Internal Standard Method

Method. A minimum of four standards of increasing concentrations are prepared in 100 ml plastic volumetric flasks using 0.0, 0.5, 1.0, 2.0 and 4.0 $\mu g\ ml^{-1}$ (ppm) Nb and make up to the mark in 50:50 glacial acetic acid and propylene carbonate solvents. A suitable internal standard of indium (In) metal is prepared in a separate plastic volumetric flask containing 5.0 $\mu g\ ml^{-1}$ (ppm). The selected internal standard must be sensitive, free from chemical, matrix, spectral interferences and close in excitation energy of the test analyte. A calibration curve is generated with the standards by measuring the emission intensity along with the internal standard for each standard. The intensity of the internal standard should be constant while the standards show corresponding increasing intensities. Solutions of sample for analysis are prepared in the same solvent mixture as that for standards and the concentration determined against the calibration curve prepared against the internal standard. A special 'V' piece described in Section 3.7.4 is used for transporting and mixing the internal standard with the standards and sample in a continuous flow mechanism. The concentration of metal in the sample is determined against the calibration curve prepared against the internal standard. Any deviation caused by the sample is corrected using the internal standard.

Table 6.1 *Results of analysis of cyananoacrylate adhesives with and without thickening agent (PMMA) using calibration curve, standard addition and internal standard method for quantitative analysis of Nb content*

% PMMA	Calibration curve [$\mu g\,g^{-1}$ (ppm) Nb]	Standard addition [$\mu g\,g^{-1}$ (ppm) Nb]	Internal standard [$\mu g\,g^{-1}$ (ppm) Nb]
0.0	1.48	1.51	1.52
2.5	1.10	1.52	1.51
5.0	0.84	1.39	1.49
10.0	0.51	1.15	1.49

Results. The results in Table 6.1 are the determination of 'Nb complex salt' added to cyanoacrylate adhesive as a tracer added at a concentration of $1.5\,\mu g\,g^{-1}$ (ppm $= 1500\,ng\,g^{-1}$) Nb using three non-destructive methods. The results obtained are similar to 0 % PMMA. Figure 6.1 shows the intensity of Nb in cyanoacrylate monomers with and without PMMA thickening agent.

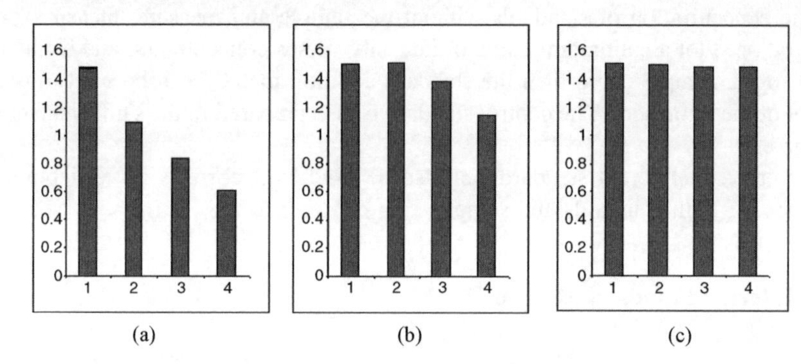

Figure 6.1 *Graphical illustration of response of intensity of Nb in cyanoacrylate monomers with and without PMMA thickening agent. (a) The results using the standard calibration curve show the decreasing of the signal response for increasing thickener concentration on the efficiency of nebulisation; (b) the standard addition method can tolerate 0.0 and 2.5% but fails above these values; (c) illustrates the reproducible results obtained using the internal standard method*

A series of samples containing increasing levels of thickening agent such as PMMA or poly(vinyl methacrylate) were also studied using the above three methods. The methods were used to study the effects of uncontrollable random errors caused by other components in the sample and the instrument. The cyanoacrylate samples were spiked with different concentrations of PMMA and signal intensity measured using the above dilution methods against the same standards as the above non-destructive methods.

6.7 Destructive Methods of Analysis for Metals Content in Cyanoacrylate Adhesives

Cyanoacrylate adhesives can be prepared for metal analysis using destructive techniques e.g. dry ashing, microwave acid digestion or oxygen bomb combustion. It is a matter of chance that most cyanoacrylate adhesives that contain metals are formulated at a level that can be detected reproducibly using dilution methods, described in Section 6.6.1. If the concentration is high it must be suitably diluted before commencing analysis. Conversely, if the concentration is low then a suitable sample preparation method must be used that will accurately determine the concentration even if a preconcentration step is required. The dilution methods are rapid, particularly when the method involves the use of an internal standard and can be applied in a quality control environment where multiple samples of variable viscosities need to be analysed in a short time. It can be automated provided the technique avoids loss through evaporation and reduces odours. Destructive sample preparation methods may be necessary for the analysis of trace levels of toxic metals, e.g. Hg, Cd, Cr, Se, Pb, Sn and Ni. Analysis of these products for toxic metals is required where health and environment certification are required, particularly if cyanoacrylate adhesives are to be used for medicinal purposes.

6.7.1 Sample Preparation Using Ashing Method

Method. Into clean platinum vessels accurately weigh approximately 2.0 g of each cyanoacryate adhesive listed in column 1 of Table 6.1. Into one series add 0.1 g of PTSA and in the second series no PTSA is added. Ash both samples using the conditions in Table 6.2. Figure 6.2 shows the ramping and hold stages for ashing cyanoacrylate adhesives.

Table 6.2 *Program for microwave ashing using ramping and holding stages*

Program	Temp. (°C)	Ramp (min)	Hold (min)
1	100	20.0	5.0
2	250	25.0	5.0
3	400	25.0	5.0
4	550	25.0	30.0

At the end of ashing allow to cool and dissolve the ash in 50:50 0.5 M HNO_3 and HCl.

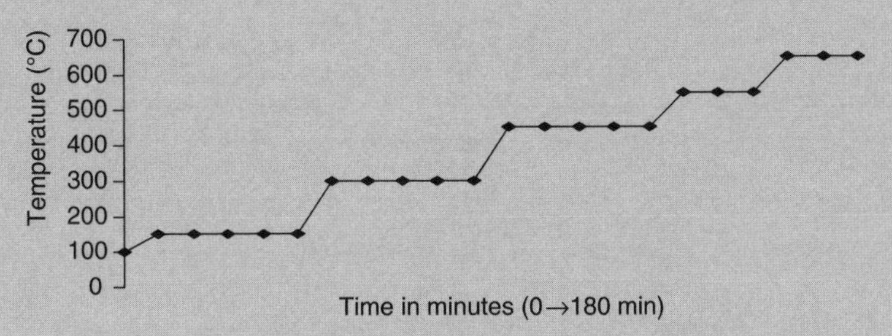

Figure 6.2 *Ramping and hold stages for ashing cyanoacrylate adhesives*

Note: The initial heating stages are low to encourage charring and reaction of metals with the PTSA. At higher temperatures the organics are burned off leaving the metal sulphate salts behind. The remaining ash is soluble 0.2 M HCl or HNO_3.

6.7.2 Sample Preparation Using Microwave Acid Digestion

Method. Weigh accurately approximately 1.0 g of each of the cyanoacrylate adhesive samples listed in column 1 of Table 6.1 into a Teflon vessel suitable for high temperature and pressure acid digestion. Add 10.0 ml HNO_3 and 5.0 ml HCl. The mixture is allowed to predigest for 1 h. A blank is prepared using the acids only. Heat to digest the sample and blank using the conditions in Table 6.3.

Table 6.3 *Program for microwave acid digestion of cyanoacrylate adhesives using ramping and holding stages*

Program	Power (W)	% power	Time (min)	Pressure (psi)	Temp. (°C)	Hold (min)
1	1200	100	10.0	250	150	5.0
2	1200	100	10.0	400	170	5.0
3	1200	100	10.0	500	180	5.0
4	1200	100	20.0	600	190	10.0

6.7.3 Sample Preparation Using Oxygen Bomb Combustion

Method. Weigh accurately approximately 1.0 g of each cyanoacrylate adhesive into a cup attached to the lid of the oxygen bomb combustion vessel. To the vessel add 5.0 ml of water and fill with oxygen to 30 atm. The bomb is fired according to manufacturer's instructions.

All samples prepared by each method are in aqueous solution and measured against an aqueous standard calibration curve prepared from a certified 1000 ppm stock standard to give 0.0, 0.5, 1.0, 2.5 and 5.0 ppm Nb in 0.5 M HNO_3.

Results of analysis of cyanoacrylate adhesives listed in column 1 of Table 6.1 using the three destructive methods are given in Table 6.4.

Table 6.4 *Results of comparison of destructive sample preparation methods for the determination of the concentration of Nb in cyanoacrylate adhesive, in µg/ml(ppm)*

% PMMA	Microwave acid digestion	Bomb combustion	Ashing without PTSA	Ashing with PTSA
0.0	1.45	1.43	~0.2	1.41
2.5	1.51	1.40	~0.2	1.48
5.0	1.49	1.46	<0.1	1.46
10.0	1.48	1.48	<0.1	1.53

6.8 Conclusion to Analysis of Cyanoacrylate Products

Analysis of cyanoacrylate adhesives was carried out using both non-destructive and destructive techniques in order to study the effect of method selection. Non-destructive methods involved dissolving the cyanoacrylate adhesive in a solvent that is compatible with the sample and the plasma torch of the ICP-OES. A suitable solvent that meets these criteria is glacial acetic acid mixed with propylene carbonate and it also reduces the strong odour effect in the laboratory environment.

The dilution methods involved are standard calibration curve, standard addition and internal standard. The results in Table 6.4 illustrate the importance of selection of the correct method. The calibration method showed a corresponding decrease in signal response with increasing concentration of PMMA. The standard addition method was satisfactory at lower levels of PMMA but showed a decrease at higher concentrations. Standard addition method only works assuming that there are no serious matrix effects i.e. non-corresponding reduction or enhancement of the analyte signal. Standard addition is difficult to automate and requires large sample volumes to prepare several samples for analysis. In statistical observations its principal disadvantage is that the extrapolation techniques are less precise than interpolation techniques. The internal standard approach is the best option providing the viscosities are not too different from the standards.

Internal standard methods are affected by matrix, line selection, sensitivity, precision, spectral interferences, and signal to background noise. Matrix effects are the most important concerns in organic samples and slight differences can cause considerable systematic errors resulting in a definite bias. Internal standard can be used to correct for most viscosity effects and the choice of internal standard is important. The internal standard must be compatible with the matix containing the analyte, free of spectral interferences, give good signal to noise ratio and be free from impurities. The disadvantage is that precise addition into standards, blanks and samples requires extreme care. Any contravention of these conditions may introduce errors.

Destructive sample preparation techniques involve ashing, microwave acid digestion and oxygen bomb combustion. The ashing method returned good results in the presence of the retaining agent PTSA and poor results in the absence of PTSA. The enclosed microwave acid digestion and oxygen bomb combustion methods returned excellent results. The disadvantage of destructive methods is that they are time-consuming and tedious, require a high degree of skill and are unsuitable for routine analysis.

With certain exceptions, cyanoacrylate monomer formulations containing additives e.g. rubbers, high-density neutral resins, silicon dioxide, etc., may hinder accurate and precise analysis using dilution methods. In such cases it may be necessary to prepare samples using destructive techniques, particularly where the levels are very low. Solvent selection for dilution of cyanoacrylate adhesive must be compatible for the entire journey of the sample solution from sample vessel to torch. Failure to do this could cause the cyanoacrylate to polymerise locally and block the entire sample transport system in ICP-OES and can cause serious damage requiring expensive replacements. The solvents suggested in the above dilution methods were found to be satisfactory.

6.9 Metals Associated with Anaerobic Adhesives [5]

Anaerobic adhesives can be formulated from 2 to 20 chemicals (raw materials) involving monomers, inhibitors, stabilisers, initiators, accelerators, colorants, rubbers, resins, fillers and plasticisers. The monomer is usually the highest concentration and is the most important chemical part of these adhesives. All adhesives are prepared by careful blending of these chemicals at controlled concentrations so that the product will function to suit a specific application. Most anaerobic adhesives are designed to stay liquid after applications so that adjustments and final fitting can be carried out on the machine parts prior to permanent bonding. Other adhesives will bond immediately so that hundreds and sometimes thousands of components are instantly assembled as part of a rapid manufacturing production line. A range of chemicals used in anaerobic adhesives are metal-based and may need to be monitored as part of quality requirements. They can be analysed using ICP-OES after appropriate sample preparation.

Most metal salts in anaerobic adhesives are present as inorganic salts or organometallic salts and can be formulated as stabilisers, initiators, fillers, or as thixotropic agents. An example is mono- or diphosphate methyacrylate monomer which increases bond strength between metal parts containing aluminium, zinc-coated mild steels and stainless steels. These metals are difficult to bond using standard anaerobic adhesives as they contain low concentrations of transition metals. All metals and salts can be easily monitored using an ICP-OES technique using dilution or destructive methods. Anaerobic adhesives are used extensively in the motor, aircraft, shipbuilding, electrical, household and spacecraft industries. There has been a 10-fold increase in the use of anaerobic adhesives in the last 10 years and demand is increasing. Other metal salts may be present as contaminants and are present at trace concentrations that may require sensitive destructive methods of sample preparation.

Anaerobic adhesives are designed to be metal-sensitive, especially with respect to the mono- or divalent transition metals, e.g. Cu, Fe, Co, Ni, Mn, etc.; these elements function by homolysing the hydroperoxide compound into free radicals which initiate polymerisation. It is important to be able to measure trace concentrations (ppb levels) of these metals as such levels may cause instability in these products. Contamination can be caused by raw materials or metal from the plant components used in the manufacturing process.

Sample preparation of anaerobic adhesives for metal content is an important step, be it by destructive or non-destructive methods. Inactive metal salts are added directly to anaerobic formulations as fillers or for thixotropic reasons. Generally, active transition metals are not added directly to anaerobic adhesives but are prepared as activators in aerosol solvents to be applied to inactive surfaces as part B of an adhesive formulation. In the majority of cases trace metal analysis of anaerobic adhesives is only required for batches with problematic stability and is best done using destructive methods.

To study methods of sample preparation for metal analysis of anaerobic adhesives the following laboratory-prepared adhesives containing the usual additives and 'spiked' with and without metals were used for comparison:

- *Type A.* Anaerobic adhesive containing monomers, stabilisers, initiators, promoters, and colorants and no fillers, rubbers or thixotropic agents.

- *Type B*. Anaerobic adhesive containing Type A with thixotropic agents, inorganic fillers and rubbers.

Note: Type of monomer used can vary from basic ethylene glycol mono- or di-methacrylates to complex structures that are beyond the scope of this book. Type of stabilisers, initiators, promoters, colorants, fillers, rubbers and thixotropic agents are propriety to adhesive manufacturing companies.

6.10 Destructive Methods of Sample Preparation for Metals Content in Anaerobic Adhesives

Dry ashing is a means of combusting the organic components in anaerobic adhesives and at the time retaining the metals for quantification by ICP-OES.

6.10.1 Ashing Method of Type A and Type B Anaerobic Adhesives

Method. Into two clean platinum dishes, weigh accurately approximately 2.0 g of each adhesive and 0.1 g of PTSA. Repeat this preparation for a second two vessels and 'spike' each with 0.5 ppm Cu, Fe, Co and Ni. Into the fifth vessel add 0.1 g of PTSA only and use this as a blank. All samples are prepared using a microwave asher (Figure 6.3).

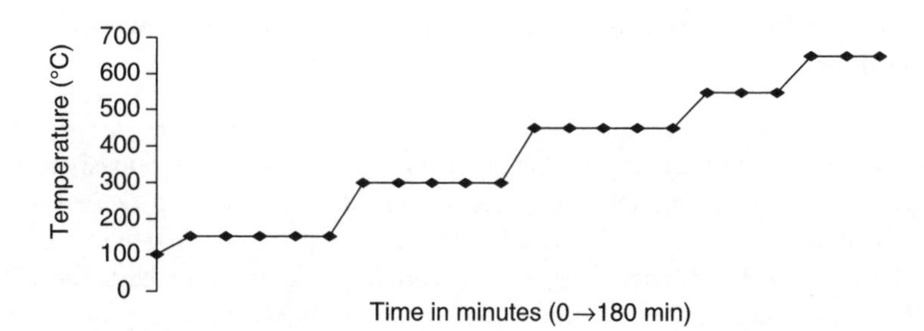

Figure 6.3 *Ramping and hold stages for ashing anaerobic adhesives*

Note: The initial heating stages are low to encourage efficient charring and to react with the retaining compound PTSA. Finally the heating is increased to higher temperatures to completely burn off the organic carbon leaving the metal sulphates behind. The remaining metals are dissolved in 0.25 M HCl.

Results. Microwave acid digestion is carried out as a means of destroying the organic components in the anaerobic adhesive product and retaining the metals of interest in the acid solution. Results of analysis of metal for metal content of Type A and Type B adhesives are shown in Table 6.5.

Table 6.5 Results of analysis of metal for metal content of Type A and Type B adhesives after sample preparation using dry ashing method

Metal	Blank	Sample + 'Spike' ($\mu g\, g^{-1}$)	Sample only ($\mu g\, g^{-1}$)
Type A anaerobic adhesive			
Fe	0.31	5.60	4.93
Cu	ND	0.88	0.18
Co	ND	0.65	0.03
Ni	ND	0.55	ND
Mn	ND	0.59	ND
Type B anaerobic adhesive			
Fe	0.27	16.1	15.32
Cu	ND	1.3	0.74
Co	ND	0.69	0.23
Ni	ND	0.49	0.09
Mn	ND	1.86	1.30

ND = None Detected

6.10.2 Sample Preparation of Anaerobic Adhesives Using Microwave Acid Digestion

Method. Weigh accurately approximately 1.0 g of each anaerobic adhesive Type A, into two Teflon vessels suitable for microwave acid digestion. The second vessel is 'spiked' with 0.5 $\mu g\, g^{-1}$ of Fe, Cu, Co, Ni, Mn. To a third vessel 10.0 ml of HNO_3 and 2.0 ml HCl is added as blank. Repeat these sample preparation methods for adhesive Type B using the same acid mixtures and vessels allowing to predigest for 1 h prior to microwave digestion. The digestion conditions in Table 6.6 are used for both adhesives and blank.

Table 6.6 Conditions for microwave acid digestion of Type A and Type B anaerobic adhesives

Stage	Power (W)	% power	Time (min)	Pressure (psi)	Temp. (°C)	Hold (min)
1	600	100	20.0	250	150	5.0
2	600	100	20.0	400	180	5.0
3	600	100	20.0	500	190	5.0
4	600	100	20.0	600	210	20.0

The digested sample should be allowed to cool to room temperature, pressure is released and the cover removed. To the residue add 2.5 ml of conc. HF and re-assemble vessel and digest as shown in Table 6.7.

Table 6.7 *Microwave acid digestion conditions for HF acid digestion of anaerobic adhesives*

Stage	Power	% power	Time (min)	Pressure (psi)	Temp. (°C)	Hold (min)
1	600	100	10.0	200	160	10.0
2	600	100	10.0	300	180	10.0
3	600	100	20.0	400	200	20.0

The digested samples in the Teflon vessels are allowed to cool to room temperature and pressure released and the cover removed. Excess acid may be reduced using a micro-vap attachment (shown in Figure 3.4) and the residue re-dissolved in 0.5 M HCl to a known volume. The digested samples are analysed for the listed metals against certified standards of 0.0, 1.0, 2.0, and 5.0 in 0.5 M HCl.

Results. Results of analysis of metal for metal content of Type A and Type B adhesives are shown in Table 6.8.

Table 6.8 *Results of analysis of metal for metal content of Type A and Type B anaerobic adhesives after sample preparation using microwave acid digestion*

Metal	Blank	Sample + 'spike' ($\mu g\, g^{-1}$)	Sample only ($\mu g\, g^{-1}$)
Type A anaerobic adhesive			
Fe	0.22	4.30	3.58
Cu	ND	0.78	0.28
Co	ND	0.55	0.05
Ni	ND	0.50	ND
Mn	ND	0.50	ND
Type B anaerobic adhesive			
Fe	0.22	18.1	17.32
Cu	ND	1.9	1.4
Co	ND	0.93	0.43
Ni	ND	0.69	0.19
Mn	ND	2.36	1.86

ND = None Detected

6.10.3 Sample Preparation of Anaerobic Adhesive Using Oxygen Bomb Combustion

Method. Type A anaerobic adhesive will combust in an oxygen bomb forming CO_2 and water, which are released and the metals retained in the aqueous solution. Combustion of Type B anaerobic adhesive will remove the organic compounds and retain the soluble metals in solution. Some non-combusted inorganic salts or fillers such as SiO_2, clays, etc., are further treated with conc. $HF/HNO_3/HF$ acids which will dissolve them after a little heat is applied. This is considered a disadvantage where several samples may need to be prepared on a routine basis, it is time-consuming, requires extreme care during the sample preparation, and all reagents must be very pure. It is not suitable for a rapid turnaround time analysis. However, it is an excellent method for analysing volatile metals, e.g. Hg, Se, Sn etc., which are present as trace toxic metals and analyses are required for environmental or health reasons.

6.10.4 Conclusion to Analysis of Anaerobic Adhesives

Trace metal analysis of anaerobic adhesives is best carried out using destructive techniques because of the high concentration of fillers and thickening agents added. These additives tend to interfere with trace metals from background noise, peak blooming, and the extent of dilution may exceed the detection limits and particle sizes may be too large to nebulise an anaerobic solution into the sample introduction orifice of the ICP-OES.

Destructive methods are the most effective as they remove the organic materials and dissolve the fillers and thickeners into non-interfering aqueous solutions. Dry ashing in the presence of a retaining agent (PTSA) offers a method of efficient removal of the organics while retaining the metals of interest. Microwave acid digestion is also a useful technique in retaining metals of interest but ultra pure reagents are required to avoid contamination. Oxygen bomb combustion is also a suitable method but a second step involving concentrated strong acids is required to dissolve some incombustible materials. These latter methods are time-consuming, tedious but are precise if extreme care is taken.

Careful examination of the results obtained in Table 6.4 shows that metals are volatile in the absence of the retaining agent PTSA for Type A and Type B anaerobic adhesives when heated to ash the organics from a sample. The possible reason for this loss in the absence of PTSA is the metals may be present as or form a volatile organo-metallic salt, due to reactions with other organic chemicals present in the anaerobic adhesive during the ashing cycle. The ramping rate of the microwave oven can also affect the rate of volatilisation in the absence of PTSA, e.g. if ramping is slow the rate of loss is greater than if the rate is fast. This seems to suggest that the rapid heating gives less opportunity for a complete volatilisation of the metallic species. The results obtained for the PTSA retained metals and the 'spiked' metals agree closely to those found by microwave acid digestion above. It can be concluded that it is possible to prepare these samples for metal analysis by ashing, provided a retaining compound is used.

6.11 Metal Analysis of Chemical Raw Materials Used to Manufacture Anaerobic Adhesives

In most adhesive chemical industries large quantities of chemical raw materials are purchased in bulk quantities to make these products. Formulated quantities of raw materials are blended using shearing stirrers, possibly under vacuum to form heterogeneous or homogeneous products. Traces of organic peroxides and active metals contaminants, particularly transition metals in methylacrylate or acrylate monomers, can cause inconsistent polymerisation rates and poor physical properties in the resultant adhesive. As part of their quality assurance programme raw materials manufacturers provide a Certificate of Analysis including metal analysis. This eliminates the need for routine analysis of every raw material and is normally sufficient assurance. However, in some cases, accidental contamination during transport or storage may occur. An added problem is that equipment used to make these products is usually metal-based and can contaminate these products with traces of Cu, Fe, Co, etc., caused by wear of machinery and vessels, particularly those used on a continuous basis.

6.11.1 Column Extraction of Metal from Liquid Monomers

In the case of certain raw materials, particularly those in the liquid form, trace levels of metals contained therein can be extracted into aqueous acid solutions and preconcentrated to improve their detection limits. A second method involves passing a solution of the raw material dissolved in a suitable metal-free solvent through an ion exchange column to retain the metals while the remaining sample is washed through minus the metals. The retained metals are then washed from the column with an acid solution to remove the metals, and their concentrations are measured against standard calibration curve prepared in the same acid solution. These methods can also be used as a preconcentrating step and the advantage is that a large number of samples can be prepared in a short time by involving several extractors or columns.

A study of column extraction of metals was carried out using pre-packed columns available from Polymer Laboratories (now a part of Varian, Inc.) [6]. Trace metals in these monomers would have a detrimental effect on the stability of anaerobic adhesive products and must be absent prior to use. The columns used are specially coated macroporous polystyrene products that are compatible with polar, non-polar, protic and aprotic solvents. They are designed to remove metals from solvents and monomers. The metal removing SPE product is approximately 45 μm and based on a mono-dispersed macroporous polymeric material.

Columns were evaluated for the extraction of transition metals from ethylene glycol dimethacrylate monomer with and without $5.0\,\mu g\,g^{-1}$ of Cu, Fe, Co and Ni in divalent state to establish the most suitable column for these materials. The macro-porous polystyrene types containing the active components have a much higher loading compared with functionalised silicas and are capable of removing charged and uncharged metal species from organic solvents. The metal sequester is further extracted using 1.0 M HCl with 25% ethanol for analysis using ICP-OES. The following columns, available from Polymer Laboratories, were evaluated:

Column type	Main functional group
Column 1 PL (Thiol MP SPE)	$-CH_2NHSNHCH_2CH_3$
Column 2, PL (Urea MP SPE)	$-CH_2NHCONHCH_2CH_2CH_2CH_3$
Column 3, PL (Thiorea MP SPE)	$-(SH)_2$

Method. Three columns containing PL-Thiol MP SPE, PL-Urea MP SPE and PL-Thiourea MP SPE (available from Polymer Laboratories Ltd) are studied to test the efficiency of extraction of metals from ethylene glycol dimethacrylate dissolved in methanol. The spiked and non-spiked monomer are dissolved in ethanol at 5% concentration and a 10.0 ml solution of each monomer is passed through the columns using a vacuum suction water pump. The columns containing the metals are washed with 1.0 ml of 0.25 M HCl and extracts made up to 25.0 ml in plastic volumetric flasks with deionised water. Extracts are analysed for metal content against standards of 0.0, 0.1, 0.25 and 0.5 ppm of Cu, Fe, Co and Ni prepared in 0.25 M HCl.

Results. Column extraction offers a new method of sample preparation and can be applied to liquid raw materials and solids that are solvent soluble. The column extraction method is promising and offers an efficient alternative for a range of materials. Excellent extractions of Cu, Fe, Co and Ni are obtained with the PL-Thiourea, while only Cu and Fe are extracted with the PL-Thiol column for this monomer (Table 6.9).

Table 6.9 *Results of extraction using metal scavenging columns available from Polymer Laboratories Ltd*

Metal	PL-Thiol		PL-Urea		PL-Thiourea	
	Mon. only	Mon.+ 'spike'	Mon. only	Mon.+ 'spike'	Mon. only	Mon.+ 'spike'
Cu	0.2	4.6	0.15	2.1	0.22	4.4
Fe	0.8	4.8	0.6	1.2	0.75	4.1
Co	<0.1	0.9	<0.1	0.8	<0.1	4.3
Ni	<0.1	2.6	<0.1	0.9	<0.1	4.7

6.12 Analysis of Metal Salt Content Dissolved in Aerosol Solvent(s)

Metal activators are formulated in aerosol solvents so that they can be applied to substrates by spray action causing the solvent to evaporate leaving the active metal(s) on the surface. The active metal enhances the anaerobic cure particularly on substrates that are slow to cure with anaerobic products. They are formulated at concentrations to suit application on difficult-to-bond substrates and for a particular adhesive.

6.12.1 Sample Preparation and Analysis of Metals in Aerosol

Method. The aerosol can containing metal (or metals) is shaken for 3 min to ensure a homogenous solution. Connect the tip of the spay nozzle to a capillary tube and place the other end under a previously weighed 100 ml plastic volumetric flask containing 50.0 ml propylene carbonate solvent. A sufficient weight of aerosol is sprayed into the flask containing the propylene carbonate liquid by holding the tip beneath the liquid. The increase in weight is noted and made up to 100 ml mark with propylene carbonate solvent. Measure the metal concentration against a standard calibration curve prepared in the same solvent used to dissolve the sample. The results are shown in Table 6.10.

Results. Good recoveries are obtained for all samples providing they are thoroughly mixed prior to sampling. This is a rapid test that could be useful in quality control laboratories where rapid analysis and accurate results are required.

Table 6.10 *Results of measured versus added concentration of listed metal in aerosol solvents. The values are the average of three sample preparations and measurements. [Note: Correction for density/weight of solvent used in calculation of true concentration averaged between $\rho = 0.77$ to 0.82 depending on solvent(s) used]*

Sample	Co added ($\mu g\,ml^{-1}$)	Co found $\mu g\,ml^{-1}$	Cu added $\mu g\,ml^{-1}$	Cu found $\mu g\,ml^{-1}$	Fe added $\mu g\,ml^{-1}$	Fe found $\mu g\,ml^{-1}$
Activator 'A'	35	33.6	5.0	5.5	—	—
Activator 'B'	5	6.1	10.0	8.9	—	—
Activator 'C'	—	—	—	—	700.0	685.0
Activator 'D'	—	—	—	—	50.0	53.3

6.13 A Study of the Effects of Anaerobic Adhesives on Metallic Substrates [5]

It is universally accepted that adhesives must establish an intimate molecular contact at the interface as a necessary requisite for success in forming strong bonds. Therefore, it is important that the adhesive spreads uniformly over the bonding area and efficiently displaces oxygen from the joint when the parts are forcibly pressed together. The theory of substrate etching (picking) by anaerobic adhesives containing acidic compounds is not as well understood as with acids in aqueous media. This study aims to study what effect different organic acids commonly used in anaerobic adhesives have on a range of metals requiring bonding. The acids are dissolved in the simplest methacrylate monomer (ethylene glycol dimethacrylate) used in anaerobic adhesives. The monomer containing the acid is brought into contact with the metal for a period of time as shown in Figure 6.4.

The chemical structure of the monomer is as follows:

$$CH_2{=}C(CH_3)CO\text{-}(OCH_2CH_2)_2\text{-}OCOC(CH_3){=}CH_2$$

To prevent inherent polymerisation of the monomer by free trace peroxy radicals and metal contaminants, the monomer is 'spiked' with hydroquinone as free peroxy radical inhibitor. The acids in Table 6.11 were dissolved in the monomer to a maximum concentration of 2.5 to 5% depending on the acid for the purpose of this study.

Table 6.11 *List of acids dissolved in polyethylene glycol dimethacrylate and pH of each solution as part of metal etching study*

Solution identity	Anaerobic adhesive containing:	% concentration	pH
1	Methacrylic acid	5.0	3.6
2	Maleic acid	2.5	4.6
3	Benzoic sulphamide	2.5	6.0
4	2-Ethylhexanoic acid	5.0	2.0
5	Hydroxypropane sulphonic acid	5.0	0.5

The following metal lapshears were contacted with the above monomer/acid solutions; stainless steel (SS), mild steel (bare) [MS (bare)], mild steel (coated with ZnbiCr) [MC (ZnBiCr)] and copper metal (Cu).

Figure 6.4 illustrates the metal monomer/acid area contact of each metal substrate.

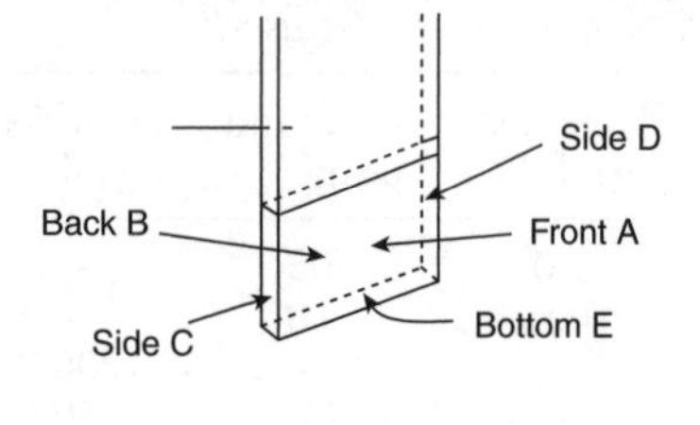

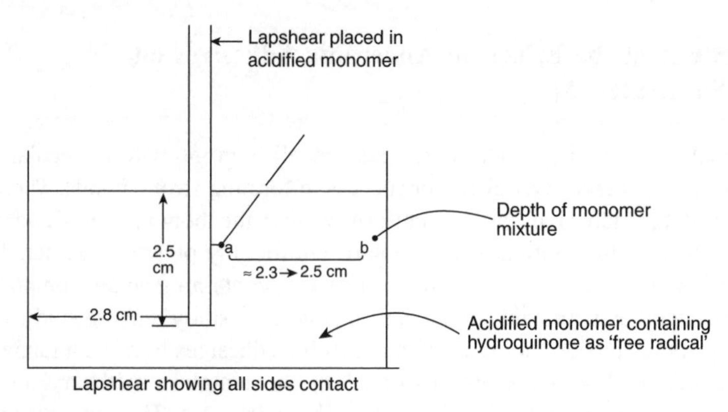

Figure 6.4 *Schematic diagram of lap shear in acid/monomer area contact study. The total contact area for each lap shear was determined by the total sum (A,B,C,D and E) of all areas of the lapshear*

All lapshears were immersed in 50 ml of each monomer/acid solution in tall plastic beakers in order to keep the solution volume low and achieve an area to depth of 2.5 cm in each case. They were left in the solution for the times (in minutes) given in Table 6.12.

Table 6.12 *Contact times (in minutes) for monomer/acid solution with each metal investigated in this study. (Note: The copper metal is softest in the study, hence shorter times)*

Solution	SS	MS (bare)	MS (ZnbiCr)	Cu	Al
1	120	120	120	30	120
2	120	120	120	30	120
3	120	120	120	30	120
4	120	120	120	30	120
5	120	120	120	30	120

Method. Analyse the concentration of 'etched metals' of each extract after diluting in 50% solution of propylene carbonate and glacial acetic acid. They are measured against a calibration curve prepared from 0.0, 1.0, 2.5, 5.0 and 10 ppm of each metal measured and prepared in the same solvents as the sample.

Results. This experiment demonstrates that metal ions are etched from metal substrates by acidified monomers. This experiment is a little more than assessment of the degree of surface etching and mobility of the metal ions throughout the adhesive joint. The results obtained are only 'apparent' when compared with the same monomer/acid mixture measured before contact with the lapshears (Table 6.13).

Table 6.13 *Results of analysis of 'etched' metal for each monomer (mon)/acid mixture. The solution identifications are listed in Table 6.11. [SS, stainless steel; MS (ZnbiCr), mild steel coated with zinc bichromate; MS (bare), mild steel no coating; Al, aluminium metal containing ~5.0% Cu and Cu metal only]*

Mon+acid	Al	Fe	Cr	Cu	Zn
Mon + methacrylic acid					
Sol 1 + SS	<0.1	0.8	<0.1	<0.1	<0.1
Sol 1 + MS (ZnbiCr)	<0.1	13.6	0.63	2.3	87.9
Sol 1 + MS (bare)	<0.1	39.6	2.7	3.4	10.5
Sol 1 + Al	18.3	1.3	0.26	119.5	6987
Sol 1 + Cu	<0.1	<0.1	<0.1	897	45.1
Mon + maleic acid					
Sol 2 + SS	<0.1	0.2	<0.1	<0.1	<0.1
Sol 2 + MS (ZnbiCr)	<0.1	2.6	8.2	<0.1	11.3
Sol 2 + MS (bare)	<0.1	4.3	2.9	1.9	0.40

(*continued*)

Table 6.13 (*Continued*)

Mon+acid	Al	Fe	Cr	Cu	Zn
Sol 2 + Al	5.3	0.44	<0.1	<0.1	84.2
Sol 2 + Cu	<0.1	<0.1	<0.1	1.6	0.87
Mon + benzoic sulphamide					
Sol 3 + SS	<0.1	0.47	<0.1	<0.1	<0.1
Sol 3 + MS (ZnbiCr)	<0.1	0.24	9.7	<0.1	4.1
Sol 3 + MS (bare)	<0.1	2.1	4.6	<0.1	<0.1
Sol 3 + Al	3.1	<0.1	0.32	<0.1	3.3
Sol 3 + Cu	<0.1	<0.1	<0.1	14.2	<0.1
Mon + 2-ethylhexanoic acid					
Sol 4 + SS	<0.1	19.3	6.7	<0.1	<0.1
Sol 4 + MS (ZnbiCr)	<0.1	649	111.9	19.0	40.1
Sol 4 + MS (bare)	1.3	2070	15.2	37.8	2.9
Sol 4 + Al	1860	2.6	0.2	2333	15.9
Sol 4 + Cu	4.9	0.6	0.15	39965	0.66
Mon + hydroxypropane sulphonic acid					
Sol 5 + SS	<0.1	34.7	66.9	<0.1	<0.1
Sol 5 + MS (ZnbiCr)	<0.1	89.3	208.6	38.5	106.9
Sol 5 + MS (bare)	0.8	4049	43.6	79.0	4.7
Sol 5 + Al	2644	1.15	<0.1	5679	2990
Sol 5 + Cu	5.7	0.91	<0.1	42390	1.9

6.14 Metals Associated with Epoxy Adhesives

Curing of epoxy adhesives can be controlled by addition of suitable curing agents and several different types are available that function at different rates. The most common curing agents in the epoxy adhesives are organic-based and function in a two-part adhesive with one part containing the active monomer resin with fillers, and the second part containing the curing agent with fillers, etc. Some epoxy adhesives are formulated with the curatives encapsulated with a suitable coating which when fractured under pressure releases the curatives. Recent development of epoxy adhesives incorporates a room temperature inactive curative that only cures with the application of heat.

A number of curing agents and catalysts used in epoxies are complex metal salts that are added to cure at room temperature or with heat. Curing agents or catalysts such as cationic dinonato (acetylacetone, etc.) complexes of Si, B, Ge, and P behave as hydrolytic activated Bronsted acid precursors, e.g.:

$$M^{-}(\text{acetylacetone})X$$

where

$$M = \text{Si, B, Ge and P}$$
$$X = SbF_6^-, AsF_6^-, PF_6^-, BF_4^-, ClO_4^-, \text{ etc.}$$

Metal curing salts function with aliphatic or aromatic epoxy resins in the presence of trace water.

Heat curing epoxy resins use a range of primary amines but are considered inferior to room temperature curing epoxies. Metal salts such as M[II] 4,4',4'',4''' -phthalocyaminetetramines where M = Cu, Co or Ni have been known to cure resins with considerable improvements in heat resistance and anaerobic-type curing characteristics compared with resins prepared with conventional primary amine curatives.

The catalysts dibutyl tin dilaurate and dioctyl tin dilaurate typically used to prepare resins for epoxy adhesives must be removed on completion of reaction in the case of epoxy adhesives to be used in computer mainframes. The vapour generated through heat releases tin vapour salts which can damage sensitive electronic components.

Inorganic fillers such as clays, $CaCO_3$, talc, silica, titanates, Al and asbestos are commonly used in epoxy adhesives, as they are cheap and readily available. Conducting epoxies can be formulated with powdered copper metal or a mixture of a blend of Sn-Pb-Bi.

The fillers are added so that the adhesives are sag-free, motionless and form shapes that suit the components to be bonded, reduce shrinkage during cure, increase thermal conductivity, improve corrosion resistance and reduce costs. The concentration of these curatives, fillers and conducting agents is usually monitored using ICP-OES.

6.14.1 Composition of Epoxy Adhesives

The most important additives in epoxy adhesives are the epoxy resins followed by fillers, colorants, curing agents, toughening agents, catalysts, etc. Resins can have viscosities ranging from 2000 cps to high melting solids and can be aliphatic or aromatic containing mono- or polyfunctional epoxy groups. The active group attached to these resins is the three-membered ring containing oxygen as follows:

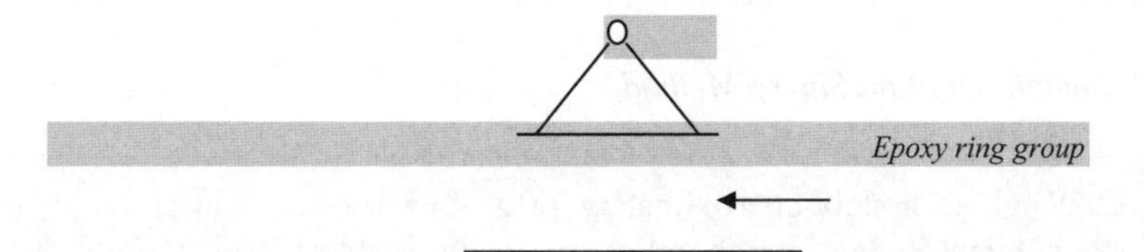

It is not the intention of this book to review the complex chemistry associated with these products or to review information other than the analysis for metallic type agents that are critical to these formulations. The metallic additives need to be monitored as part of quality assurance and for environmental and health requirements.

6.14.2 Preparation of Epoxy Adhesive 'Spiked' with $Ge(AcAc)BF_4$

There are many metal-type additives in epoxy adhesives and for the sake of analytical input to these products for metal content the following formulations were prepared to study the effect of different sample preparation methods as applied to these products. Four typical epoxy formulations containing active monomer/resin, colorants, curatives and fillers were prepared in the laboratory as part of a study of sample preparation methods for the determination of the concentration of the $Ge(AcAc)BF_4$ additive. The four preparations were formulated as shown in Table 6.14.

Each of the formulations (Table 6.14) contains 0.1% $Ge(AcAC)BF_4$ which equates to 279 ppm germanium and 42 ppm boron, respectively. Both elements can be used to quantify the concentration of this curative and the laboratory prepared samples using a high shear mixer for 3 h to ensure homogenous products.

Table 6.14 *Laboratory preparation of epoxy adhesive formulations containing the curative Ge(AcAc)BF_4 in a typical epoxy adhesive formulation*

Epoxy adhesive formulation	Additives
1 + 0.25% Ge(AcAc)BF_4	No fillers
2 + 0.25% Ge(AcAc)BF_4	6% SiO_2 + 6% $CaCO_3$
3 + 0.25% Ge(AcAc)BF_4	12% SiO_2
4 + 0.25% Ge(AcAc)BF_4	8% SiO_2 + 4% Talc

6.14.3 Determination of the Concentration of Ge(AcAc)BF_4 in Epoxy Adhesives Using Non-Destructive Methods

A study of the quantitative analysis of epoxy adhesives for the measurement of the concentration of Ge(AcAc)BF_4 was initially carried out using dilution methods. They contain particles of insoluble fillers that are usually less than 5 μm so that they can flow through narrow needles in rapid production lines. These particles sizes are suitable for a slurry method of analysis using the ICP-OES with a V-groove cross flow nebuliser. The sample solution is first stirred with a magnetic stirrer and then pumped to the nebuliser using a peristaltic pump. The sample inlet orifice in the 'V' nebuliser is large enough to allow the slurry solution to pass into the spray chamber and eventually to the plasma for excitation. The particle sizes are also suitably small for maximum efficiency of the sample in the spray chamber. The stirring of the slurry is carried out for the duration of analysis.

6.14.3.1 Sample Dilution/Slurry Method

Method. Weigh accurately approximately 1.0 g of each epoxy adhesive into a 100.0 ml plastic volumetric flask and dissolve in 10.0 ml of chloroform. Make up to 100.0 ml mark with 50:50 glacial acetic acid and propylene carbonate. The solution is shaken to ensure homogeneity and to give potential concentrations of 0.42 ppm B and 2.8 ppm Ge, respectively. The sample is stirred using a magnetic stirrer prior to the peristaltic pump and transferred to the nebuliser using the pump to the torch for excitation. The B is more sensitive than the Ge but the Ge is at a higher concentration. This means that both elements can be detected with ease and used for compound quantification*. Standards of 0.5 ppm B+ 2.0 ppm Ge and 1.0 ppm B + 4.0 ppm Ge, respectively, are prepared in 10.0 ml of chloroform and 50:50 glacial acetic acid and propylene carbonate.

*There is a possibility that the SiO_2 or $CaCO_3$ could be contaminated with B but is unlikely to be contaminated with Ge.

6.14.3.2 Standard Addition Method

Method. Separately weigh accurately approximately 1.0 g of each epoxy adhesive listed in Table 6.14 followed by 10.0 ml of chloroform to dissolve the samples. Add no metal standard into the first flask and add 0.5 ppm B and 2.0 ppm Ge into the second flask. Into the third flask add 1.0 ppm B and 4.0 ppm Ge and into the fourth flask add 1.5 ppm B and 6.0 ppm Ge. Each flask is diluted to 100 ml with 50:50 glacial acetic acid and chloroform. A similar series of standards are prepared for both metals in the same solvent system without the samples. The intensity of standards and samples are measured with stirring on a magnetic plate and extrapolate to the negative concentration line to read the concentration for each metal. The standards prepared without the sample should pass through the origin and if a parallel line is obtained similar to the sample, the concentration obtained for the sample is acceptable.

6.14.3.3 Internal Standard Method

Method. A minimum of three standards of increasing concentrations are prepared in a 100 ml plastic volumetric flask and dissolved in 10.0 ml of chloroform. Make up to mark with 50:50 glacial acetic acid and propylene carbonate solvents. A single internal standard of yttrium metal is prepared in a separate plastic volumetric flask with the same solvent mixtures. A calibration curve is generated using the emissions intensity from standards along with the internal standard intensity for each standard. The intensity of the internal standard should be constant while the standards show corresponding increasing intensities. The special 'V' described in Section 3.7.4 is used to mix the internal standard with the standards and samples prior to nebulisation. The samples listed in Table 6.14 are dissolved in the same solvent as that for standards and stirred prior to the pumping and nebulisation. The concentrations are determined and corrected for viscosity differences using the internal standard for both metals against the calibration curves prepared.

Results. Table 6.15 gives results of analysis of epoxy adhesives 'spiked' with $Ge(AcAc)BF_4$.

Table 6.15 *Results of analysis of epoxy adhesives 'spiked' with $Ge(AcAc)BF_4$. Both germanium and boron elements were measured at the same time. The results show the effect of fillers on signal intensities. The presence of boron as a contaminant in the talc rules this element out as a means of quantifying the concentration of $Ge(AcAc)BF_4$ in epoxy adhesives*

	Dilution only		Standard addition		Internal standard	
	Ge	B	Ge	B	Ge	B
Epoxy, no filler	223	33.4	245	39.0	265	44.1
Epoxy, 6% SiO_2 + 6% $CaCO_3$	201	25.6	225	35.3	270	44.8
Epoxy, +9% SiO_2	211	23.4	240	31.6	270	45.6
Epoxy, 8% SiO_2 + Talc	140	>1000	190	>1000	250	>1000

6.14.4 Determination of the Concentration of Ge(AcAc)BF_4 in Epoxy Adhesives Using Destructive Methods

A study of the effect of ashing of epoxy adhesives for metals content was carried out as follows.

6.14.4.1 Ashing Method for the Analysis of Ge and B in Epoxy Adhesives

Method. Accurately weigh approximately 2.5–3.0 g of into four clean platinum dishes of each epoxy adhesive listed in Table 6.14 along with 0.15 g of PTSA. The samples are ashed using a stepwise 'heating and holding' ramping to a maximum of 650 °C as shown in Figure 6.3. On completion of ashing the residue is transferred to plastic beakers and 'spiked' with boron and germanium. The mixture is dissolved in an acid mixture of 5.0 ml of 1 M HNO_3 and 10 ml of 0.1 M HCl. The resulting solution is filtered through a fast flow filter paper and the clear solutions analysed against a standard calibration curve of 0.0, 0.5, 2.5 and 5.0 ppm for each metal prepared in the same acid mixtures.

Results. The results of analysis of epoxy adhesives using the ashing method are shown in Table 6.16.

Table 6.16 *Results of analysis of epoxy adhesives using the ashing method*

	Ge	B
Epoxy, no filler	41	36.4
Epoxy, 6% SiO_2 + 6% $CaCO_3$	102	79.9
Epoxy, SiO_2	165	35.7
Epoxy, 8% SiO_2 + Talc	236	>1000

6.14.4.2 Microwave Acid Digestion of Epoxy Adhesives for Ge and B Content

Method. Accurately weigh approximately 0.8 g of each of epoxy adhesive listed in Table 6.14 into special Teflon beakers suitable for microwave acid digestion. Add 10.0 ml of conc. HCl and 5.0 ml of conc. HNO_3 and allow the mixture to predigest for 2 h prior to microwave digestion. A blank containing the acids only is also prepared under the same conditions. The digestion is carried out according to the parameters shown in Table 6.17.

Table 6.17 *Conditions for microwave acid digestion of epoxy adhesives*

Stage	Power (W)	% power	Time (min)	Pressure (psi)	Temp. (°C)	Hold (min)
1	600	100	20	150	150	10.0
2	600	100	15.0	200	160	10.0

Table 6.17 (Continued)

Stage	Power (W)	% power	Time (min)	Pressure (psi)	Temp. (°C)	Hold (min)
3	600	100	10.0	350	180	10.0
4	600	100	10.0	400	190	10.0
5	600	100	10.0	500	210	20.0

The samples prepared in Table 6.14 were allowed to cool to room temperature and 5.0 ml of conc. HF was added to each vessel and the residue digested using the parameters shown in Table 6.18.

Table 6.18 *Conditions for second digesting of SiO_2, $CaCO_3$ and Talc in epoxy formulations*

Stage	Power (W)	% power	Time (min)	Pressure (psi)	Temp. (°C)	Hold (min)
1	600	100	5.0	200	190	10.0
2	600	100	5.0	300	200	10.0
3	600	100	5.0	400	210	10.0

On completion of digestion the vessels are allowed to cool and the excess acids reduced using the CEM micro-vap accessory. The remaining reduced acid solutions are diluted to mark with deionised water to 50.0 ml volume. Each solution is analysed against Ge and B standards prepared in de-ionised water from certified standards.

Results. The results of the analysis of epoxy samples after sample preparation using microwave acid digestion are shown in Table 6.19.

Table 6.19 *Results of analysis of epoxy samples after sample preparation using microwave acid digestion*

	Ge	B
Epoxy, no filler	272.0	38.0
Epoxy, 6% SiO_2 + 6% $CaCO_3$	274.0	39.0
Epoxy, 12% SiO_2	276.0	34.0
Epoxy, 8% SiO_2 + 4% Talc	271.0	>1000

6.14.4.3 Analysis of Epoxy Adhesives for Ge and B Content Using Oxygen Bomb Combustion Method

Method. An accurate weight of approximately 0.5 g of sample followed by 0.25 g of paraffin oil is added to a suitable platinum dish attached to the holder in the lid of the bomb. To this add 5.0 ml of deionised water to the bottom of the vessel away from the dish. The bomb is charged with 30 atm of oxygen and placed under water to check for visible leaks. Assuming no leaks are detected the bomb is fired according to manufacturer's instructions. On completion of combustion filter the contents through a fast flowing filter paper. The clear filtered liquid is diluted to 50.0 ml with deionised water. The sample is analysed against standards prepared from certified standards using deionised water.

Results. The epoxy monomer resins containing metal salts can be analysed for metals content after diluting in suitable solvents providing the concentration is high enough for detection (Table 6.20). The maximum concentration of epoxy products that ICP-OES can tolerate is usually 0.05–1.0% depending on the product. Trace metal content in these products requires ashing, microwave acid digestion or bomb combustion as sample preparation methods and in some cases may require the use of hydride generation or ultrasonic sample nebulistion to measure low levels accurately.

Table 6.20 *Results of analysis of epoxy adhesives using the oxygen bomb combustion method*

	Ge	B
Epoxy, no filler	275.0	48.6
Epoxy, 6% SiO_2 + 6% $CaCO_3$	157.0	46.5
Epoxy, 12% SiO_2	93	44.7
Epoxy, 8% SiO_2 + 4% Talc	21.0	>1000

Laboratory-prepared epoxy adhesives containing fillers that are representative of most epoxy products were 'spiked' with 0.25% Ge(AcAC)BF_4 curing agent. A list of prepared epoxy products is presented in Table 6.14. The samples were analysed using dilution methods and the best was the internal standard method. The exception was boron that is probably contaminated by the addition of talc. Ashing and microwave acid digestion destructive methods also gave excellent results and, again, the formulation containing talc gave higher results. The oxygen bomb reported lower values probably due to the insolubility of the oxide states of the metals.

6.14.5 Conclusion of Metal Analysis of Epoxy Adhesives

Metal catalysts used in epoxy products are usually present as complex salts and are reactive when initiated by heat and/or water when applied to substrates. Due to their high sensitivity and reactivity they need to be monitored to ensure that they are formulated correctly so as to enhance the performance of these products and at the same time be stable during storage. If the concentration of the curatives is low, it may give rise to slow bonding properties and if too high it will make the product unstable. Lewis acid type catalysts are restricted due to the rapid curing of epoxy products and poor physical properties of cured resins. However, Lewis acids are frequently modified by formation of complexes with amines or glycols.

The salt Ge(AcAc)BF_4, described in Section 6.14.2, was used to study the importance of selecting the correct sample preparation method and the effect it can have on the final results. Both destructive and non-destructive methods were studied. The non-destructive dilution method can only be applied if the sample is soluble in a solvent that is compatible with ICP-OES and a suitable internal standard is used to correct for viscosity effects. The results in Table 6.15 clearly illustrate the importance of selecting the correct sample preparation method. In the case of the destructive methods only the microwave acid digestion reported results that were acceptable with and without fillers. The bomb combustion method seems to form very volatile metal gases that are lost on releasing the excess CO_2 and water. The ashing method was not studied here as it is well known that the salt volatises when heated in an oven. Other non-volatile metals salts which are added as fillers such as $CaCO_3$, clays, SiO_2, chalks, etc., are not important because of their unreactive properties in epoxy adhesives and generally are not quantified using ICP-OES methods. A simple dry ashing method would be sufficient here. However, identification may be necessary by analysing for the type of metal filler.

6.15 Metals Associated with Phenolic Adhesives

Phenolic adhesives can contain metal salts of Ca, Mg, Zn, Pb, Cd, Co, Ni and Cu as acetals, halides, hydroxides or sulphonates as activators. These adhesives are limited to room temperature functionality but the use of a specific catalyst such as copper 8-quinolinolate can be used to increase durability at higher temperatures.

6.15.1 Preparation of Typical Phenolic Adhesives Containing Calcium and Copper Sulphonate Salts

Four laboratory samples of phenolic adhesives were prepared with and without fillers, as shown in Table 6.21, to study methods of sample preparation and analysis for metal content.

Generally, analysis for the presence of metals in phenolic adhesives is carried out using dilution methods against standard calibration curves and is possible because of the high concentrations of metal salts present. Sample preparations using ashing, microwave acid digestion and bomb combustion are similar to methods applied for epoxy adhesives but are only required if concentrations are lower than quantification detection limits after dilution or for the determination of trace metals as part of health and environmental requiremental. The most suitable method for routine analysis of metal catalysts is the internal standard method.

Table 6.21 *Composition of typical phenolic formulation with 0.1% each of* $Ca(SO_3H)_2$ *and* $Cu(SO_3H)_2$ *and listed levels of fillers. This is equivalent to 198 ppm Ca and 274 ppm Cu*

Phenolic adhesive	Additives
1 + 0.1% $Ca(SO_3H)_2$ + 0.1% $Cu(SO_3H)_2$	No fillers
2 + 0.1% $Ca(SO_3H)_2$ + 0.1% $Cu(SO_3H)_2$	3% SiO_2 + 3% TiO_2
3 + 0.1% $Ca(SO_3H)_2$ + 0.1% $Cu(SO_3H)_2$	5% TiO_2
4 + 0.1% $Ca(SO_3H)_2$ + 0.1% $Cu(SO_3H)_2$	2% SiO_2 + Talc

6.15.2 Non-Destructive Methods of Analysis of Phenolic Adhesives

6.15.2.1 Internal Standard Method

Method. Prepare a minimum of three standards of increasing concentrations of Ca and Cu in 100 ml plastic volumetric flasks and make up to the mark to 100 ml. These standards are dissolved in 10.0 ml of chloroform followed by 90.0 ml of 50:50 glacial acetic acid and propylene carbonate solvents. A single internal standard of yttrium metal is prepared in a separate plastic volumetric flask using the same solvent mixture as sample. A calibration curve is generated with the standards by measuring the emission intensity along with the internal standard intensity for each standard. The intensity of the internal standard should be constant while the standards show corresponding increasing intensities. The special 'V' described in Section 3.7.4 is used to mix the internal standard with the standards and samples prior to nebulisation. The samples listed in Table 6.21 are dissolved in the same solvent as that for standards and stirred prior to pumping and nebulisation. The concentrations of Ca and Cu are determined and corrected for viscosity differences using the internal standard for both metals against the calibration curves prepared.

Results. Phenolic adhesives are usually manufactured with little or no fillers because of their applications. They are soluble in solvents and dilution methods can be readily applied to the analysis of these products but if trace analysis is required for health or environmental reasons destructive methods may be necessary. Excellent results can be obtained for the determination of calcium and copper sulphonate salts added to products using the internal standard method. These low density adhesives can be analysed without resorting to destructive methods (Table 6.22).

Table 6.22 *Results of measurement of the concentrations of Ca and Cu after dilution in chloroform, propylene carbonate and glacial acetic acid. The identities **1, 2, 3** and **4** are the original phenolic formulations 'spiked' with fillers listed in Table 6.21*

Sample	Additives	Ca	Cu
1 + 0.1% $Ca(SO_3H)_2$ + 0.1% $Cu(SO_3H)_2$	No fillers	188	279
2 + 0.1% $Ca(SO_3H)_2$ + 0.1% $Cu(SO_3H)_2$	3% SiO_2 + 3% TiO_2	>1000	269
3 + 0.1% $Ca(SO_3H)_2$ + 0.1% $Cu(SO_3H)_2$	5% TiO_2	179	277
4 + 0.1% $Ca(SO_3H)_2$ + 0.1% $Cu(SO_3H)_2$	10.0 % SiO_2	>1000	283

6.16 Metals Associated with Polyurethane Adhesives

Polyurethane adhesives employ metal catalysts as part of their functionality and the most popular are listed in Table 6.23. Catalysts such as the borate or phosphate salts of Sb, Ge, Mo and W are also widely used in polyurethanes containing a high concentration of

Table 6.23 List of common organometallic catalysts used in polyurethane products

Catalyst	Use	Metal(s)
Dibutyl tin octoate	Flexible foams	Sn
Dibutyl tin dilaurate	Moulding foams	Sn
Dibutyl tin mercaptide	Hydrolysis resistant	Sn
Dibutyl tin thiocarboxylate	Controlled action	Sn,S
Dioctyl tin thiocarboxylate	Resilient foams	Sn,S
Phenyl mercuric propionate	Delayed reaction	Hg
Lead octoate	Chain extender	Pb
CH_3COOK	General catalyst	K
$NaHCO_3$	General catalyst	Na
Na_2CO_3/Ca_2CO_3	Catalyst + filler	Na,Ca
Ferric acetylacetonate	Catalyst	Fe
Bi^{2+} salts	Bi	Bi
Silanes	Bonding aid	Si
Phosphines	Bonding aid	P

soluble rubbers and/or fillers. These adhesives are used extensively where wide-gap bonding is required and bonding of glass windscreen to metal on automobiles, aircraft, ships, etc. Organometallic salts are added to polyurethanes for the reasons stated in Table 6.23 and are present in these products to suit the application and reaction rates.

6.16.1 Preparation and Analysis of Polyurethane Adhesives Containing Organometallic Catalysts

For the purpose of studying the effects of sample preparation and analysis of polyurethanes for organometallic catalysts, the metal salts listed in Table 6.24 were added at 0.1% each.

Preparation of polyurethane samples for metal analysis requires destructive techniques because of their high molecular weights and high concentration of rubber and/or fillers. These products are not very soluble in glacial acetic acid, propylene carbonate, kerosene or white spirit. They will dissolve in chloroform xylene, toluene but these solvents are unsuitable for ICP-OES nebulisation because they flicker and give rise to noisy plasmas and quench the plasma during analysis.

6.16.1.1 Ashing Methods for Metal Analysis of Polyurethane Adhesives

Method. An accurate weight of approximately 2.0 to 3.0 g of each polyurethane listed in Table 6.24 is placed in a clean platinum dish with 0.15 g of PTSA. The samples are ashed using a program of 'ramping/holding' stages similar to methods described for cyanoacrylate adhesives to a maximum temperature of 650 °C. The resulting ashes are contacted with 10.0 ml of 0.1 M HCl and the suspensions transferred to a Teflon beaker. To each add 2.0 ml of conc. HF and heat to ~150 °C on a hot plate to dissolve the SiO_2. The clear solutions are made up to mark in a plastic volumetric flask with deionised water.

Table 6.24 *List of basic polyurethane (PU) formulations 'spiked' with 0.1% organometallic catalysts*

PU + organometallic[a]	Chemical formula	M^{n+}	Metal (ppm)
Dibutyl tin dilaurate	$CH_3(CH_2)_3Sn[OCO(CH_2)_{10}CH_3]_2$	Sn	187
Phenyl mercuric propionate	$C_9H_{10}HgO_2$	Hg	572
dichlorodimethylsilane	$(CH_3)_2SiCl_2$	Si	218
Ferric acetylacetonate	$Fe(C_5H_7O2)_3$	Fe	158

[a]Organometallics are not necessarily present in all polyurethanes.

6.16.1.2 *Microwave Acid Digestion Method*

Method. Accurately weigh approximately 0.8 g of sample of each polyurethane listed in Table 6.24 and add into separate Teflon flasks suitable for microwave acid digestion. To each flask add 10.0 ml of conc. HNO_3 and 5.0 ml of conc. HCl. The mixtures are allowed to predigest for 2 h prior to microwave heat pressure digestion. A blank containing the acids only is also prepared under the same conditions. The digestion is carried out according to the conditions in Table 6.25.

Table 6.25 *Microwave conditions for digestion of polyurethane adhesives*

Stage	Power (W)	% power	Time (min)	Pressure (psi)	Temp. (°C)	Hold (min)
1	600	100	20	250	180	10.0
2	600	100	15.0	400	200	10.0
3	600	100	20.0	600	220	10.0

After completion of digestion allow the vessels to cool to room temperature, add 2.0 ml of conc. HF to each vessel and further digest as shown in Table 6.26.

Table 6.26 *Conditions for second digestion of SiO_2 additive in polyurethane formulations*

Stage	Power (W)	% power	Time (min)	Pressure (psi)	Temp. (°C)	Hold (min)
1	600	100	5.0	200	190	10.0
2	600	100	5.0	300	200	10.0
3	600	100	5.0	400	210	10.0

The acid volumes were reduced using a CEM micro-vap evaporation apparatus and the remaining reduced liquid made up to mark in a plastic volumetric flask with deionised water.

6.16.1.3 Oxygen Bomb Combustion Method

Method. Accurately weigh approximately 0.5 g of sample followed by 0.25 g of paraffin oil and add to a suitable platinum dish attached to the holder in the lid of the bomb. Add 5.0 ml of deionised water to the bottom of the vessel. The bomb is charged with 30 atm of oxygen and placed under water to check for visible leaks. Assuming no leaks are detected the bomb is fired according to manufacturer's instructions. On completion of combustion the resulting suspension is transferred to a Teflon beaker. Then 2.0 ml of conc. HF is added and heated for 2–3 min to dissolve the SiO_2. The solutions are diluted with deionised water to mark in plastic volumetric flasks. This is repeated for all samples including a blank without sample. The solutions are analysed against certified standard solutions prepared in deionised water.

Results

Table 6.27 *Results of analysis of polyurethane (PU) adhesives spiked with 0.1% organometallic catalysts using destructive methods; ashing, microwave acid digestion and bomb combustion methods of sample preparation*

PU + organometallic	Ashing	Microwave digestion	Bomb method	Added conc.
Dibutyl tin dilaurate (Sn)	66	183	191	187
Phenyl mercuric propionate (Hg)	31	566	493	572
dichlorodimethylsilane (Si)	210[a]	211[a]	222[a]	218
Ferric acetylacetonate (Fe)	98	183	~755	158

[a]HF acid was added to suspensions to dissolve the SiO_2.

6.17 Conclusion to Metal Analysis of Phenolic and Polyurethane Adhesives

There are a range of metal catalyst/activators used in phenolic and polyurethane adhesives that can be monitored using ICP-OES. In the metal analysis study of phenolic adhesives, the salts $Ca(SO_3H)_2$ and $Cu(SO_3H)_2$ were formulated as listed in Table 6.21, with and without fillers. Several metal salts are employed in polyurethane adhesives for the reasons listed in Table 6.23. As part of the sample preparation study of these products metal salts were formulated into typical polyurethane products consisting of volatile and non-volatile metal salts, listed in Table 6.24, showing the metals of interest and the concentrations expected.

Phenolic adhesives are usually manufactured with little or no fillers because of their applications. They are soluble in solvents and dilution methods can be readily applied to the analysis of these products. However, if trace analysis is required for health or environmental reasons destructive methods may be necessary. Excellent results can be obtained for the determination of calcium and copper sulphonate salts added to products using the internal standard method (Table 6.22). These low density adhesives can be analysed without resorting to destructive methods.

Polyurethane adhesives are high density, low solvent soluble adhesives that require destructive methods of sample preparation for metal analysis. Three methods were studied using laboratory prepared polyurethane adhesives 'spiked' with catalysts (Table 6.23). The ashing method gave poor results even in the presence of a retaining agent (PTSA); the microwave acid digestion and oxygen bomb combustion gave good results. The exception is the Fe content using the bomb, which gave higher results, probably due to attack of the stainless steel vessel making it a poor method for metals present in the bomb material.

References

[1] Harthorn, S.R. (Ed.) (1986) *Structural Adhesives, Chemical and Technology Topics in Applied Chemistry*, New York: Plenum Press.

[2] Coover, H.W. and McIntire, J.M. (1973) 'Cyanoacrylate Adhesive Composition', US Patent 3,728,375, Eastman Kodak [CA 78,44509].

[3] Krieble, R.H. (1962) 'Anaerobic Curing Sealant Having Extended Shelf life', US Patent 3,043,820, Rocky Hill, Connecticut.

[4] Christe, K.O., Dixon, D.,A., McLemore, D., *et al.* (2000) On quantitative scale for Lewis acidity and recent progress in polynitrogen chemistry, *Journal of Fluorine Chemistry*, **101**(2), pp151–153.

[5] Brennan, M.C. (1992) *Novel Electrochemical and Atomic Spectrometric Techniques in the Characterization of Anaerobic Adhesives*, PhD Thesis, Cork: University College Cork.

[6] Stratospheres™ SPE Resins, mini high load scavenger resin columns', Polymer Laboratories (now a part of Varian Inc.).

7

Hyphenated and Miscellaneous Techniques Used with ICP-OES

7.1 Introduction

As analytical techniques improve and capabilities become more sophisticated, more and more multi-component systems are appearing on the market. They are linked together with a greater degree of compatibility that improves speed, accuracy and, in some cases, can be automated. Combined techniques linked together are called 'hybrids' or hyphenated techniques and several readily available examples are used in chromatography techniques, e.g. GC/MS, HPLC-MS, GC-IR, GC-AED, GC-MS-MS, LC-MS-MS, etc. These techniques are the combination of two independent instruments with the purpose of enhancing information associated with measurements that is poor or non-existent using a single instrument.

Hyphenated techniques associated with ICP-OES include ultrasonic nebuliser, hydride generator, cold vapour trap (mercury), graphite furnace, DIN (direct injection nebulisation), GC/AED and HPLC. The hyphenated attachment gives an extra dimension to capability of selectivity, increase in sensitivity, accuracy, isotopic ratio studies and so on, which under normal measurements may be difficult to detect. With the advance of multi-array detectors, multi-component analysis is feasible since each improvement is designed to supply more information associated with the analysis. An example is the inclusion of HPLC to ICP-OES that can be used as a means of a preconcentration step in improving quantitative detection of a series of elements in samples, e.g. oxidation states of metals and metal-ligand complexes. Computer control and data handling have the power to enhance the capabilities of most complex systems and for reproducible control of the combined systems. The ideal hyphenated system for the future will detect and quantify smaller sample sizes at lower concentration, have greater speed and better accuracy and will be automated.

A Practical Approach to Quantitative Metal Analysis of Organic Matrices Martin Brennan

An analytical method must be capable of delivering signals that are free from interferences and report true values. The differences between the analyte and interferant is associated with the extent to which a method can determine the accurate quantity of analyte in a complex matrix, particularly signals affected by volatilisation, viscosity, matrix salts, etc. Modern methods are designed by combining several measurement principles and introducing their own selectivity to the complete operation in order to enhance the above stated requirements. Setting up a hyphenated instrument to an ICP-OES often requires a considerable degree of trial and error before accepting it as an official technique. More often than not the effort would be worth it for the sake of the information obtained and confidence in the results reported.

Speciation studies of metals are also an important requirement in some metal complexes and a method for selectively separating and detecting these species is becoming very important. An example is the successful application of a hyphenated technique in the speciation study of the toxic compounds associated with arsenic compound mixtures and the determination of the four forms of arsenic metal salts which can exist – As(III), dimethylarsenic acid (DMAA), monomethylarsonic acid (MMAA) and arsenate (V), using the combination of LC-ICP-OES. A similar example is the chromatographic separation of a mixture of monomethyltin trichloride (MMT-TCl), dimethyltin dichloride (DMT-DCl), diethyltin dichloride (DET-DCl) and trimethyltin chloride (TMC-Cl); each tin metal complex can be detected by GC/ICP-OES. Hyphenated systems such as reverse phase liquid chromatography and ICP-OES/MS detect the separation and quantitative detection of sub-trace levels of cationic species [2] of Hg (Hg^{2+}, $MeHg^+$, $EtHg^+$ and $PhHg^+$) and Pb [Pb^{2+}, $(Me)_3Pb^+$ and $(Et)_3Pb^+$].

The following hyphenated and miscellaneous techniques with ICP-OES will be discussed in this chapter:

(i) ICP-OES-FIA
(ii) Internal standard analysis
(iii) ICP-OES-IC
(iv) ICP-OES/GC-AED
(v) ICP-OES/ETA
(vi) ICP-OES/Laser ablation
(vii) thickener content using ICP-OES
(viii) pharmaceutical products
(ix) antibiotics
(x) cancer drugs
(xi) organometallic compounds
(xii) forensic support
(xiii) metals in health supplements
(xiv) metals associated with foods.

7.2 Coupling of Flow Injection Analysis with ICP-OES

Flow injection analysis (FIA) is a continuous flow method in which highly precise sample volumes are introduced into a stream that is segmented or non-segmented. The flow injection technique began to appear in the 1970s [1] in which samples were injected into a flowing system and carried by a peristaltic pump to a detector. The detector signal output in the form of transient peaks serves as a basis for quantitative analysis. Rizicka and Hansen [2] improved the technique by dispensing with a mixing chamber and using flow induced sample dispersion to provide contact between analyte and reagent. The method avoids the excess sample dilution that accompanied the mechanical stirring in the earlier procedures. These ideas were tried out successfully using atomic spectroscopy as a detector and can be adopted as part of routine analysis. The technique can be applied to the analysis of organic matrices for the analysis of major and trace levels of metals. The first attempt to combine FIA with ICP-OES was made as early as 1981 [3]; since then mathematical models have been set up to include standard addition and internal standard methods involving FIA/ICP-OES. Usually solutions are sprayed into the plasma using manual operations or in the case of a large number of samples with the aid of an auto-sampler. Continuous spraying of liquid (aqueous or organics) into the plasma is quite feasible using a flow injection technique and the attempt to combine FIA with ICP-OES has been successfully carried out.

7.2.1 Theory of Flow Injection

Flow injection signal response depends on the dispersion of the sample zone within the carrier stream, together with the dynamic characteristics of the detector reading the signal intensity. Zone dispersion results from the hydrodynamic process taking place in the tube caused mainly by convection and diffusion. The relationship between dispersion and residence time is important for the optimisation of the system and must be jointly considered when designing an ICP-OES-FIA system. In the normal flow of sample solution through a tube the sample plug is a result of countless repositioning of the elements of the fluid in the axial and radial direction caused by the twin processes of conversion and diffusion. The axial mixing depends on the turbulent or laminar flow. A flowing liquid in a narrow tube would show an increasing axial dispersion as it flows faster in the centre than at the walls of the tube. The effect of a sample plug flowing in a tube can be demonstrated by observing the behaviour of loop distances from the detector as shown below in Figure 7.5(A) and Figure 7.5(B) respectively.

The effect illustrated in Figure 7.5 was observed with the element Mo using the FIA method. The signal obtained in 'A' is for the loop at 12 cm (nearest attainable due to instrument design) from the detector forming a perfect single signal with no evidence of a tailing peak while the signals obtained for 'B' is for a loop 50 cm from the detector giving rise to a reduced single signal and tailing second peak for the same sample solution. The second broad peak in 'B' suggests that the tail end broke away forming a second but smaller plug a short distance from the main plug. FIA is made possible by the existence of forces that promote radial dispersion since they allow the repositioning of the sample from the original streamline. The forward movement of the sample is retarded when it moves away from the central streamline and is accelerated when it moves towards it. Hence, if repositioning occurs at random, the axial dispersion is reduced and the more intense radial movement will be compared with the forward convection motion; hence, the lower the dispersion of the sample will be per unit length travelled.

7.2.2 Configuration of ICP-OES/FIA System

A liquid sample containing the element(s) of interest is injected into a continuous stream of carrier liquid and is transported to the plasma jet via the spray chamber for atomisation and excitation. On its way to the detector the sample plug is mixed with the carrier stream and is partially dispersed. The degree of dispersion depends on the distance of injection point from the plasma, the volume of sample, flow rate, viscosity of the sample solution and inner diameter of the tubing. These parameters have to be optimised for samples, solvents and carrier liquid and in a given procedure experimental conditions have to be kept constant for both samples and standards. Flow injection systems are characterised by short response times during which analytical signals are obtained within 2–3 s and lead to high sample throughput.

A schematic diagram of the automated ICP-OES-FIA system is shown in Figure 7.1. It consists of an auto-sampler, FIA device, the ICP-OES and a microcomputer that controls the ICP-OES/FIA system. The carrier liquid is transported using a multi-roller peristaltic pump at a rate predetermined to suit the particular analysis of interest. Suitable tubing (0.25–1.2 mm int. diam.) is used where appropriate and silicone tubing is applied for organic liquids due to solvent compatibility at the roller heads. It will also accommodate roller pinching to force the liquid through the tubes, and samples can be introduced singly or using an auto-sampler through a loop. The size of the loop can vary from 10 μl to 1.0 ml and is adjusted to suit the concentration of analyte in the sample and standards.

A precision-controlled flow injection valve is necessary to keep the dispersion to a minimum by keeping the volume and tube length as short as possible. A sixport valve* was found to be the best for this purpose. A schematic flow diagram of the valve configuration for filling and injecting is shown in Figure 7.2, illustrating two ports are used for the loop, one for the inlet and one for the outlet of the carrier stream. The fifth

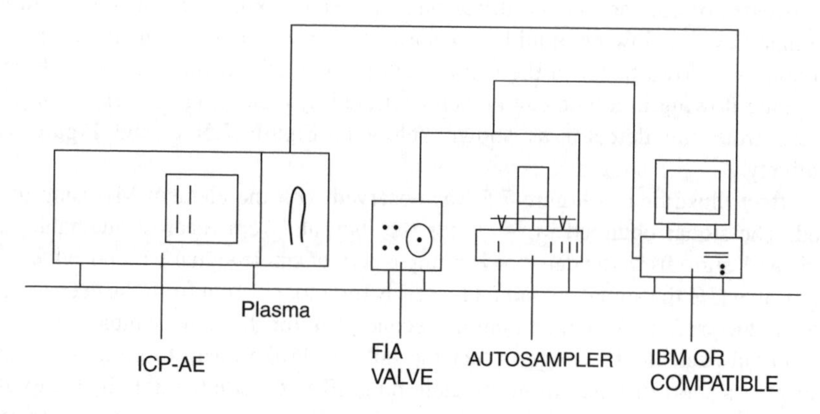

Figure 7.1 *Schematic diagram of the automated ICP-OES-FIA analyser. (Reproduced with kind permission from PS Analytical, Orpington BR5 3HP, UK)*

*Available from PS Analytical, Orpington UK, Cat. No. PSA 60.043.

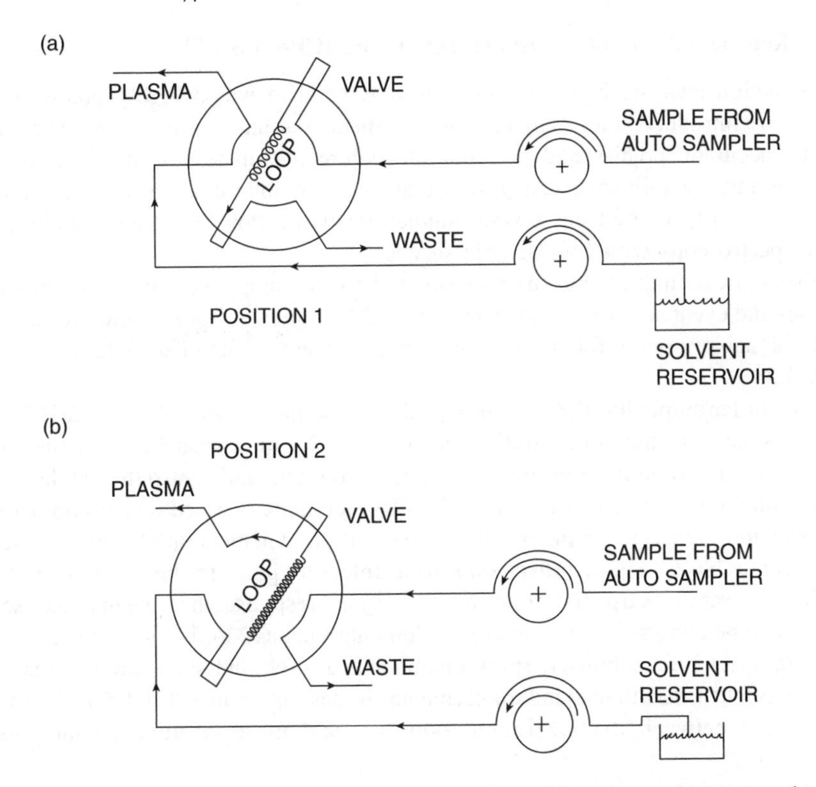

Figure 7.2 *Schematic diagram of flow injection valve in filling (a) and injection (b) configurations. (Reproduced with kind permission from PS Analytical, Orpington BR5 3HP, UK)*

port is used for carrying the plug to the ICP-OES detector and the sixth is for excess waste. The valve is controlled and in sync with the ICP-OES using the microcomputer.

7.2.3 Signal Acquisition and Data Management

The system consists of a suitable computer with a RS232 interface, an enhanced graphics adaptor screen, and a chart recorder. The touchstone software used to operate the system is available from PS Analytical. The data acquisition program is designed to operate with a standard computer using MS dos or better. The program is user-friendly and follows a step-by-step guide through its facilities. It can collect, store and process data generated as line intensities in the spectrometer. Data are transferred continuously, through the serial asynchronous communication interface. When measurements for standards are completed, be it either by standard addition or standard curve, a curve will be generated on the screen. Statistical analysis can also be carried out with data generated for quality assurance purposes. In atomic emission spectroscopy, the detector responds linearly and instantly to the injected sample and does not differentiate between peak height, peak area or peak width as they all give useful information. Peak height is the simplest and most popular as it is directly related to the detector response. Peak area is also popular in certain cases as it involves the entire signal measurement while passing through the detector. Peak width is proportional to the logarithms of the concentration and is not widely used as it is less precise.

7.2.4 Reproducibility of Measurements Using ICP-OES/FIA

Flow injection analysis is a continuous flow method in which highly precise sample volumes are introduced into a stream using segmented or unsegmented flow. The method must be accurate, precise and reproducible before it can be considered as a useful technique and the following test proves that this technique does meet all the requirements. Tyson [3], carried out several studies involving flow injection techniques and atomic spectroscopy with considerable success.

In the application of atomic spectroscopy to FIA the sample plug is carried first to the nebuliser and eventually to the plasma source for excitation and atomisation for detection to give signal responses for the corresponding concentrations of analyte, as shown in Figure 7.3.

The recorder output has the form of a peak with height H, width W, or area A, each of which is related to the concentration of analyte. The time span between the sample injections is important in determining the peak maximum and is also the residence time during which the sample is travelling. The FIA technique is based on a combination of three principle steps: (i) sample injection; (ii) controlled dispersion of sample zone; and (iii) reproducible timing of movements from injection point to the detector. It should also have a rapid sharp and reproducible signal response that means that several samples can be analysed in a short time. Elements such as Mo, B and W tend to 'stick' in the transport line, nebuliser, spray chamber and torch and such samples may need longer washout times than standard elements. In designing an ICP-OES-FIA system a fine balance between the maximum loop size and most sensitive signal must be established.

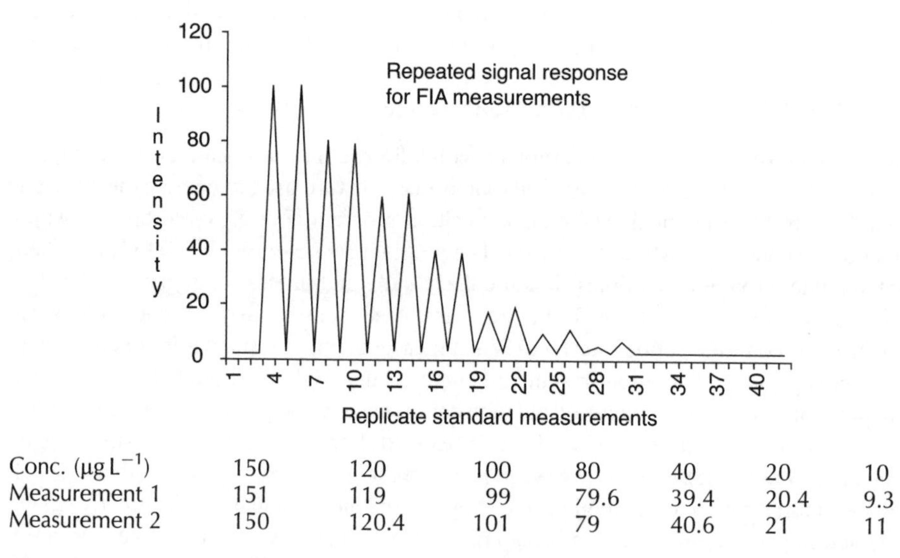

Conc. (μg L^{-1})	150	120	100	80	40	20	10
Measurement 1	151	119	99	79.6	39.4	20.4	9.3
Measurement 2	150	120.4	101	79	40.6	21	11

Figure 7.3 *Typical duplicate measurements of decreasing concentrations (150–10 μg L^{-1}) of Cu using ICP-OES/FIA. Actual response readings are shown*

7.2.5 Dispersion and Diffusion of 'Sample Plug' in a Carrier Stream [4]

The signal response of the FIA peak is a result of a physical process of zone dispersion as the sample in the carrier stream is not a homogeneous mix but a dispersion of concentration gradient, as shown in Figure 7.4.

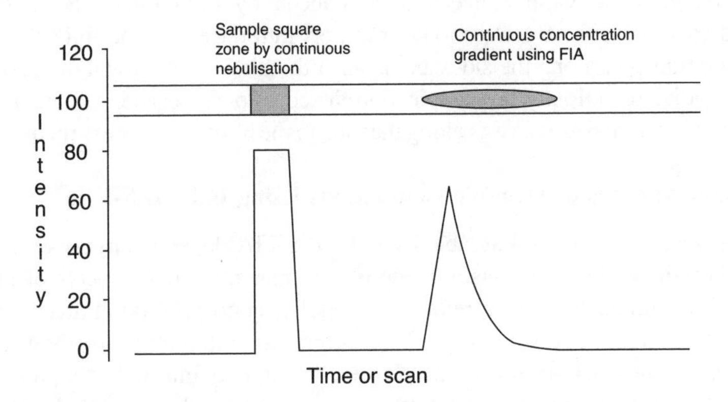

Figure 7.4 Diagram of sample square zone by continuous nebulisation and continuous concentration gradient using FIA

When a homogeneous sample of concentration c is measured, the signal intensity would be continuous as in Figure 7.5(a); however, if the sample is a slug in a carrier stream then the signal obtained is as shown in Figure 7.5(b). The latter signals show the change in analyte concentration within the slug and illustrate the importance of distances between the sample loop and the injector. The biggest challenge to the FIA method of sample introduction is keeping the peak as sharp and reproducible as possible to achieve maximum sensitivity and rapid sweep of measurement so that analyses can be carried out in a shorter time.

The signal shows the various concentration gradients of the injected plug using an FIA method as it passes the detector plug reaching a maximum. This type of scan shows the extent of dilution by the carrier stream, and to achieve a maximum signal the sample loop

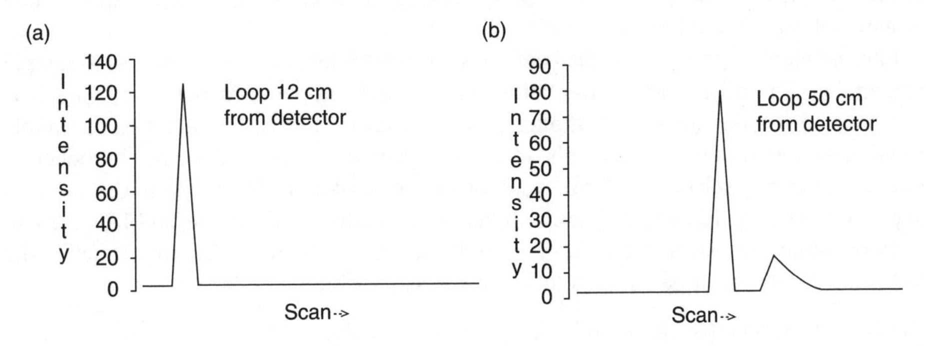

Figure 7.5 Effect of loop distance from detector. The nearer the loop to the detector (a) the less tail end effect (b)

must be as close to the detector as possible, hence a higher sharper signal response. Several factors contribute to the transport of a substance in a flowing liquid and two of the most important are diffusion and convection. Other factors include migration, ionic mobility, concentration effects, adhesion to tube wall and compressibility of the carrier solvent. The first two – diffusion and convection – are the most important as diffusion is caused by a concentration gradient while convection is induced by fluid flow. The distribution is determined by a balance of convectional transport introduced by the fluid flow in a tube and diffusion transport along the sides and against the flow (longitudinal or axial diffusion) and in the radial direction. The sample introduced into the carrier stream is gradually lengthened and distorted as it passes along the tube to the transport process mentioned above.

7.2.6 Metal Analysis of Organic Compounds Using ICP-OES-FIA

The accuracy and precision obtainable by ICP-OES-FIA depends largely on the way the sample is introduced into the plasma. One of the most attractive aspects of introducing the sample as a liquid lies in its relative simplicity, good reproducibility and speed of analysis. For routine measurements FIA offers an alternative method of sample introduction to direct nebulisation. The dissolved sample is injected as a plug into either a segmented or non-segmented stream of a carrier liquid and transported to the plasma. The nebulisation efficiency in ICP-OES is lower by a factor of one-fifth compared with AAS.

Analysis for metals dissolved in organic liquids using atomic absorption methods can often increase the sensitivity of analysis. This is partly due to the highly reducing nature of the plasma, and partly to the ease of evaporation of the solvent dissolving the metal leading to a higher concentration of metals free of the solvent in the flame, and the slightly improved efficiency of nebulisation. These, along with the combined effects of lower density, low surface tension and high vapour pressure, all contribute to better sensitivity. In the case of ICP-OES and FIA, solvents do not enhance signal sensitivity. The introduction of the sample as a plug into the carrier stream causes a transient signal in response, which soon decreases to the background level caused by the carrier stream and should ensure that a constant nebulisation is maintained over a longer period of time. This signal compares well with the direct nebulisation of signals obtained by transporting and nebulising larger volumes of a sample over an extended period of time. The major advantage of FIA is that the continuous nebulisation of the carrier stream into the plasma cleans the plasma transport line reducing memory effects and allowing a more rapid turnaround time in analysis.

In using atomic spectroscopy analysis the sample introduction is an extension to sample preparation. To understand the limitations of practical sample introduction systems it is necessary to reverse the train of thought, which tends to flow in the direction of sample solution → nebulisation → spray chamber → excitation → atomisation. An introduction procedure must be selected that will result in a rapid breakdown of species in the atomiser to give reproducible results irrespective of the sample matrix. In designing an FIA system to carry out atomic emission and to generate efficient free atom production for excitation the following criteria must be adhered to as closely as possible:

(i) suitable acceptable dropsize;
(ii) optimum solvent loading;

(iii) maximum analyte mass loading;
(iv) appropriate gas flow for effective plasma penetration; and
(v) maximum height to allow suitable residence time in the plasma.

7.2.7 Effect of Loop Size on Signal Response [5]

The volume of sample transported to the plasma by the carrier solvent plays an important role in determining the shape of the peak and, to study this, loop sizes of 100, 200, 300, 400, 500, 600, 700 and 800 µl were connected in an FIA line. Samples of 1.0 µg ml^{-1} of Cu were passed through each loop and the peak height measured and plotted on a chart recorder. A peak height for continuous measurements of the sample Cu solution and plotted alongside the FIA peaks as a comparison are shown in Figure 7.6

Table 7.1 shows the measured peak heights compared with loop size and continuous nebulisation of the sample solution.

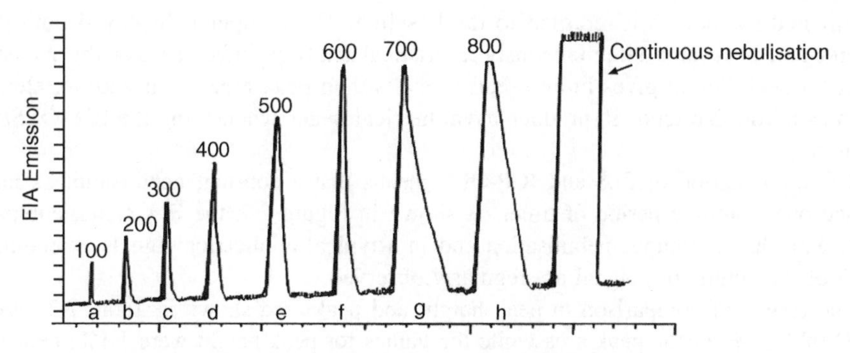

Figure 7.6 Results of the peak height obtained for variable loop sizes from (a) 100 to (h) 800 µl by injecting 1.0 µg ml^{-1} of Cu dissolved in propyl alcohol using glacial acetic acid as a carrier solvent

Table 7.1 Results of measured peak height for each loop size after injecting 1.0 µg ml^{-1} of Cu dissolved in propyl alcohol using glacial acetic acid as a carrier solvent

Loop size (µl)	Peak height ×1000	Continuous nebulisation ×1000
100	22	127
200	34	127
300	51	129
400	72	126
500	93	128
600	118	126
700	116	128
800	118	127

7.2.8 Comparative Measurements of Peak Height and Peak Area

It has been suggested that improved precision could be achieved if peak area rather than peak height was measured. To test this hypothesis a comparison was made by measuring both peak height and peak area simultaneously using a computer peak monitoring program facility. The element boron was used as an example because of its tendency for a minute fraction to 'stick' in the transport line and appear probably in the subsequent injection resulting in a slightly higher peak. An explanation for this suggests that the front of the plug is constant while a fraction of the tail could break away forming a smaller plug giving rise to a trace peak. To demonstrate this effect, a sample solution containing 5.0 μg ml^{-1} boron was dissolved in isopropyl alcohol and injected ten times into a glacial acetic acid carrier stream and the peak height and peak area measured for boron content against 0.0, 5.0 and 10.0 μg ml^{-1} standards prepared in isopropyl alcohol. The results are shown in Table 7.2 with the mean and standard deviation.

Synchronisation of sample introduction and initiation of peak integration is achieved by using the PSA FIA valve. The integrator has a built in mechanism whereby the integration of the peak commenced immediately prior to the initial rise of the peak and terminated as soon as it returned to the baseline. The computer displayed both peak height and peak area simultaneously. Statistical analysis indicates that the measurement of peak height gives more precise results than peak area, even with an element such as boron that tends to produce asymmetrical peaks when using the ICP-OES/FIA system.

The combination of FIA and ICP-OES means that a constant nebulisation is maintained over a longer period of time. As shown in Figure 7.7, the FIA signal compared well with the continuous nebulisation and in a typical application signals extending to 94% of the continuous signal are regularly observed.

The results of comparison in peak height and peak area shown in Table 7.2 give an RSD of 12.8% for the peak area while the values for peak height were 1.1%; hence the peak height is more accurate than peak area.

Table 7.2 *Results of comparison of peak height and peak area*

Injection no.	Peak area signal counts	Peak height (mm)
1	840	121
2	806	123
3	1100	120
4	870	122
5	1106	122
6	828	123
7	896	120
8	1100	119
9	823	123
10	1002	122
$\bar{x}$	937.1	121.5
δ	119.8	1.3
%RSD	12.8	1.1

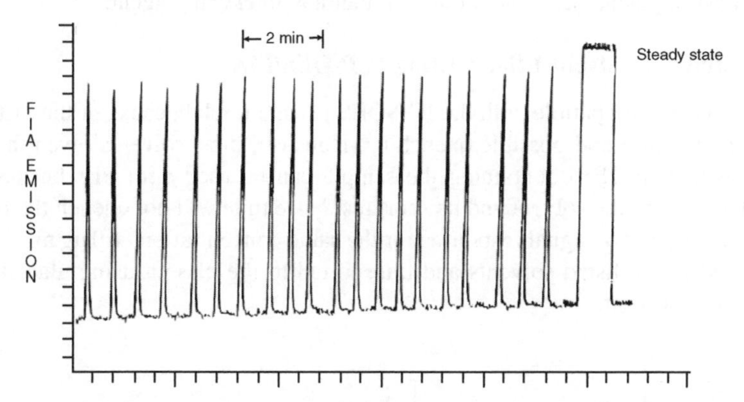

Figure 7.7 Reproducibility of signals obtained with 20 injections of a 5 μg ml^{-1} Cu standard in ethanol using a fixed loop size of 600 μl and their comparison with the steady state signal. Mean peak height of injected sample is 90% of the steady state signal with an RSD of 1.6%

7.2.9 Effect of Viscosity Using ICP-OES/FIA

It is well established in atomic spectroscopy that viscosity affects the efficiency of nebulisation. In this study the effect of viscosity can be illustrated by diluting a high viscosity motor oil (Mobil DTE-18 ~200 cps dynamic viscosity) in the solvent mixture 50:50 tetralin ($C_{10}H_{12}$) and glacial acetic acid at concentration levels of 0.0, 2.0, 5.0, 10.0, 15.0 and 20% with each solution containing 1.0 μg ml^{-1} Fe. All solutions were injected into a 600 μl loop using the standard instrument flow rate. Plots of results in Figure 7.8 show that with increasing viscosity the peaks get broader and lower. It appears from the plots that the effect of viscosity between 2 and 5% gives almost 100% response while the remaining solutions show a corresponding reduction in peak heights. It is possible to analyse for high viscosity samples using a standard addition calibration approach but it is tedious and time

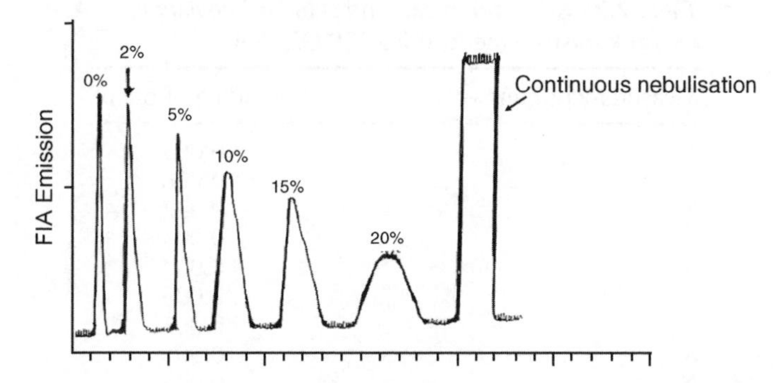

Figure 7.8 Effect of viscosity of 1.0 μg ml^{-1} Fe prepared with increasing concentration of 200 cps Mobil DTE-18 oil. The oil and standards are prepared in 20% kerosene in glacial acetic acid and the samples were measured against 0.0, 0.5 and 1.0 ppm Fe standards prepared in the same solvent mixture

consuming. The continuous nebulisation shown in Figure 7.8 is the same concentration of metal dissolved in glacial acetic acid only without any thickening agent.

7.2.10 A Study of Solvent Effects Using ICP-OES/FIA

Not all solvents are compatible with the ICP-OES plasma torch because of the instability caused by signal noise and possible quenching. However, most solvents (even those not compatible with ICP-OES) containing the sample can be used with FIA because it is carried and diluted by the solvent and its short residence time will not quench the plasma. Figure 7.9 shows a list of signal responses for the same concentration (1.0 μg ml^{-1} Cu) of metal prepared in the listed solvents and transported to the plasma using glacial acetic acid as a carrier solvent.

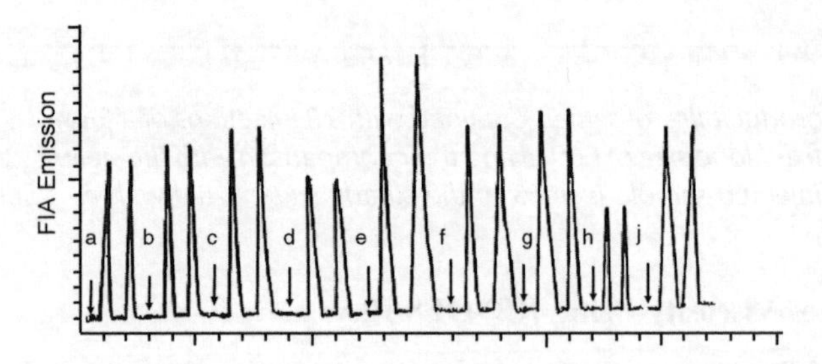

Figure 7.9 *Signal response for 1 μg ml^{-1} Cu prepared in (a) carbon tetrachloride; (b) chloroform; (c) ethanol; (d) glacial acetic acid; (e) methyl isobutyl ketone; (f) tetralin; (g) toluene; (h) water; (j) xylene all using frequency 40 MHz, power 1.2 kW and 850 V PMT*

7.2.11 Determination of Limit of Detection and Quantification

The limit of detection was determined by measuring a blank solution of glacial acetic acid 10 times against a calibration curve prepared from standard solutions of 0.5, 1.0, 2.0 and 4.0 μg ml^{-1} Cu in glacial acetic acid. The results are shown in Table 7.3.

Table 7.3 *Results of mean and standard deviation of blank measurements using ICP-OES-FIA*

Measurement number	Reading of blank
1	0.005
2	0.006
3	0.007
4	0.003
5	0.002
6	0.006
7	0.007
8	0.006
9	0.005
10	0.004
$\bar{x}$	0.0052
SD	0.001508

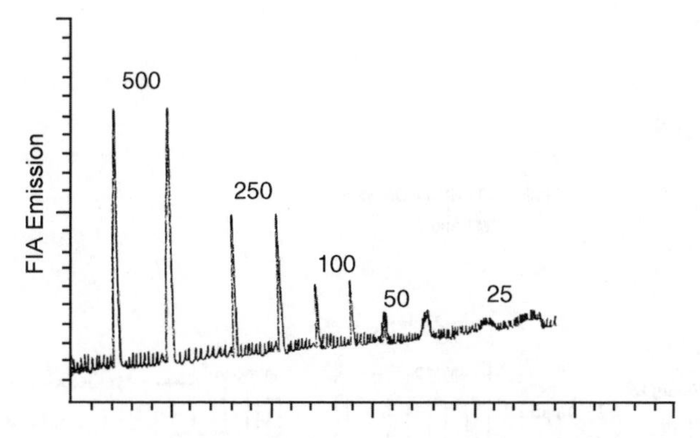

Figure 7.10 *Duplicate measured signals for 500, 250, 100, and 25 ng ml^{-1} Cu. The lowest measured peak is just 10 ng ml^{-1} above the calculated the limit of detection level of 15 ng ml^{-1} based on three times the standard deviation of the baseline noise. These values are close given the experimental errors associated with actually measuring the lowest level*

The recommended method for calculating the limit of detection is based on three times the standard deviation of the noise of the blank and limit of quantification is based on calculating it as ten times the standard deviation of the noise of the blank. Therefore the limit of detection estimated from the values in Table 7.3 is 0.0045 μg ml^{-1} (4.5 ppb) of Cu and limit of quantification is 0.015 μg ml^{-1} (15.0 ppb) of Cu using the ICP-OES/FIA method. Evaluation of the results is based on calculation of the differences between peak height and the average of the background noise in the vicinity of the peak. It must be emphasised that the background noise tends to creep between recalibration points (when the zero is readjusted to the middle of the noise range). This can be seen clearly in Figure 7.10, which shows the response obtained in duplicate with solutions of various concentrations of copper. This may vary from metal to metal but it is expected that ICP-OES-FIA would have a higher limit of detection than that for direct nebulisation.

7.2.12 Conclusions of Analysis Using ICP-OES-FIA

The ICP-OES-FIA technique allows a rapid and routine method of analysis for both major and trace levels of metals in aqueous and non-aqueous solutions in most samples provided that the sample is in solution form. The flow injection method can be used to correct for baseline drift that may originate from uncontrollable thermal and electronic noise during analysis. However, these errors can be corrected if the peak obtained is measured over at least three points, i.e. immediately before the peak, at the peak and immediately after the peak and the height or area is integrated over these points. The elaborate time consuming correction procedures required for batch operations are not required for FIA methods and the baseline is defined by the emission obtained from the carrier liquid and is reproduced between each sample injection. A typical FIA analysis of signals for standards and samples is shown in Figure 7.11 for triplicate injections of variable concentrations of boron.

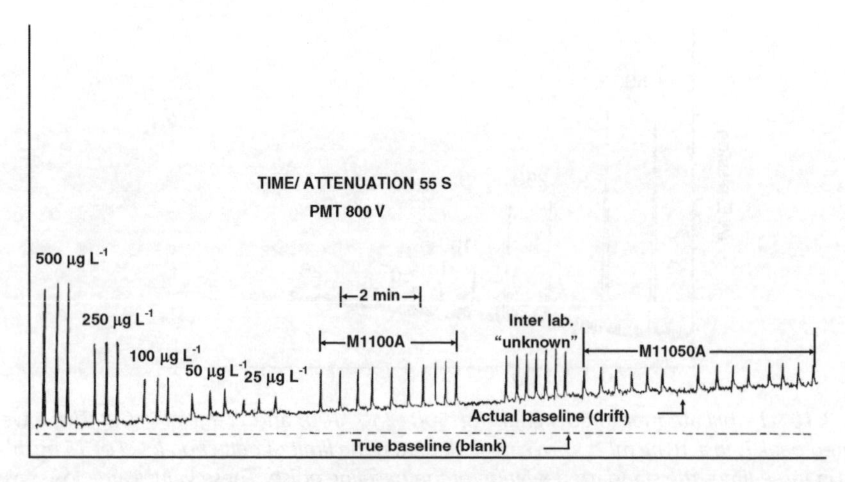

Figure 7.11 *Reproducibility of signals for triplicate measurements of standards and samples using the optimised ICP-OES-FIA conditions. The baseline drift as seen is common for most FIA techniques particularly at trace levels. It does not present a problem in reporting precise analytical results because each signal starts integration before, at and after the peak*

Automation is especially advantageous if a large number of samples need to be analysed on a routine basis. There is little doubt that sample introduction using flow injection is superior to other solvent delivery techniques. Analysis using flow injection techniques with ICP-OES offers a number of advantages:

(a) Can analyse small volumes of sample solutions using smaller loops (provided that the concentration of analyte can be quantitatively detected) and allows rapid routine analysis of multiple samples.
(b) Generation of a calibration curve from a single standard solution using a standard diluting computing program; the exponential decay of the signal effectively provides a calibration with an infinite number of points.
(c) Cost of instrument and operator's time, the setting up and operation of the ICP-OES-FIA instrument makes the technique relatively inexpensive.

The technique improves the analysis for most metals in most sample solutions in terms of precision of results, ease of sample handling, less physical interference, higher sample throughput and versatility towards physical and chemical properties of reagents. The technique allows several elements to be measured at the same time using a simultaneous CCD. Disadvantages are loss of sensitivity compared with continuous nebulisation and may not be suitable where high salt contents are present in solutions.

Many examples are known where the FIA technique is used for sample transport only and an example of this is where sample contains a concentration of interfering matrices. These samples can be injected in very small volumes (10 to 100 µl) into a carrier stream to minimise these interferences due to excessive dilution. Standard addition and internal standard methods can equally be applied to FIA techniques to reduce matrix, spectral and other potential interfering effects. Ion exchange columns connected in the sample feed line containing Chelex-100 resins have been used to successfully preconcentrate metals prior to FIA measurements.

7.3 Use of Internal Standard(s) with ICP-OES

The concept of measuring the ratio of signals of analyte and internal standard is well known in atomic spectroscopy where the internal standard is an added metal of known concentration and not present in the sample. Internal standard addition can be performed manually or by use of an automated flow injection technique at the nebuliser during sample introduction. The advantage of an internal standard in metal analysis is that it can be used where different samples with different viscosities require analysis against the same calibration curve. Kahn [6] showed that the response for Fe decreased with increasing concentration of NaCl, but could be corrected by using Mn as an internal standard. Similarly, he showed that the concentration of Pb was corrected when he used Cu as an internal standard in a NaCl solution. Later, this theory was extended to correct for fluctuations in sample transport effects, instrumental drift, and electronic and plasma noise. The element selected as an internal standard must be similar in chemical behaviour and excitation energy as the analytes of interest.

Signal enhancements can be achieved by several means without affecting the precision of analysis. For optimisation with respect to the higher signal response, parameters such as the power of the ICP-OES, nebuliser gas flow, burner geometry, correct peak height and correct horizontal peak position of the plasma in relation to the optics of the instrument must be studied for each analyte and sample. Increasing the nebuliser gas flow rate helps the droplet size decrease due to collisions in the spray chamber introducing higher sample efficiency to the plasma torch. However, this may be at the expense of residence time, flow mechanics and plasma temperature. These improvements in signal enhancement of the analyte will also affect the internal standard present in the same solution.

An additional problem is that known elements in the periodic table, e.g. boron (B), tungsten (W) and molybdenum (Mo) tend to 'stick' in the transport line, nebuliser, spray chamber and torch causing memory effects in atomic spectroscopy. Elements that 'stick' cause problems with quantification, and detection limits and it is important that methods of reducing these are rigorously applied in analysis of these elements. However, memory effects in ICP-OES are not as pronounced as they are with graphite furnaces for refractory elements but they are present to some extent, and must be reduced or removed.

A study was carried out on solutions of organic compounds containing the known 'sticking elements' B, W and Mo and compared for signal intensity and memory effects with and without thickening agents using FIA-ICP-OES. The study was carried out in an organic medium using low viscosity oil with and without poly(vinyl acrylate) as thickening agent.

Mannitol and phosphoric acid have been suggested as memory-reducing additives for these elements and are also known to enhance sensitivity of these elements. A study of the effect of this memory-reducing additive was carried out to test this theory using an automated internal standard and sample mixer, as shown in Figure 7.12. The metals W, Mo and B were studied and the calibration curves obtained are shown in Figure 7.13. The curves in Figures 7.14 and 7.15 show graphically the effect of the memory-reducing additives with and without the thickening agent.

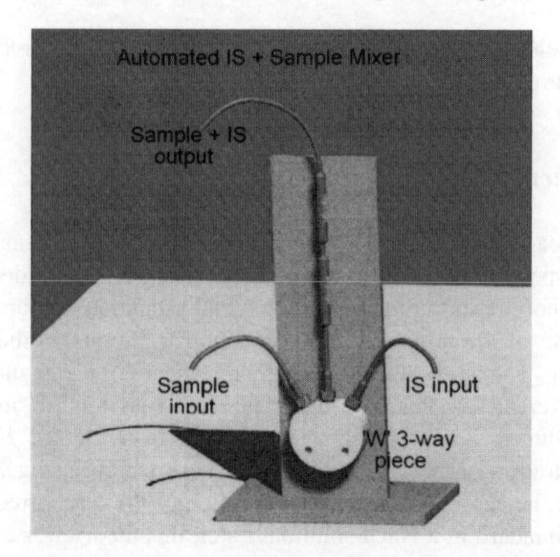

Figure 7.12 Automated internal standard and sample mixer showing the sample input and internal standard input channels using a three-way 'V' or 'W' piece (author's own invention)

General Method. The method involves analysing samples against a calibration curve generated using an automated internal standard method. A calibration curve is prepared by adding to four 100 ml grade 'B' plastic volumetric flasks 0.0, 0.25, 0.5 and 1.0 ml of B, W and Mo from a 100 $\mu g\,ml^{-1}$ multi-element stock solution. The standards are diluted with 50:50 glacial acetic acid and propylene carbonate to give 0.0, 0.25, 0.5 and 1.0 $\mu g\,ml^{-1}$ of each metal, respectively. An internal standard solution is prepared by diluting 0.10 ml of 1000 $\mu g\,ml^{-1}$ scandium (Sc) to give 1.0 $\mu g\,ml^{-1}$ Sc metal prepared separately in 50:50 glacial acetic acid and propylene carbonate. The samples containing the metals are pumped into one side of the 'V' or 'W' using an automatic internal standard mixer as shown in Figure 7.12. The internal standard is pumped into the second side. The two solutions are mixed in the 'V' or 'W' piece and transported to a spiral-mixing chamber directly above the 'V' or 'W' piece and eventually to the plasma torch using a suitable peristaltic pump. The results are shown in Table 7.4.

Table 7.4 *Results of intensities for B, W and Mo metal measured against Sc internal standard. (Note: There is excellent constant reading for the Sc internal standard)*

Conc. Int. Std Sc ($\mu g\,ml^{-1}$)	Int. Std. Sc (intensity)	Conc. B, W and Mo Stds ($\mu g\,ml^{-1}$)	B Std. (intensity)	W Std. (intensity)	Mo Std. (intensity)
1.0	16 400	0.0	160	120	270
1.0	16 490	0.25	5860	4750	7800
1.0	16 390	0.5	11 720	9350	15 400
1.0	16 440	1.0	23 500	18 300	30 870

Using the automated mixer described in Section 3.7.4 the calibration curves for B, W and Mo using Sc as internal standard shown in Figure 7.13 were obtained.

The calibration curves in Figure 7.13, showed excellent correlation; greater than 0.99985 for the three elements. The benefit of internal standard(s) for analysis can be

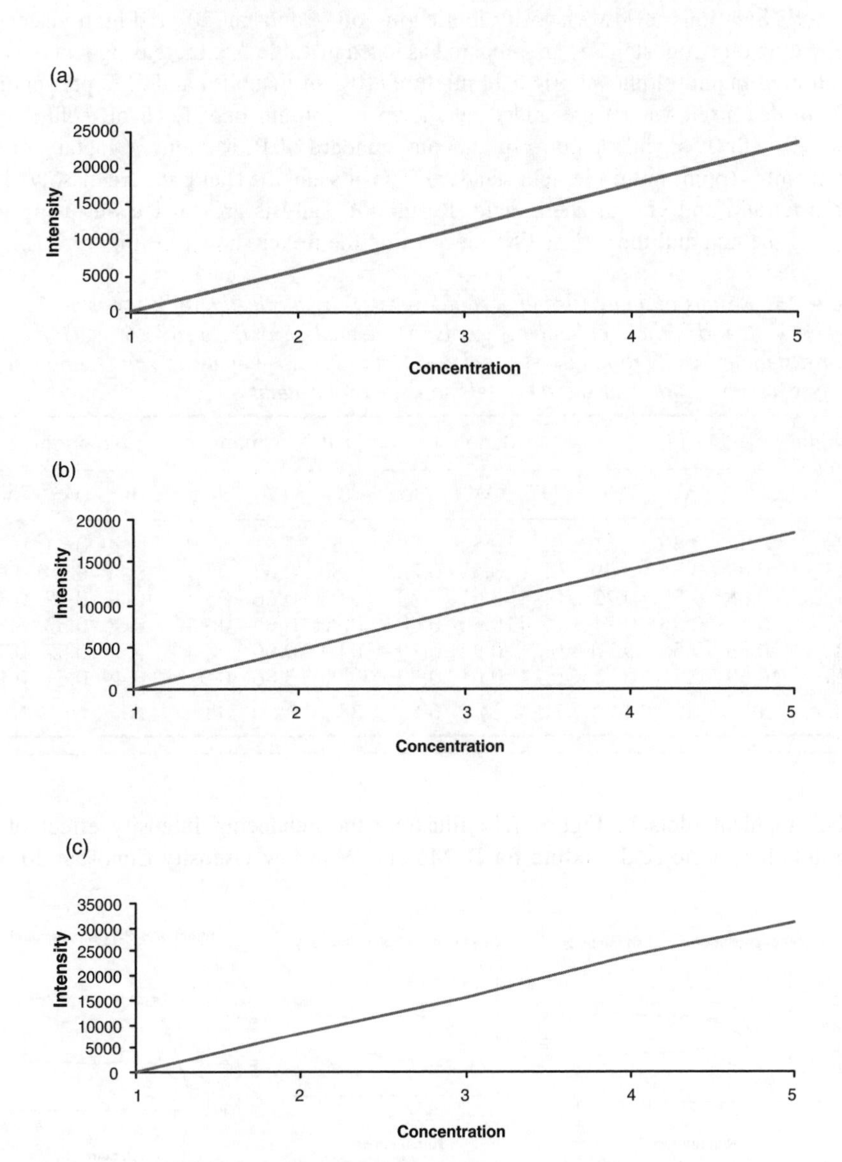

Figure 7.13 *Calibration curves for boron, tungsten and molybdenum from 0.0 to 4.0 µg/ml (ppm) using scandium 'Sc' as an internal standard. The units on the 'x' axis are flask numbers of standards 0.0, 1.0, 2.0, 3.0 & 4.0 ppm metal respectively*

compared by studying the effects of a low viscosity lubricating oil (e.g. Conostan 20) with those of a high viscosity lubricating oil (e.g. Conostan 75). The same study can be used to illustrate the improved intensities achieved by the addition of mannitol and/or phosphoric acid in the analysis of these refractory elements.

Method. Solutions of low viscosity lubricating oil (Conostan 20) and high viscosity lubricating oil (Conostan 75) are prepared as listed in Table 7.5. Increasing volumes of solution of mannitol/phosphoric acid mixture (10% of mannitol and 25% phosphoric acid in deionised water) are added as shown in column one. Each oil solution is 'spiked' with 0.5 μg ml^{-1} (ppm) of aqueous standard of B, Mo and W metal from a 100 μg ml^{-1} (ppm) multi-element standard. The oils and the standards are dissolved in 75:25 tetralin and glacial acetic acid. Results of analysis are carried out using the internal method and the 'V' or 'W' piece using the mixer shown in Figure 7.12.

Table 7.5 *Results of analysis of effects of low viscosity lubricating oil (Conostan 20) 'spiked' with and without enhancing agents for the analysis of 0.5 μg ml^{-1} of each metal against standard calibration curves, similar to Figure 7.13. All analyses were carried out using scandium as internal standard. NI, no increase in signal*

Volume (ml)	H_3PO_4			Mannitol			H_3PO_4 + mannitol			No additive		
	B	W	Mo	B	W	Mo	B	W	Mo	B	W	Mo
0.0	0.48	0.49	0.51	0.49	0.52	0.50	0.51	0.51	0.49	0.49	0.51	0.50
0.25	0.63	0.54	0.69	0.71	0.58	0.76	0.88	0.65	0.93	0.50	0570	0.47
0.50	0.68	0.59	0.72	0.75	0.61	0.79	0.93	0.66	0.94	0.48	0.49	0.51
0.75	0.67	0.58	0.71	0.74	0.60	0.80	0.92	0.65	0.94	0.50	0.53	0.51
1.0	0.68	0.58	0.70	0.73	0.62	0.79	0.94	0.66	0.95	0.53	0.52	0.50
1.25	0.69	0.57	0.71	0.74	0.61	0.79	0.92	0.66	0.92	0.49	0.51	0.52
% Inc.	38	18	44	50	24	60	88	32	88	NI	NI	NI

The graphical plots in Figure 7.14 illustrate the enhancing intensity effect of the mannitol/phosphoric acid mixture for B, Mo and W in low viscosity Conostan 20 oil.

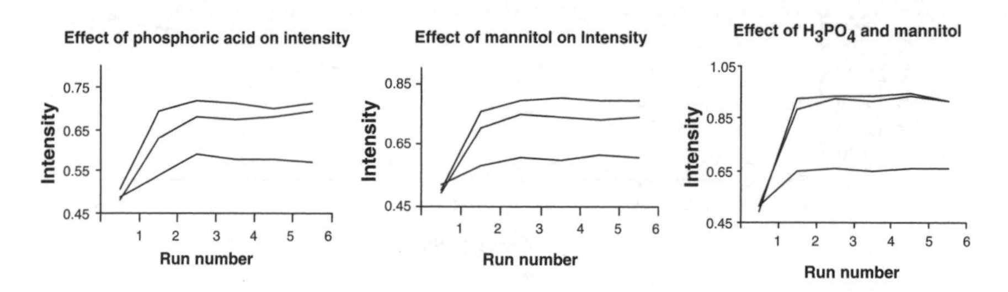

Figure 7.14 *Plot of B, Mo and W containing mannitol and phosphoric acid listed in Table 7.5. The top line is Mo, the middle line is B and the bottom line is W. All solutions contained 5% poly(vinyl acrylate) as thickener*

A similar list of samples was prepared for high viscosity lubricating oil (Conostan 75), as shown in Table 7.6.

Table 7.6 *Results of analysis of effects of high viscosity lubricating oil (Conostan 75) 'spiked' with and without enhancing agents for the analysis of 0.5 μg ml^{-1} of each metal against standard. NI, no increase in signal*

Volume (ml)	H_3PO_4			Mannitol			H_3PO_4 + mannitol			No additive		
	B	W	Mo	B	W	Mo	B	W	Mo	B	W	Mo
0.0	0.51	0.50	0.52	0.49	0.51	0.50	0.51	0.51	0.49	0.49	0.51	0.50
0.25	0.57	0.50	0.63	0.66	0.57	0.66	0.66	0.59	0.71	0.52	0.50	0.51
0.50	0.58	0.55	0.68	0.68	0.56	0.70	0.67	0.63	0.73	0.49	0.49	0.52
0.75	0.60	0.56	0.67	0.69	0.59	0.69	0.70	0.60	0.72	0.50	0.53	0.50
1.0	0.59	0.56	0.67	0.68	0.58	0.71	0.68	0.61	0.75	0.53	0.51	0.52
1.25	0.60	0.57	0.68	0.67	0.59	0.70	0.69	0.63	0.74	0.49	0.50	0.51
% Inc.	20	14	36	38	18	40	40	26	50	NI	NI	NI

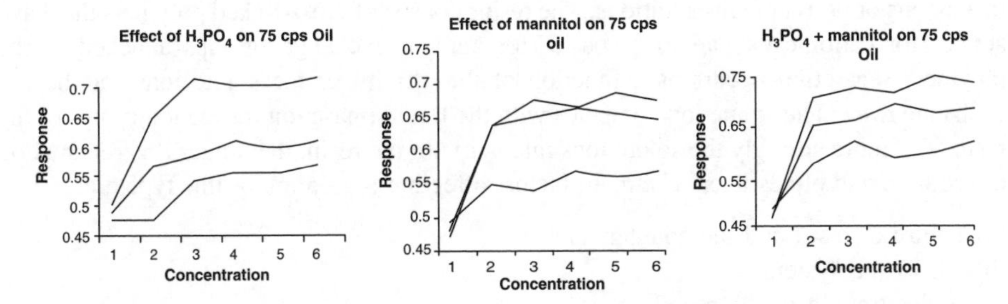

Figure 7.15 *Plot of B, Mo and W containing the enhancing compound listed in Table 7.6. The top line is Mo, the middle line is B and the bottom line is W. No thickener added*

The graphical plots in Figure 7.15 illustrate the enhancing effect of mannitol/ phosphoric acid on the intensities of these elements in high viscosity Conostan 75 oil.

7.3.1 Conclusion to Internal Standard(s) Study

The application of internal standards in the analysis of a range of metals can be successfully used in overcoming various matrix effects caused by some samples, e.g. density, viscosity and surface tension. The droplet size can affect the efficiency of analyte introduction, hence the sensitivity of measurements. Similarly, increasing concentrations of acids may have the effect that a lowering of sensitivity occurs and is most noticeable with highly concentrated acid solutions, e.g. sulphuric and phosphoric acids which also have high densities and high viscosities. Viscous organic solutions, e.g. crude oils, intermediate reaction products, paints or organic compounds containing thickening agents of different concentrations, may affect the fluctuation in sample transport, instrument drift and electronic and plasma noise. These properties can also affect the accuracy and precision of analysis that may be corrected using a suitable internal standard. Selection of an internal standard is important as it must be as close as possible to the behaviour of the analyte(s) of interest. In some cases it may be feasible to run more than one internal standard, particularly where multi-elemental analysis is required.

Signal enhancement is also a useful technique associated with ICP-OES particularly where trace analysis is required. The study was extended to observe the effect of increasing concentrations of these memory-reducing and signal-enhancing compounds by analysing low viscosity and high viscosity oil 'spiked' with these elements. The results in Tables 7.5 and 7.6 are shown for H_3PO_4 and mannitol and a mixture of H_3PO_4 and mannitol. This study has also shown that by using mannitol and phosphoric acid and their mixtures, the signals increased from 14% for high viscosity samples to 88% for low viscosity samples. (It is important that these chemicals come from boron-free containers, e.g. plastics, etc.)

7.4 Coupling of Ion Chromatography with ICP-OES

Ion exchange chromatography is a variation of adsorption chromatography in which the stationary phase is an ion-exchange resin used for the separation of ionic solutes, usually in aqueous or non-aqueous solutions. The resins consist of cross-linked polymers that have acidic (for cationic exchange) or basic (for anionic exchange) groups attached to the polymer. Separation occurs as a function of the affinity of the solute ions for the ion exchange resin. The solute ion competes with the liquid phase ion for the ionic sites of the resin. The more strongly the solute ions interact with the resin, the longer it is retained on the column and elutes later. The main factors affecting separation of this type are:

(a) nature of resin and the counter ion;
(b) pH of the system;
(c) ionic strength of the mobile phase;
(d) temperature.

Prepared samples can be passed through an ion-exchange column attached to an HPLC system to selectively retain metals present in the samples while the remaining sample is eluted from the column. The separation of organometallic or inorganic ions is favoured by ion exchange columns and the operating conditions are different from the analysis of non-metallic compounds particularly for retention and separation of trace elements.

Mixtures of organometallic or inorganic ions are often analysed using ion chromatography equipped with single or combined UV or a conductivity detector. Columns used may involve a variation of ion-exchange or ion pair types that are suitable for the selective separation of metals. The principle of operation is the conversion of the metal ions into ionic or ionisable acid or base form that can be separated by these columns. The surface of the packing in the column is coated with an ion-pair agent resulting in a net charge for which it can behave as an ion-exchange column and the charged groups are covalently bonded to the surface of the packing material. Displacing the counter ions associated with the ionic groups bound to the particle surface retains the ionised metal species in acids or bases. The retention of the ions M^{n+} onto the cation exchange column can be represented as follows:

$$M^{n+} + (S^{-}Na^{+}) \Leftrightarrow Na^{+} + (S^{-}M^{n+})$$

The M^{n+} and Na^+ refer to ions in the mobile phase with Na^+ ions being the counter ions and an ionic group on the particle surface of the packing in the column being represented by S^-. Retention of a sample containing ions X^- in an anionic exchange column, e.g. SO_3^- as counter ion, is given by:

$$X^- + (S^+SO_3^-) \Leftrightarrow SO_3^- + (S^+ + X^-)$$

Considering these reactions, it is apparent that an increase in the mobile phase counter ion concentration (Na^+ or SO_3^-) will proportionately decrease the retention of the sample ion. The pH of the mobile phase is very important as it can affect the relative ionisation of acids or bases, i.e. a higher pH leads to increased ionisation of acids on anion exchangers while a lower pH favours increased ionisation and retention of bases on cation exchangers.

Depending on the type of metal analysis required and the nature of the matrix in which the metals are present, either ion exchange or ion pair HPLC may be used. Ion exchange columns are favoured for separating inorganic ions and large organometallic compounds but require specific columns. In the case of ion pair columns used in HPLC, reversed phase columns are also used for specific applications. Ion exchange requires aqueous mobile phases while ion pair can operate with a controlled gradient mixture of organic and aqueous phases. Retention of species of interest using ion exchange columns is usually controlled by varying the concentration of the salt or buffer in the mobile phase, i.e. concentrated salt solutions are similar to strong solvents and vice versa for dilute solutions. Changing the pH or concentration of the ion pair agent or type of ion pair alters the retention of ionisable solute. For both ion exchange and ion pair chromatography, compounds with different pK_a values can be separated by change in mobile pH, salt concentration and temperature. Increasing–decreasing solvent ratios are the primary variables used for band spacing of signals.

Analysis of metals using ion exchange chromatography utilises a complexing eluent that enables metal ions to be separated by anion or cation exchange. The common mobile phase employed for these metals is pyridine 2,6 -dicarboxylic acid (PDCA) which complexes the metals to be separated by anion and cation exchange. The PDCA forms a strong complex with metals forming anionic metal complexes, hence anionic exchange predominates. Columns with special CS12A packing (available from Dionex) are mainly used for alkali and alkaline metals (Li, Na, K, Mg, and Ca) and columns containing special CS5A packing (also from Dionex) are used for transition metals, e.g. Fe^{3+}, Cu, Ni, Zn, Cd, Mn and Fe^{2+}. Ion exchange chromatography is the preferred technique for separating inorganic ions at low concentrations in aqueous solutions and for metals occurring at variable oxidation states.

The method of separation is initially carried out using a 'separator' column containing a dilute salt solution of a mobile phase. This salt is subsequently removed by a second high capacity ion-exchange column (stripper) as a result of neutralization of the counter ions, e.g. cation analysis. The mobile phase can be a dilute acid solution and the stripper column is an anion exchange column containing hydroxide counter ions. The mobile phase containing the sample ions is passed into the ICP-OES nebuliser/spray chamber and to the plasma torch for atomisation and excitation. For increased sensitivity the mobile phase can be nebulised using an ultrasonic nebuliser and an axial plasma arranged ICP-OES/MS for greater sensitivity.

The advantage of using HPLC-ICP-OES for metal analysis is when direct nebulisation of solutions of samples can cause matrix interference on ICP-OES. An important use of this technique would be the detection of variable oxidation states of elements and it can also preconcentrate trace elements on a column which can be eluted from the column and nebulised using ICP-OES friendly solvents. See schematic diagram 7.16 showing an anion and cation HPLC coupled with ICP-OES.

Factors which affect selectivity, mobile phase, temperature and nature of solute can be shown by the following selective coefficient equation:

$$K_x = \frac{(X^-)_s}{(X^-)_m}$$

where subscripts s and m denote solute and mobile phase, respectively.

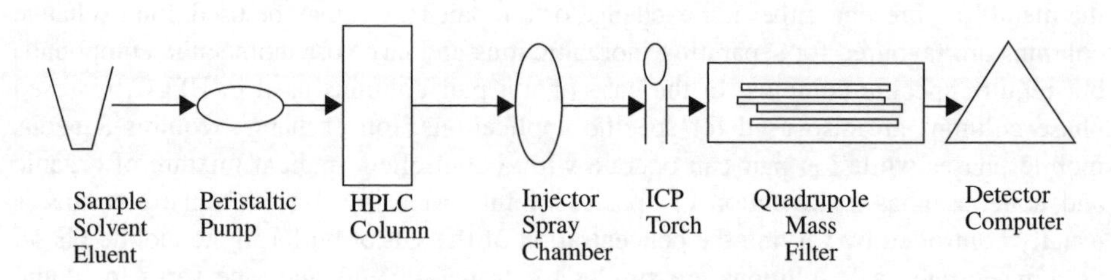

Figure 7.16 *Schematic diagram of HPLC anion and cation chromatography system used with ICP-MS*

7.4.1 Preconcentration of Metals Using Ion Chromatography

The coupling of ion chromatography with ICP-OES offers the potential of analysing organic samples without tedious sample preparations. Liquid organic compounds containing added or contaminated metals may be extracted by aqueous liquids and injected onto a suitable column that will bond to the surface using the ion exchange chemistry described above. Large volumes of extracting liquid can be passed through the column collecting the metals in the sample. A small volume of a suitable reagent is then passed through the column to displace the metals previously bonded and collected in a suitable container. Analysis is carried out by pumping it to the nebuliser and plasma for detection. This method is a useful technique for preconcentrating the metals content in the sample, particularly where the metal(s) are present at very trace levels. A good application of this is the analysis of organic samples for trace levels of As, Pb, Hg, Cd, Se, Cr, Ni, etc., which would have a cumulative toxic effect with long-term contact. Large samples can be extracted into a small volume of solvent and this solvent is further reduced after passing through a column. The disadvantage of this method is that different column packing may be required for different metals and another packing may be required for other elements. An example is the IonPac CS5A column by the Dionex Corporation, a high resolution ion exchange column for the determination of transition and lanthanide metals. This column also can detect the two common oxidation states of iron, i.e. Fe^{2+} and Fe^{3+}. Figure 7.17 is a typical chromatogram of these metals detected by UV and swept from the column using oxalic acid. The sample solution can also be analysed using ICP-OES along with

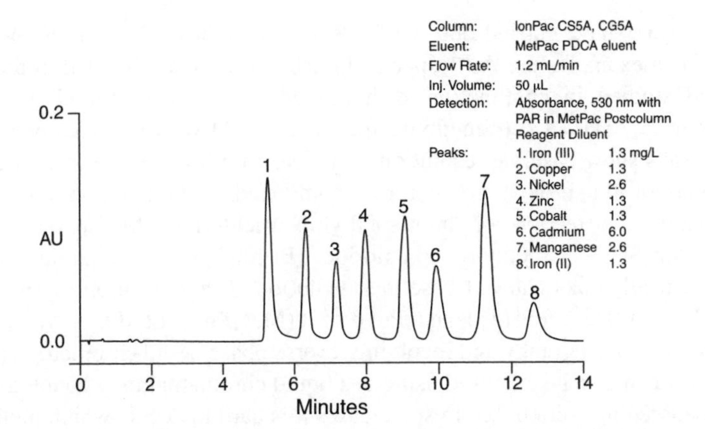

Figure 7.17 *Separation of transition metals by anion exchange using MetPac PDCA eluent concentrate as a complexing agent. (Reproduced by kind permission of Dionex Corporation)*

other metals not separated or detected by the UV detector. It is not possible to detect the oxidation states of iron (Fe^{2+} and Fe^{3+}) using direct ICP-OES but if separated using an appropriate column they can be individually detected and quantified in the presence of other metals not detected with the UV detector.

7.4.2 Analysis of Lanthanide and Transition Metals with ICP-OES/IC

Similar to the transition metal scans shown in Figure 7.17, metals and lanthanides can also be preconcentrated on columns using ion chromatography. The metals in Figure 7.18 can be detected with a single injection; this illustrates the power of ion exchange methodologies for ultra trace reproducible and precise analysis.

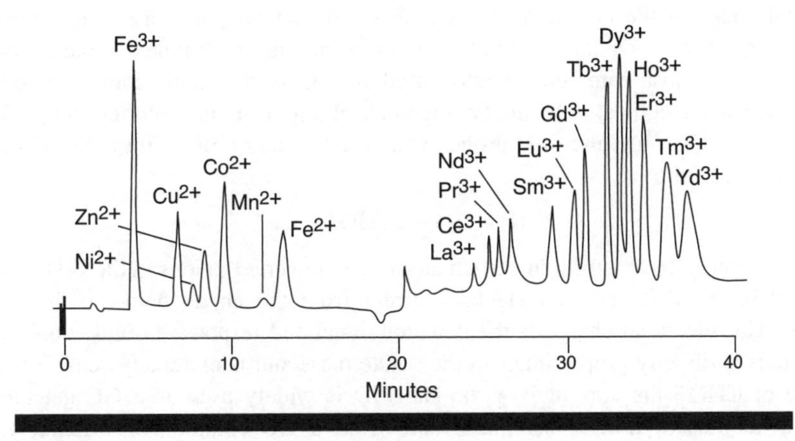

Figure 7.18 *Scan of transition elements using the CS5A column, oxalic acid as eluting solvent and UV detector. Note the Fe^{2+} and Fe^{3+} oxidation states. (Reproduced by kind permission of Dionex Corporation)*

Several workers have carried out metal speciation studies using ion chromatography and well known examples are As^{n+} species in fish tissues and urine, Cd in food, Pb and Se in blood studied in conjunction with ICP-MS. Khan [6] studied As as As(III) compounds, e.g. AsDMAA (dimethylarsinic acid), MMAA (monomethylarsinic) and As(V); all peaks show baseline resolution with detection limits close to $1.0\,\mu g\,L^{-1}$. The superior nature of ion chromatograms for tin compounds and other componds in organic matrices reported detection of monomethyltin trichloride ($MeSnCl_3$), dimethyltin dichloride (Me_2SnCl_2), diethyltin dichloride (Et_2SnCl_2) and trimethyltin chloride (Me_3SnCl) and all peaks showed baseline resolution. The separation of species of Hg (Hg^{2+}, $MeHg^+$, $EtHg^+$, $PhHg^+$) and Pb $[Pb^{2+}, (Me)_3Pb^+, (Et)_3Pb^+]$ was carried out using ion pair chromatography and involving reverse phase liquid chromatography. These Pb compounds can also be analysed using gas liquid chromatography coupled with AED which is discussed in Section 7.5. In some cases it is hard to decide which method to use for trace analysis; the Dionex ion chromatography methods are superior for rapid and trace analysis of selected elements and, in most cases, require very little sample preparation. Sample preparation for analysis using ICP-OES may be simple or tedious but its main advantage over ion chromatography is its ability to analyse other elements not detected by ion chromatography.

7.5 Coupling of Gas Chromatography with ICP-OES or Atomic Emission Detector

The much-improved resolving power of modern capillary columns used in gas chromatography (GC) for the analysis of volatile organic compounds makes this technique/instrument the most popular in modern analytical laboratories. The combinations of most compounds that make complex formulations can be separated using capillary columns that have lengths from 1 to 100 m wound into a coil of multiple turns to give the desired length for separation of metals of interest. The column is placed in a temperature control oven and coupling the heat and the number of theoretical plates associated with that column determines whether individual compounds in a complex mixture can be separated. The most common detector used in GC is the flame ionisation detector (FID) that can detect most individual compounds eluting from the column. In the FID the eluate is burned in a mixture of hydrogen and air producing CHO^+ from the CH radicals in the flame as follows:

$$2CH + O_2 \rightarrow 2CHO^+ + 2e^-$$

Approximately 10 in a million carbon atoms are converted to ions in the FID. The ions produced in the FID carry an electric current from the anode flame to the cathode detector. This electrical charge is the detector signal and response to individual organic compounds is directly proportional to the solute mass and is extremely sensitive. In the absence of CHO^+ the current is zero. The FID is widely used as a GC detector and responds to most hydrocarbons that eluate from a GC column. Unfortunately, it is insensitive to hydrocarbons, H_2, He, N_2, O_2, CO, CO_2, H_2O, NH_3, NO, S, H_2S, SiF_4 and a host of volatile metallic compounds in gasoline, diesel, distillates, crude oils, natural gas, environmental pollutants and volatile organics in water samples.

More and more complex mixtures can be successfully separated on GC columns as their chemistries become more sophisticated. Therefore, the need to differentiate between sample components using FID is limited purely to hydrocarbons. An ICP-OES or atomic emission detector (AED) has been developed to detect volatile organometallic compounds by taking advantage of the atomisation and excitation of metals and some non-metals, depending on the gas used to generate the heat source (Figure 7.19). The metals eluted from a GC column are passed into a microwave plasma source to atomise and excite atoms of interest. It will operate with helium, argon or nitrogen gas. Helium gas is the most commonly used inert gas because of interference from compounds that form with argon or nitrogen.

The strength of this technique lies in its ability to simultaneously atomise, excite, selectively detect and determine with a photodiode array detector the elements that emerge as eluants from a GC column. The principle of operation is that the compound containing elements of interest emerging from the column is passed into a microwave power-induced plasma cavity where atoms of metals are atomised and excited by the energy of the plasma. The light emitted by the excited atoms is separated using diffraction grating (echelle or holographic) with associated optics to focus the dispersed spectral atomic lines which finally pass to the photodiode array for detection.

The photodiode array detector is connected to a computer fitted with a suitable package to control the method, calculate and report the results. The photodiode array detector consists of 1000 or more silicon photodiodes arranged side by side on a single small silicon chip and absorption of electromagnetic radiation by a pn-junction causes promotion of electrons from the valence bands to the conduction bands and thus the

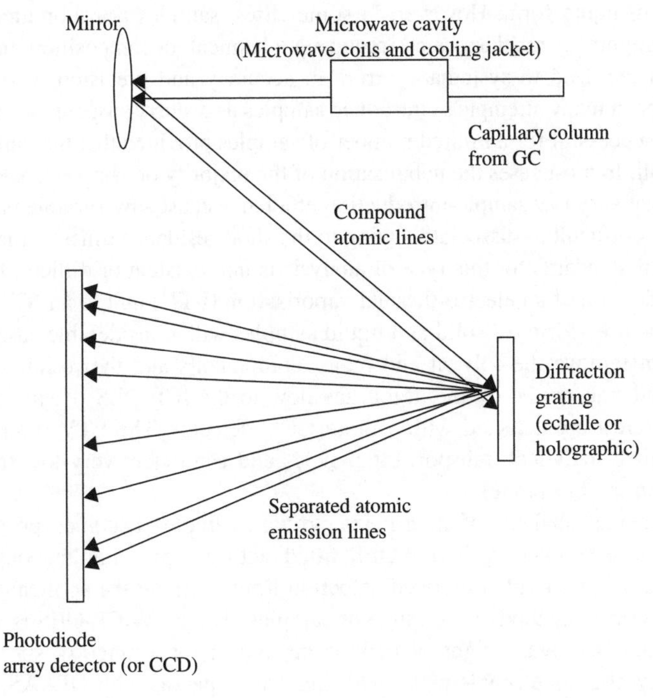

Figure 7.19 *Schematic diagram of a gas chromatographic atomic emission detector*

formation of electron hole pairs in the depletion region. By combining two or more diode array detectors along the focal plane of the monochromator, all wavelengths of interest can be detected simultaneously. Most elements can be detected using this technique provided that they can be separated using the capillary column on the GC. Excited atoms produce emission lines characteristic of the elements of interest.

This technique is suitable for metals that are in the gaseous state in a sample after passing through a GC column, and can be used to study the compound nature of metals as well as the oxidation states provided that they are separated by the columns. The determination of trace organic, inorganic and organometallic compounds is of major interest to most scientists. Biological effects on metals and metalloids can cause speciation changes that are of major concern to toxicologists, biologists and pharmacologists and they need to be monitored in their original form in order to understand their behaviour. The combination of gas chromatography and ICP-OES can provide a sensitive method for the detection of a variety of these analytes at sub-trace levels and improves detection limits and reduces interferences. A considerable amount of work has been carried out using this technique for trace metal studies as part of research and development support but it is not suitable for routine analysis.

7.6 Metal Analysis Using ICP-OES Coupled with Electro-Thermal Vaporisation

Samples for trace metal analysis by AAS or ICP-OES must be presented to these instruments in liquid form. However, in some cases, samples are submitted as powders or solids (chippings, residues, etc.) requiring chemical decomposition prior to metal analysis that can lead to systematic errors in accuracy and precision of measurements. There have been many attempts to introduce samples as a slurry suspension and these were found to be successful for a limited number of samples provided that the particle sizes are suitably small. In most cases the nebulisation of the majority of samples analysed this way has shown that very low sample-introduction efficiency caused by variable particle sizes is in some cases difficult to dissociate, owing to the short residence times in the plasma. The availability of standards for this type of analysis is non-existent or difficult to obtain.

The introduction of an electro-thermal vaporisation (ETV) unit to an ICP-OES plasma source can be used for most solid and liquid samples with considerable ease. Drying and pyrolysis can remove the solvent and major components and the residual analytes are vaporised and transported by the argon gas flow to the ICP-OES plasma source where metals of interest are detected with a rapid CCD detector. The ETV sampling/analysis provides higher analytical transport efficiencies and can detect very low trace levels of metals (i.e. in the ppt range).

Multi-elemental analysis of micro-size difficult samples becomes possible using a graphite furnace coupled with ICP-OES fitted with a rapid simultaneous CCD. This modern system can provide improved detection limits without the application of tedious sample preparation methods for a range of samples. The ETV-ICP-OES is very much in its infancy and is showing signs of making an impact for a range of specific samples. Unfortunately, the same problems lie with this technique, as with GFAAS, i.e. memory effects, reproducibility and precision. Automation of this technique can reduce the reproducibility problems. The complete apparatus is expensive and time consuming and requires a high degree of skill by operator.

The ETV-ICP-OES has been applied using graphite cups, graphite furnaces, and tungsten filaments for the analysis of dry solids, suspensions, organic liquids and solid compounds. It has been shown that the number of elements that can be excited in most sample matrices at 2500–3000°C occurs with 60 or more elements in the periodic table. Direct sample injection allows the direct analysis of used oils, plastics, paint chippings, blood, fingernails, hairs, and volatile elements in refractory matrices. An interesting series of analyses was carried out on hair samples as part of forensic support measuring ultra trace levels of Cu, Zn, Mn, Fe, Cd and Pb by preparing a slurry with the addition of poly(tetrafluoroethylene) (PTFE) as a chemical modifier for the improvement of vaporisation characteristics of the analytes. This method gives detection limits as low as 0.03 ppb for Cu to 2.5 ppb for Pb. These results are almost as good as those determined using ICP-OES/MS that require the hair sample to be prepared by tedious conventional microwave acid digestion methods. The selection of the chemical modifier as part of the total metal analysis is important, when using ETV furnaces, and is similar with direct GFAAS. The modifier poly(vinylidene difluoride) (PVDF) is used for the determination of difficult elements such as Ti, Zr, V, Mo, Cr and La. The combined use of KNO_3 and $Pd(NO_3)_2$ in C_6H_{12} (cyclohexane) as a chemical modifier improves the detection limits and efficiency for the determination of Ag, Pb, Cu and Ni ranging from 86% to 96%.

The design of the combined ICP-OES/ETV instruments allows the vapours from the ETV unit to be swept into the plasma fitted with an axial or radial torch, with which simultaneous measurements can be obtained using a rapid CCD detector.

Good calibration curves are difficult to achieve with the ICP-OES/ETV system. The latter necessitates a simultaneous detector with a time-resolved measurement of the transient signals for the analyte line and background as required for trace analysis. The use of a CCD for detection is extremely important, as it is rapid and reproducible. In some cases, samples containing variable particle sizes (must be suitably small) can be analysed readily using this technique, as evaporation is the same from small particles and bigger particles because of the rapid high heating cycle available with the ETV. In the analysis, where there is a large difference in volatility between analyte and matrix, trace–matrix separation can be readily performed in the oven, resulting in a reduction of interferences, provided that the sample is thoroughly mixed prior to analysis. Analyses are usually carried out using the standard addition approach where sample and standards are both taken through the same heating cycle, i.e. dried, charred, ashed, vaporised and fired together in the same cup to give reproducible results.

Difficulties with reproducible sample injection using manual methods can be overcome by using an automatic sample injector. The method is labour intensive and requires a considerable level of skill, making it unsuitable for routine analysis [7].

The use of ETV with ICP-OES fell out of favour because of the high level of skill required to use it, its expense, slow analysis time and the introduction of the more sensitive modern ICP-MS. The precision associated with ETV-ICP-OES is difficult to maintain and the high cost of replacing parts does not make the technique popular. However, the technique is very good for dealing with biological samples that require little or no sample preparation and very small sizes can be introduced into the graphite furnace for excitation and measurements. Most metals (except refractory) give reasonable recoveries provided that they are analysed by a skilled analyst. A wide range of neat samples can be analysed using this technique, e.g. blood, urine, some pharmaceutical products, etc.

7.7 Surface Analysis Using Laser Ablation with ICP-OES

Laser ablation is a process for removing material from a solid or liquid surface by irradiating it with a laser beam (Figure 7.20). The surface is struck by the laser at one spot or more causing heat in that area from absorbed laser energy and the material heated is vaporised or sublimed when using low laser flux. At higher laser flux, the material is removed with a pulsed laser and transported to the plasma for detection. The depth over which the laser energy is absorbed depends on the material and laser wavelength. Laser pulses (can vary from femtoseconds to milliseconds) and can be precisely controlled. The laser is used to generate minute fractions of the sample in a finely dispersed form and they are immediately carried to the plasma torch for atomisation and excitation using the argon gas.

Laser ablation can be carried out on any material without special sample preparation. The laser beam can be directed onto a defined spot of the sample or moved to different parts to analyse over a defined area. It can be moved in an XYZ plane using a stepper motor and driven in translational motions on which the cell is mounted and with more expensive models can be turned for analysis in other parts of the sample. Lasers can operate in UV, visible, and IR regions of the spectrum and a recent development in laser technology uses neodymium:yttrium aluminium garnet (Nd:YAG) which gives high repetition rate at a comparatively low power. This method of analysis is suited to bulk analysis of solid materials and the amount of volatility varies from sample to sample. The size of the laser spot can vary from 10 to 250 μm and little or no sample preparation is required. Errors are greatly reduced because of the simple sample preparation, and the fact that no solvents are required reduces interferences.

Typical materials analysed using this technique include glass, uranium oxides, steel, platinum, geological samples, ceramic materials and plastics (Figure 7.21). Excellent information is obtained where surface composition is important, e.g. catalysts, presence of toxic elements, surface analysis in the application of adhesives, etc. It can be used as a quick qualitative test for the analysis of semiconductor components as specified as part of WEEE and RoHS compliance.

Manufacturers of electrical and household goods must comply with the WEEE and RoHS regulations for the presence of toxic materials. This is defined by a European Union directive that from August 2005 companies selling all types of electrical goods in Europe must conform to WEEE and from July 2006 the same companies also have to conform to RoHS. As part of these directives, products containing higher than the specified levels of toxic metals such as Pb, Cd, Hg, and hexavalent Cr are banned for sale within the EU. The latest edition of ICP-OES and laser ablation techniques have been used to analyse most of these products rapidly by integrating a large spot using UV Nd:YAG laser ablation with a simultaneous ICP-OES system incorporating a large CCD format detector. A wide range of goods is affected by this legislation from computers and telecommunication equipment to domestic appliances and electronic tools, toys and automatic dispensers. The solid sampling technique has a number of advantages over traditional dissolution techniques. These include high sample throughput and the elimination of additional mixed waste, typically generated by dissolution methods, and both bulk and microstructure chemical analysis can be performed and data reported.

Detection limits can be as low as 2.0 ng to 100 ng depending on the sample and ease of etching because the sample measured is 100% not diluted. It can be used as a semi-quantitative analyser but is difficult to quantify due to lack of available standards. However, for certain samples it may be possible to carry out quantitative analysis if special solids standards are prepared containing increasing concentrations of metals blended into similar blank samples under tests. In such cases the unknown sample and standard blends must be ablated under identical conditions.

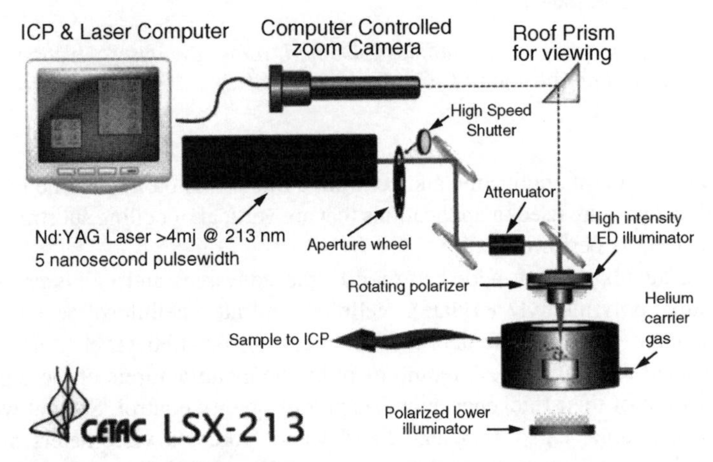

Figure 7.20 *Schematic diagram of laser ablation. (Reproduced by kind permission of Cetac Technologies Ltd)*

The power of laser ablation can be extended as a popular method for trace and bulk analysis in conjunction with ICP-OES and is an invaluable tool in the study of surface behaviour particularly where sensitive surfaces are important. The common area for surface knowledge is in environment, medicines, adhesives, powders, slurries, oil-based samples and liquids. It finds application in the analysis of metallurgical samples, non-conductive polymers, ceramic materials, surface mapping, elemental migration, depth profiling, thin film coatings, biological and clinical specimens, forensic, paint chips, inks, bullets, fabrics, etc.

7.8 Determination of Thickener Content of Paints, Pharmaceutical Products and Adhesives Using ICP-OES

The addition of thickening agents to paints, pharmaceutical products and adhesives has many functions and the most important is for flow behaviour, gap filling and wetting ability for paints and adhesives. They are added at variable concentrations to pharmaceutical

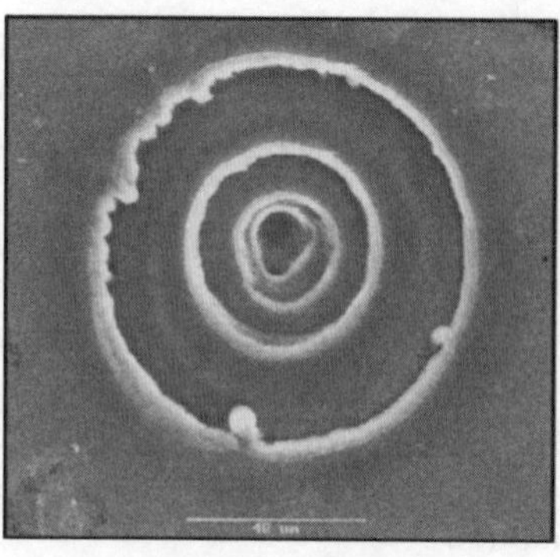

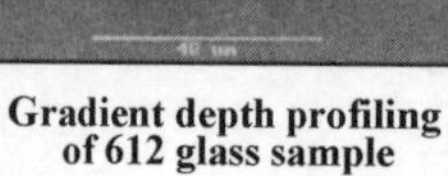

Gradient depth profiling of 612 glass sample

50 μm laser shot in steel coating

Figure 7.21 *Depth profiling of samples obtained using the laser ablation technique. (Reproduced by kind permission of Cetac Technologies Ltd)*

products for ease of oral application and controlled rate of fusion into the body system. In adhesives, thickeners are used in applications that are vertical or ceiling substrates and hold while the bonding is performing.

Thickening agents that are added to modify the adhesives and paints are poly(alkyl cyanoacrylate), poly(methyl acrylate), cellulose nitrate, cellulose acetate butyrate, chlorosulphonated polyethylene, acrylic elastomers, nitrile rubbers, etc., and are mixed by heat and shear force stirring. A requirement by the manufacturers of these products is the determination of their thickener power as part of quality control. Several well known methods are available, e.g. Canon and Fenske and Norcross viscometers are used to measure the thickening strength of these products. The following method describes an atomic spectroscopy method that is quick and gives a qualitative estimate of the level of thickener present. It uses the principle of measuring the signal response for increasing concentrations of thickener added to a monomer.

Method

Sample preparation

A free flowing monomer, ethylene glycol dimethacrylate (EGD) stabilised with 0.1% hydroquinone is thickened with 0.0, 2.0, 4.0, 6.0, 8.0, 10.0, 12.0, 14.0 and 16.0% of methyacrylate butadiene styrene (MBS). Two further samples of the same monomer are thickened with similar increasing concentrations of cellulose nitrate (CN) and silicone dioxide (SiO_2) for comparison. An accurate weight of 10.0 g of each thickened monomer is dissolved in 25% n-propanol and 75% glacial acetic acid. Then 0.25 ml of 1000 ppm Zn metal stock standard solution is added to each mixture. These solutions are also spiked with 0.5 ml of 1000 ppm indium (In) metal as internal standard. All mixtures are diluted to mark with the 25% n-propanol/glacial acetic acid. The mixtures contain 2.5 ppm Zn and 5.0 ppm In per ml.

Preparation of Standard Calibration Curve

Standards containing 0.0, 0.1, 0.25 and 0.5 ml of the 1000 ppm Zn metal are diluted to 100 ml of 25% n-propanol and 75% glacial acetic acid to give 0.0, 1.0, 2.5 and 5.0 ppm Zn, respectively. The prepared samples above are measured for Zn content for each sample with a 2 min washout time for each standard and samples.

Results. Table 7.7 gives the results for 5.0 ppm Zn with increasing concentration of thickener.

Table 7.7 *Results of the same concentration of zinc with increasing concentration of thickener*

Zn added to each solution (ppm)	Sample, EGD + % Thickener	Peak ht ×1000 + % MBS	Peak ht ×1000 + % CN	Peak ht ×1000 + % SiO_2
5.0	0.0	121	144	177
5.0	2.0	109	139	161
5.0	4.0	87	121	144
5.0	6.0	65	102	128
5.0	8.0	43	80	109
5.0	10.0	38	64	92
5.0	12.0	22	59	86
5.0	14.0	29	53	77

Figure 7.22 shows the graphical trend for the results in Table 7.7. The curve appears to decrease in linearity. The trend-line plotted for each slope is reasonably linear which means such a line can give a practical guide to the level of thickener added to the monomer. This differs for other thickeners added to the same monomer, and vice versa for other monomers

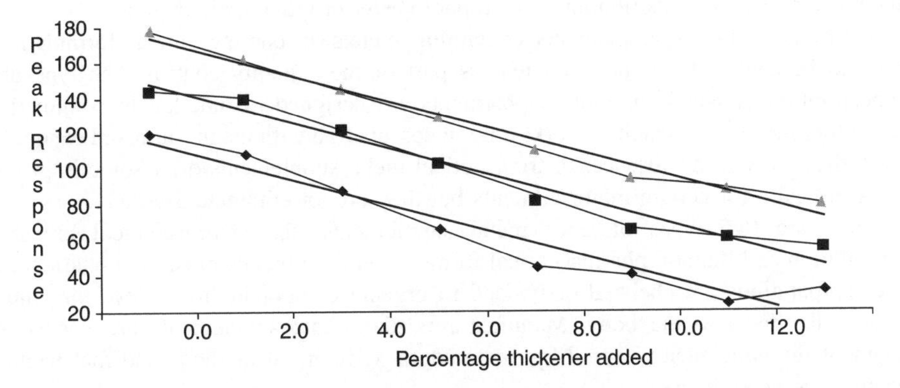

Figure 7.22 *Graphical trend for increasing concentration of thickener MBS, CN and SiO_2 added to EGD monomer and response determined by measuring the same level of zinc in all samples*

added to the same thickener and only works for the monomer/thickener slope. However, it only works if there is no partial polymerisation of the monomer itself and can give an indication of the level of thickener present using the ICP-OES method.

7.9 Metal Analysis of Metallo-Pharmaceutical Products

The selectivity and sensitivity offered by atomic spectroscopy techniques can be used for direct and indirect determination of metals in a range of pharmaceutical preparations and compounds. Metals can be present in pharmaceutical preparations as a main ingredient, impurities, or as preservatives which can be prepared for analysis using non-destructive (direct or solvent dilution) or destructive methods (microwave acid digestion, bomb combustion, extraction, etc.) and the metal of interest measured against standards of the metal prepared in the same solvents as the sample. Methods associated with some pharmaceutical products are already described in the international pharmacopoeias and must be used in order to comply with regulations associated with these products, e.g titration techniques are carried out according to methods that are the same for all pharmaceutical products.

Metal analysis of pharmaceutical products using ICP-OES methods has gained widespread interest because of the selectivity, sensitivity and rapid analysis. Metals in pharmaceutical products are added for beneficial reasons or are present as contamination. It is known that the oxidation state of some metals is important as one state may be beneficial while the other is toxic. An example is chromium: Cr^{3+} is necessary for diabetics as part of the normalisation of the sugar cycle in the body while Cr^{6+} is carcinogenic to the lungs. To determine the states of this metal it would be necessary to separate them using a HPLC column or by chemical reactions and by analysing them individually using ICP-OES.

Sample preparation of pharmaceutical products is an important step in the analysis methodology and must be carefully carried out to avoid contamination, loss of metal(s) or addition of interferences that could result in errors in the measurements. Modern versions of the international pharmacopoeias contain methods involving atomic emission methods, and some replace tedious titration, spectrophotometer or gravimetric methods.

Nutritional vitamin supplements containing metals as pharmaceutical formulations need to be analysed for metal content as part of their quality control. The type and concentration of metals in some supplements are designed to function in conjunction with other metals, e.g. calcium works only in conjunction with magnesium and cannot be used effectively in its absence. Pharmaceutical metal supplements offer some improvement in health for certain minor ailments but they do not eliminate disease.

There are two types of trace metal supplements: the pharmaceutical prepared formulation and the non-pharmaceutical formulation. The metals present in pharmaceutical preparations are chelated or bonded to organic compounds to achieve maximum beneficial effects to the body. Manufacturers claim that supplements are needed to augment the deficiencies caused by our modern agricultural methods and that healthy people also need them.

The non-pharmaceutical supplements are extracts from natural products such as seaweeds and sea shale, available in ancient mines and caves. Manufacturers claim that naturally extracted supplements do not contain other beneficial additives, such as omega-3 fatty acid that is required for the treatment of the nervous system and mental illness. Claims have also been made that naturally extracted supplements are not as well absorbed or metabolised by our bodies as synthetic supplements and are not chelated with suitable organic compounds to give the maximum benefit. Neither theory has been proven scientifically.

Metals other than those in supplement formulations are used extensively in pharmaceutical products and are added at various concentrations as an aid to health benefits, or as fillers or encapsulants. Magnesium hydroxide is used as a gentle laxative while aluminium present as hydroxide, lactate, salicyclates, acetates and alums is used frequently in a number of dermatological products. Metals in creams, lotions and powders are also part of the pharmaceuticals range. The metals formulated into pharmaceutical products are salts of p-block metal ions and the following is a brief outline of their use in medicine:

(i) *Barium.* Patients are orally given barium sulphate solution as part of an X-ray examination of a problematic gastrointestinal tract. This solution is non-toxic while in the body for a short duration and it is easy to monitor this part of the body without surgery and most of it will be excreted over a short time. Analysis for residual barium in urine, blood and semen can be analysed directly after digestion in alkaline solution and dilution in deionised water prior to analysis using ICP-OES.

(ii) *Calcium.* Calcium compounds (e.g. ascorbate, lactate, carbonate, chloride) are used as therapeutic agents for calcium deficiency and as anti-anaphylactic and antacid agents. These salts are readily soluble in dilute hydrochloric acid solutions and separated from other organic and insoluble inorganic fillers by filtering and measuring against a certified standards calibration curve using ICP-OES. The advantage of ICP-OES over AAS methods is that the signals are not affected by low molecular weight soluble organic compounds such as gelatines, acids, bases, etc. However, where large molecules are concerned it may be necessary to digest or ash the sample prior to analysis.

(iii) *Copper, cobalt, iron, manganese, gadolinium, gold, vanadium and molybdenum.* These metals, which are used in creams and ointments and multi-vitamin tablets, are analysed after dissolving the organic compounds in chloroform and extracting the metal with 10% nitric acid solution. Analysis for the presence of iron as carbohydrates in pharmaceutical formulations or as EDTA (chelator), as sulphates, chlorides and phosphates is also required as part of product specification. Other metals present in the preparations do not interfere with analytical protocol. Chelators can be used to remove toxic metals that are poisonous and to be sure that the treatment is successful body fluids must be analysed.

Thalassaemia is an inherited disorder characterised by an abnormal production of haemoglobin. This results in low haemoglobin production, and excessive destruction of red blood cells. Monitoring the metal Fe can provide important information as part of diagnostic control of this disease. Wilson's disease is an inherited disorder where there is excessive copper (Cu) in the body's tissue. This can cause a variety of effects, including liver disease and damage to the nervous system. Painful rheumatoid arthritis can be reduced by the use of gold and gadolinium salts in controlled doses. Similarly, with diabetic patients recovering from infections, the salt vanadyl sulphate can be used to cure lingering tiredness and apathy.

(iv) *Mercury.* The metal Hg can be used as a fungicide while other forms can cause mental disorder. Organic mercury compounds are used commonly at concentrations of 10–30 μg ml^{-1} as a preservative in eye drops, injection solutions, vaccines, etc., which are applied for short-term use to prevent micro-organism growth during their application. The modern method of analysis of these products for mercury content is by the cold vapour trap method attached to the ICP-OES. The mercury compounds in Table 7.8 are used in pharmaceutical preparations such as pharmaceutical powders, tablets, gels, injections, tinctures, suspensions, ointments and ophthalmic solutions.

Table 7.8 *Mercury compounds used in pharmaceutical formulations*

Compound	Function
Thiomersal	Antiseptic and antifungal agent
Mersalyl	Diuretic with antiviral properties
Merbromin	Topical antiseptic, ineffective by FDA
Nitromersal	Antiseptic for skin and mucous membrane, disinfectant
Sodium mercaptomerin	Treatement of kidney disorder
Phenyl mercury borate	Antiseptic, treatment of periapical periodontitis and preservative in ophthalmic solutions
Phenyl mercury benzoate	Banned antiseptic
Phenyl mercury acetate	Fungicide, unclassified herbicide

Mercury salts that are present in the formulations in Table 7.8 are used in low and controlled concentrations and must be analysed for their concentration prior to use. Therefore, sample preparation is usually carried out in closed vessels such as microwave acid digestion or bomb combustion and diluted in deionised water to a known volume.

(v) *Tin, titanium and zinc used in pharmaceutical formulations.* Dibutyl tin dilaurate (DBTDL) is a drug added to chicken and turkey feeds to remove round worms, caecal worms, tapeworms and to protect them from coccidiosis and hexamitiasis. The DBTDL is extracted from the feeds using chloroform and the suspension of feeds/chloroform is filtered through a fine fast flowing filter paper. The clear and separated chloroform is evaporated leaving the tin salt that is re-dissolved in glacial acetic acid (GAC) for analysis against tin standards prepared in GAC using ICP-OES.

Titanium metals are used in sun protection creams and function effectively in conjunction with an iron complex that is present at 100 and 150 ppm, respectively. The samples are digested for analysis using a conc. HNO_3/HCl acid mixture in a microwave oven and analysed against calibration curves prepared with standards of the same metals after filtering to remove high concentrations of silicon fillers from the digested solution through a fast flow filter paper.

Zinc salts such as stearate, naphthenate and oxides are formulated mainly in ointments and pastes for external use only. Zinc salts are also present in calamine powder, calamine lotions and dusting powders and are determined by extracting the organic components with chloroform and digesting the dried insoluble product in concentrated acids using a microwave acid digester. The metal is analysed against standards prepared in the diluted aqueous solutions using the ICP-OES. Mouth washes and some eye drops also contain zinc complexes as sterile reagents and are simply diluted in deionised water and analysed for metal content.

(vi) *Manganese and lanthanide salts*. Manganese salts are used extensively in magnetic resonance imaging (MRI) to observe internal abnormalities without the use of invasive surgery. Lanthanide complexes are used in X-rays to identify structural abnormalities assisting in the diagnostic decision of whether or not operations are necessary.

7.9.1 Metallic Type Antibiotic Drugs

Pharmaceutical compounds used in formulations need to be monitored as part of quality assurances and this equally applies to the presence of added metals or contamination of metals. Strict conditions are applied to all medical compounds used. All drugs used in medicine are tested, both biologically and pathologically, as well as for purity prior to releasing to the public.

Techniques involving AAS, ICP-OES, ICP/MS are also applied to support the quality of a range of metal related pharmaceutical drugs and for metal contamination.

Various workers in the field have developed several atomic spectroscopy methods used in the international pharmacopoeias. The following is a short list of methods that are still used today:

- Bohme and Lampe [8] and Minamikowa and Matsumura [9] developed an atomic spectroscopy method for detecting drugs such as chlorprothixene and noscapine using the Reineckes salt ammonium tetrathiocyanatodiamminochromate[III] monohydrate, $NH_4[(Cr(NH_3)_2(SCN)_4)]H_2O$, through the chromium metal complex. Recoveries close to 100% can be obtained for each of the drugs which makes the method an excellent technique for the detection and confirmation of their presence. Alkaloids such as atropine, codeine, emetrine, narcotine, procaine, quinidine and strychnine can also be quantified using their chromium complex provided that it is the only alkaloid present in the pharmaceutical product.
- An atomic spectroscopy method for the analysis of vitamin B_1 in the presence of other vitamins such as B_2, B_6, B_{12}, nicotinamide, and vitamin C can be carried out after reacting it with a Pb^{2+} salt in a basic solution (NaOH). The test involves measuring the unreacted lead using ICP-OES in solution after centrifugation. The sulphur in the vitamin is quantitatively precipitated as PbS after heating to 85°C. The difference between the unreacted lead in solution and the precipitated PbS can be used quantitatively to determine the level of vitamin B_1 in pharmaceutical preparations or in natural products. This technique was studied extensively by Hassan [10] and compared favourably with other techniques giving recoveries close to 100%.

- Benzylpencillin can be analysed using ICP-OES after reacting it with excess tris(1,10-phenanthroline)Cd salt and the complex is extracted into dimethyl sulphoxide (DMSO) and analysed against organo-metallic standards prepared in DMSO. The structure of the complex formed using this reagent is Cd(penicillin)$_2$ (phen)$_3$ and is stable and trace levels of this antibiotic can be detected. Other penicillins such as ampicillin, epicillin, phenoxymethyl penicillin and sodium penicillin G can be determined after desulphurization with alkali metal plumbite using ICP-OES for the excess Pb. The Pb metal reacts 1 mol of PbS with 1 mol of penicillin and this method was also studied extensively. Sodium and potassium metals can form cyclic complex salts that are beneficial as antibiotics.

7.9.2 Platinum and Palladium Drugs for Cancer Treatments

Platinum salts used as anti-tumour agents [*cis*-dichloroplatinum(II) diammino and platinum(IV) chlorides] are mostly used in chemotherapy as anti-tumour agents. The site of action is thought to be the DNA, where the nitro strand binding the nitrogen of the adjacent guanine resides in a prominent lesion. Cisplatin is thought to act by cross-linking DNA in several different ways, making it impossible for rapidly dividing cells to duplicate their DNA for mitosis. The damaged DNA sets off a repair mechanism, which activates apoptosis when repair proves impossible. The salt complex was found to possess a pronounced activity against tumours in mice with low animal toxicity. This chemotherapeutic drug is an important treatment for many types of solid tumours such as sarcomas, some carcinomas (e.g. small cell lung cancer and ovarian cancer), lymphomas and germ cell tumours, in particular non-small cell lung cancer. A series of Pt(IV) drugs has been shown to improve the cure rates of certain cancers and reduction potential correlated well with cytotoxicity. The complex containing a chloro axial ligand was demonstrated to have a better potency while the hydoxy ligand was the least effective. Platinum drugs retained the ability to evoke apoptosis in quiescent cells. It is believed that Pt(IV) drugs retain cytotoxicy potency under resistance inducing tumours in microenvironmental conditions, and may be used as an alternate to the current platinum ammonium type drugs.

Cisplatin and its salts have a number of side effects that can reduce the use of this treatment and each person's reaction to chemotheraphy is different. Some people have very few side effects while others have more severe reactions which may be further complicated if taking other drugs. The following is a short list of side effects:

- kidney damage (nephrotoxicity)–must be checked using nerve conduction studies before use;
- sickness–can be controlled by use of anti-sickness drugs;
- hearing loss–no known treatment for this;
- alopecia (hair loss)–not considered a problem with patients;
- electrolye disturbance–can be caused by low serum magnesium secondary to cisplatin and not caused by this drug.

On completion of the use of the platinum salts to damage the cancerous cells they must be removed from the body as soon as possible because if they are present for a prolonged period of time they would be toxic. Therefore they must be removed from the body within 24–48 h after treatment using natural bodily excretion methods. Analysis of samples of tissues, albumin, blood plasma, urine, excrement and saliva for platinum content is an essential part of the treatment process.

Direct analysis using atomic emission methods will only detect the total platinum content after appropriate sample preparation and the information is useful for diagnostic support. It can be used to indicate if the platinum has been removed from the body after use. This analysis alone does not reveal any information of the effects of pharmaceutical studies associated with this salt as it fails to distinguish between the different forms of platinum complexes that may result after the treatment. It does not discriminate between the drug and other metabolites or breakdown of the platinum species. To gain further information about the states of platinum, HPLC apparatus is connected to the ICP-OES prior to the nebuliser and each platinum species can be separated on a column using a strong base anion-exchange column containing Partisil-10 Sax. Identification of each species is obtained from prior knowledge of separations by that column.

Microwave acid digestion of the tissue, blood, serum, etc., can be used to prepare samples for metal analysis. The ICP-OES method is useful for monitoring the distribution of platinum compounds in the body but the information alone is not sufficient to support rigorous pharmacokinetic studies required to fully understand the total functionality as a cancer killing drug.

7.10 Metal Analysis of Infusion and Dialysis and Bio-Monitoring Solutions

Electrolytic solutions used for extra-renal infusion and dialysis contain metal chlorides of Na, K, Ca and Mg salts at concentrations that are critical for effective treatment. These solutions also contain dextrose, citrate and lactate additives as part of this special formulation. The analysis for these metals must be precise and accurate and this can be achieved with ICP-OES using yttrium or scandium as internal standard to correct for matrix affects. The method of standard addition may also be used with similar success but is a more tedious method. The ability to dilute the sample several fold due to the high concentrations of metals reduces/eliminates the effect of EIE* (easily ionised elements) caused by other elements in the same solution. The dilution and the ease of detection and corrections with an internal standard using the multi-element capability make this an excellent method.

*Easily ionised elements have many influences:

(i) They can decrease excitation temperature due to consumption of energy for ionisation.

(ii) They may shift the ionisation equilibrium for partly ionised elements as they influence the electron number through their easy ionisation.

(iii) They may cause changes in plasma volume as a result of ambipolar diffusion. (Ambipolar diffusion is differences in positive and negative particles in a plasma due to their interaction via the electrical fields. It is closely related to the concept of quasi neutrality. In plasmas, the force acting on the ions is different from that acting on electrons, so one would expect one species to be transported faster than the other, whether by diffusion or convection or some other process. If such a differential transport has divergence, then it will result in a change in the charge density, which in turn creates an electric field that will alter the transport of one or both species in such a way that they become equal.)

Metal analysis of bio-monitoring samples, such as blood, urine, saliva, semen, skin, internal and external body parts outside clinical testing, is often required to support toxicological and other studies. Screening of samples for metals analysis may be used to expose the presence of toxic metals in water, air pollution or foods consumed by the public. Some elements that are essential nutrients at low levels can be toxic at higher levels and some are toxic at any level.

Analysis of these samples must be carried out with extreme care particularly during the sampling and preparation stages because many trace metals are prone to analytical interferences or matrix effects from the other biological specimens. The official methods accepted by regulatory bodies within the medical/pharmaceutical bodies have been developed to report accurate, precise and rapid determination of a wide range of major and trace levels of metals using ICP-OES. In the case of ultra trace analysis the use of an ultrasonic nebuliser, hydride generator, cold vapour trap or ICP/MS would greatly improve their detection limits.

Analysis of urine samples for metals content is a useful way to study the presence of toxic metals in humans. The metal content can give an indication of the performance of kidneys in regulating the body electrolyte, water metabolism and rate of excretion of metals from the body. ICP-OES can be used to measure the level of heavy metals in urine of both healthy and pathological cases. Sample preparation for analysis of these samples must involve an acid digestion in a microwave oven or bomb combustion to destroy the interfering organics present. Metals such as Pb, Cd, Tl, Se, Sn and Hg are the usual metals requiring analysis.

Teeth are a good indication of past exposure to metals because of their physical stability. The tooth material is digested in a Teflon vessel using bomb combustion at elevated temperature and pressure followed by diluting in deionised water to a known volume and analysed against a standard calibration curve for metals of interest.

Human eye tissue is usually analysed for the presence of zinc salt (sterilising agent) content after acid digestion in a microwave oven at elevated temperature and pressure. This solution can also be used for analyses of other metals for medical management and post mortem studies.

7.11 Organometallic Compounds

The application of organometallic compounds in medicine, pharmacy, agriculture and industry requires the accurate determination of these metals as part of their application. Most π complexes characterised by direct carbon-to-carbon metal bonding may be classified as organometallic and the nature and characteristics of the π ligands are similar to those in the coordination metal-ligand complexes. The π-complex metals are the least satisfactorily described by crystal field theory (CFT) or valence bond theory (VBT). They are better treated by molecular orbital theory (MOT) and ligand field theory (LFT). There are several uses of metal π-complexes and metal catalysed reactions that proceed via substrate metal π-complex intermediate. Examples of these are the polymerisation of ethylene and the hydration of olefins to form aldehydes as in the Wacker process of air oxidation of ethylene to produce acetaldehyde.

In most cases the determination of organometallic complexes by atomic spectroscopy techniques is the only acceptable method because the analysis is selective, accurate and precise. Analysis of these complex salts may only involve a simple dilution in a solvent or destruction methods depending on the matrix it is formulated into. The presence of some sample matrices containing organometallic complexes can be severely restricted by the matrix material to achieve accurate detection and quantification of these salts.

Analysis of initial, intermediate and final stages of most reactions involving metals used as catalysts, activators, etc., needs to be monitored at each stage to ensure that the process in which the metal salt is used is effective. In certain reactions it may be necessary to carry out analysis to determine if the metals have been effectively removed, if the process so requires. All metal catalysts can be readily monitored using atomic spectroscopy techniques after appropriate sample preparations.

7.12 Metals and Metalloid Analysis in Support of Forensic Science

Forensic science is the application of a broad spectrum of science used to answer questions of interest to the legal profession. This may be defined as the 'application of science to law'. This science can be applied in solving criminal cases or used in civil action cases. The use of analytical science in the legal profession is primarily concerned with the analysis of trace materials in which the crimes have taken place. The principle is that every contact leaves a trace that will offer potential evidence to link a suspect with the scene of the crime, victim or weapon.

Forensic chemistry is concerned with the application of the principles of chemistry and related sciences to the examination of physical evidence collected at scenes of crimes, e.g. blood stains, paint fragments, bomb residue, clothes, drug samples and hair. The samples could provide conclusive evidence linking a suspect to a crime. Results from chemical analysis are usually conclusive and can be used to support eye witness testimony especially when trials take place many months or even years after the offence.

Metal analysis can give valuable information as part of investigations carried out by forensic scientists. The number and type of samples that require chemical analysis is very large and the methods required for their analysis can vary from simple visual inspection to the use of most advanced mass spectrographs that can be a part of gas–liquid chromatography, HPLC or ICP-OES. The concentrations of suspect analyte can vary from percentage levels to parts per trillion (ppt) levels, and detection and quantification must be precise and accurate. The sample collection, transportation, storage, etc., must be carried out with extreme care so as not to contaminate the sample and so undo any potential conclusion required by the courts. All forensic analyses are carried out by proven procedures because any break in the procedural chain would give rise to doubts concerning the admissibility of evidence. Most forensic analysis is repeated to confirm the results using the same method or an alternative method or scientist. New, advanced and more sensitive methods are being developed in the forensic laboratory but must be validated and verified by the scientific community to be registered by the courts. Without this they will not be accepted as valid proven methods suitable for the proof of evidence

Table 7.9 *List of some of the samples analysed by forensic laboratories as part of criminal investigations*

Crime	ICP-OES metal and metalloid support in forensic analysis
Arson	Metals in paints, petrol, kerosene, magnesium ribbons, etc.
Assault	Trace metals after knife, sword, hammer or metal weapon attack
Robbery	Same as robbery including gunshot residues, etc.
Blood	Metal from pharmaceutical drugs, poisons, contamination, etc.
Plasma	Same as for blood
Urine	Same as for blood
Hair	Same as for blood
Foods	Contamination, excess metal(s), toxic metals from containers, etc.
Beverages	Same as for foods
Metals	Containing toxic metals as part of composition
Poisons	Metal poisons, e.g. As, Pb, Hg, Se, Sn, Cd, etc.
Tissues	Same as for blood
Bones	Identification purposes, etc.
Teeth	Same as for bones, poisons, etc.
Metal-proteins	Metals reacting with proteins, etc.
Gunshot residues	Metal scan for Ag, Ba, Cr, Cu, Mn, Pb, Sb and Sn
Cotton swabs	Metal scan

in a criminal trial. Table 7.9 shows some of the areas of criminal investigation in which ICP-OES can provide evidence.

The common elements associated with forensic analysis are Al, Ag, As, B, Ba, Be, Bi, Cd, Co, Cr, Cu, Fe, Ga, Ge, Hg, Li, Mn, Mo, Ni, Pb, Pd, Pt, Rb, Sb, Se, Sn, Sr, Tl, U and Zn, and are readily determined using ICP-OES.

Several elements can be analysed together using the simultaneous or sequential ICP-OES and an internal standard can be included to correct for matrix effects. The internal standards yttrium, scandium, gold or rhodium are the common elements used provided that they are not already present in the sample being analysed. Sufficient samples must be provided for analysis as duplicate or triplicate analysis may be required prior to court evidence. Concentrations of metals that may need to be detected can be as low as $0.5\,ng\,ml^{-1}$ to 100%. Some elements may not be detected using conventional direct nebulisation and may require hyphenated techniques such as hydride generation, ultrasonic nebulisation or the cold vapour trap method for mercury, etc., to improve their detections. Axial torches are 5–10 times more sensitive than the corresponding radial torches.

Sample preparations associated with forensic evidence can vary from simple dissolution to solvent extraction with or without the use of a complexing agent, e.g APDC. Some samples may require destructive techniques, i.e. ashing, microwave acid digestion or oxygen bomb combustion. It is recommended that the sample preparation also include a repeat test involving ‘spiking’ the sample with a known concentration of metals of interest provided that sufficient sample is available. A blank must also be prepared containing all reagents without the sample.

The correct choice of sample preparation and analysis will give excellent confidence in the reported results. Microwave acid digestion methods can be used to prepare almost all forensic samples because of the closed, inert, microwave-transparent digestion vessels using strong acid mixtures at elevated temperatures and pressures. Preparation by the oxygen bomb combustion method has the advantage in that the samples are usually dissolved in a weak solution of a base or in deionised water. Unfortunately, this method only allows a single sample preparation to be carried out at a time making it time consuming and erroneous if care is not taken.

Prolonged ingestion may be required to achieve toxic proportions of most essential metals. Human toxicity from heavy metals such as Pb, Hg and Cd are most commonly associated with accidental or intentional consumption rather than true natural occurrence. An example is the mercury contamination in Minimata Bay in Japan shortly after the Second World War causing the 'Minimata' disease and this was sourced to the disposal of mercury in the bay in the 1950s. This disease was caused by the long-term consumption of fish and water by the local populace. This metal in the seawater was eventually sourced to waste effluent from the Cisso Chemical Corporation [9]. This chemical company was manufacturing acetylene, acetaldehyde, acetic acid, vinyl chloride and octanol and using a range of metals as accelerators and catalysts and disposing them after use into the local bay. The company was disposing of the metals Pb, Mn, As, Se, Tl, Cu and Hg and forensic investigation discovered that hair samples taken from local people contained these metals.

Local scientists concluded that symptoms of poisoning such as ataxia, numbness, weakness, insanity, paralysis, coma and death were associated with the cumulative effect of the metal mercury. It has since been proven that mercury can give rise to a neurological syndrome caused by severe mercury poisoning which affects foetuses in the womb causing abnormal babies or even early miscarriages. More than 50 years later, legal cases are still being fought in Japanese courts.

Heavy metals are important with regard to human toxicity because the body possesses only inactive mechanisms for their excretion and low level intake can lead to chronic toxic proportions over time. Treatments are not effective except for symptomatic relief as no effective means has been discovered to increase their excretion.

Residues from explosive devices are commonly analysed for metals. Explosive materials can be analysed for metals such as potassium in potassium nitrate, sulphur and sodium in sodium nitrate. Traces of metal content can also be detected on clothes, hands and hair after using the appropriate sample preparation method. The development of a reliable, accurate procedure for identifying gunpowder residues on hands, gloves and clothing of suspects has been a problem for forensic scientists in the past. When a user of a gun fires the bullet, the metal components of the handle, barrel and bullets leave traces of Sb, Pb, Cu, Fe, Ba and S that can be detected using ICP-OES coupled with an ultrasonic nebuliser and axial torch. These metals can be detected after washing the hands with swabs of cotton wool soaked in 1.0 M HCl. Clothing and gloves can be soaked in 1.0 M HCl to dissolve the metals. A blank of each cloth must also be prepared to rule out any presence of these metals in the original cloth.

7.13 Non-Prescription Nutritional Dietary Supplements

Nutritional dietary supplements are intended for inadequate diets. According to the FDA (Food and Drug Administration) of US Dietary Supplement Health and Education Act

1994, supplements are foods, and not drugs. Unlike pharmaceutical companies, manufacturers of supplements are not required to prove the safety or effectiveness of their products but the FDA can take action if these supplements proved to be harmful. The purity and quality of different brands of supplements are not necessarily controlled, but where supplements are consumed by the vulnerable population (i.e. infants and invalids) they are controlled for their quantity and quality.

The European Union Food Supplements Directive requires that supplements are demonstrated to be safe both in quantity and quality. Supplements that have been proved to be safe may be sold without prescription 'over the counter'. It also requires that supplements *should not* claim to be drugs but can claim to be beneficial to health.

A random sample of a supplement containing 12 minerals (metals) was analysed using ICP-OES for their concentrations. The elements are listed in Table 7.10 along with the approximate concentration of each element in the tablet.

Table 7.10 *List of metals present in health supplement tablets available from the local supermarket*

Element	mg per tablet	μg per tablet
Ca	160	160 000
P	125	125 000
Fe	3.5	3500
Mg	100	100 000
I	0.1	100
Zn	5.0	5000
Cu	1.0	1000
Mn	1.0	1000
K	40	40 000
Cr	0.06	60
Mo	0.03	30
Se	0.03	30

The following sample preparation procedure was used:

Method. The average weight of a tablet was first established by weighing 10 tablets in grams to four decimal places and calculating the mean.

Results

1.4252; 1.4223; 1.4060; 1.4314; 1.4034; 1.4184; 1.4071; 1.4026; 1.4329; 1.4276

$$\text{Mean}\,\bar{x} = (14.1769/10) = 1.4177\,\text{g}\,(\text{spread} \pm 0.0303\,\text{g}) \quad (4)$$

Ten tablets containing metals listed in Table 7.10 and insoluble gelatine, starch, sucrose, gum, thickener, hydrogenated vegetable oils, vitamins, etc., were filtered through a GF/C Whatman filter paper and the clear solution was analysed as follows:

Method. The tablets are crushed to a fine powder using a mortar and pestle. Approximately 1.0 g of the crushed tablets is accurately weighed into a 100 ml grade B plastic volumetric flask followed by 5.0 ml of conc. HNO_3, 2 ml of conc. HCl and approximately 60 ml of deionised water. The mixture is stirred on a stirring plate for 1 h and made up to 100 ml mark with deionised water and shaken to form a homogeneous solution.

(a) The clear filtered solution is further diluted to 100 ml with deionised water using a 100 ml plastic volumetric flask to be analysed for the elements Ca, P, Fe, Mg, Zn, Cu, Mn and K using the multi-element facility of the ICP-OES. This solution is analysed against 0.0, 5, 10 and 20 $\mu g\,ml^{-1}$ standards of the same metals prepared in 5.0 ml conc. HNO_3 and 2.0 ml HCl and diluted to 100 ml with deionised water. A radial mode of a dual axial/radial plasma torch for metals at high concentrations is used for these metals.
(b) The original solution containing lower levels of metals is used to analyse for I, Cr, Mo and Se against 0.0, 0.5, 1.0 and 2.0 $\mu g\,ml^{-1}$ standards of the same metals prepared in 5.0 ml conc. HNO_3 and 2.0 ml conc. HCl diluted to 100 ml with deionised water. An axial mode of a dual axial/radial plasma torch for metals at low concentrations is used for these metals.
(c) A blank solution containing 5.0 ml conc. HNO_3 and 2.0 ml conc. HCl only is diluted to 100 ml with deionised water.

Figures 7.23 and 7.24 show the results of scans for metal content of dietary supplements using the radial and axial modes, respectively.

Results. The results of analysis of dietary supplements are shown in Table 7.11.

Table 7.11 *Results of analysis of dietary supplements for metal content of diluted filtered solutions containing high levels of metals using the radial mode of the plasma torch and axial mode for undiluted filtered solutions for the lower levels of metals. (The results show excellent agreement with label quantity)*

Element	mg per tablet (Spec.)	μg per tablet (Spec.)	μg per tablet (Found)
Radial			
Ca	160	160 000	164 000 (164 mg)
P	125	125 000	119 000 (119 mg)
Fe	3.5	3500	3700 (3.7 mg)
Mg	100	100 000	99 390 (9.9 mg)
Zn	5.0	5000	5300 (5.3 mg)
Cu	1.0	1000	920 (0.92 mg)
Mn	1.0	1000	1010 (1.01mg)
K	40	40 000	40 600 (40.6 mg)
Axial			
Cr	0.06	60	55 (0.055 mg)
Mo	0.03	30	27 (0.027 mg)
Se	0.03	30	21 (0.21 mg)
I	0.1	100	88 (0.088 mg)

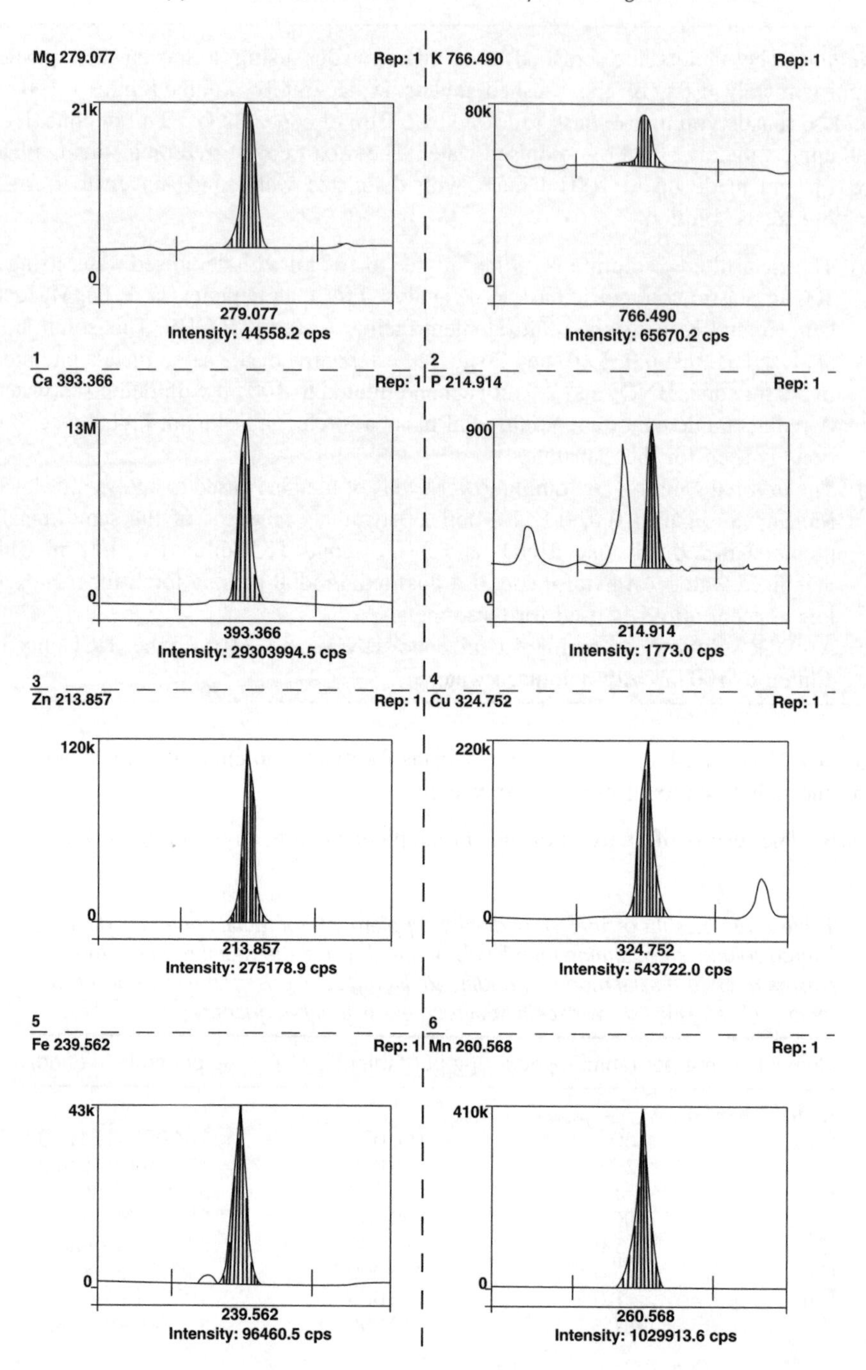

Figure 7.23 *Analysis for Mg, K, Ca, P, Zn, Cu, Fe and Mn using the radial mode of the dual radial/axial ICP-OES system. (The final solution was between 0.5 and 10.0 ppm metal.) Results obtained using a Perkin Elmer 2100DV ICP-OES. (Reproduced with kind permission: copyright © 1999–2008, all rights reserved, Perkin Elmer, Inc.)*

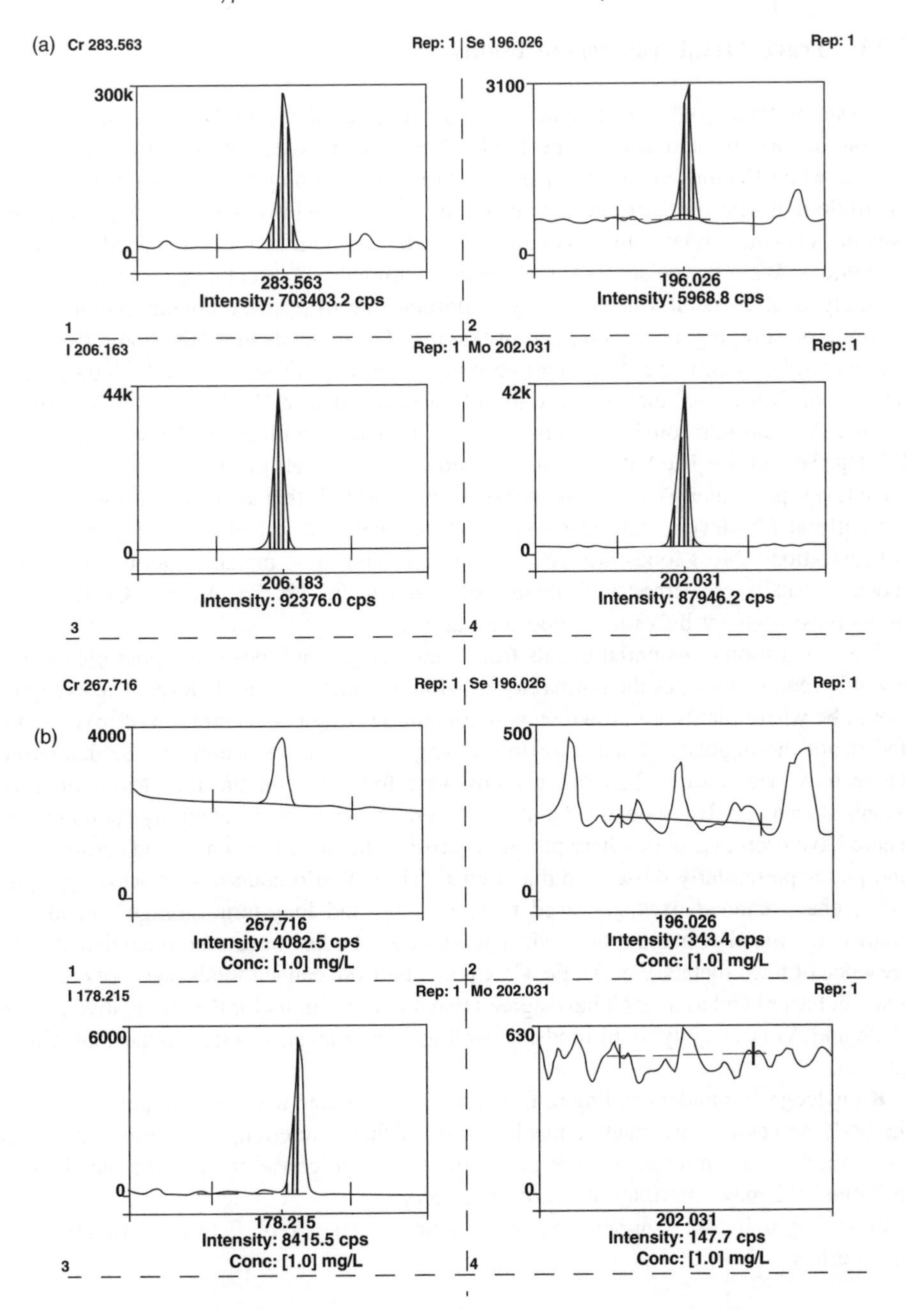

Figure 7.24 *Comparative analysis for trace levels of Cr, Se, I and Mo using the axial mode (a) and the same solution measured using the radial mode (b) of the dual radial/axial ICP-OES. They were scanned to illustrate the advantage/disadvantage of the two viewing methods. (The final solution was between 0.02 and 0.08 ppm metal.) Results obtained using a Perkin Elmer 2100DV ICP-OES. (Reproduced with kind permission: copyright © 1999–2008, all rights reserved, Perkin Elmer, Inc.)*

7.14 Trace Metal Analysis of Foods

Analysis of Napoleon's and Beethoven's hair revealed that they had excessive concentrations of arsenic and lead, respectively. Napoleon was definitely poisoned with the arsenic while Beethoven suffered from lead toxicity that may have been responsible for his lifelong illness that affected his personality and caused his death. Evidence for these poisons became known only recently with the introduction of advanced analytical techniques. Were the poisons administered intentionally or accidentally via foods?

Analysis of foodstuffs is extremely important due to possible contamination caused during crop spraying, processing and packaging. Trace metals in foods should be at very low levels, i.e. $<100\,\mu g\,g^{-1}$ and can be divided between those that are toxicological or nutritional. Nutritional elements are usually associated with Co, Cu, Fe, I, Mn, Mg and Zn and the non-nutritional elements are those associated with Al, B, Cr, Ni, Sn, As, Pb, Cd, Hg, Se and Sb. The latter metals are known to have detrimental effects even at very trace levels particularly over a long period of time. Metals that are essential for health are normally at low levels and have emetic actions when absorbed into the body at high concentrations. Most foods are free of toxic and non-toxic metals. Routine analysis of foods is usually concerned with metals such as Cu, Fe, Pb, Sn, As, Hg, Cd for which limits have been set by various food regulations.

Toxic ingestion of essential metals from naturally grown foods is not possible because it would require 15 times the normal level before it comes near toxic level. The exception would be where plants are grown on soil containing a high concentration of toxic metals and if not thoroughly cleaned prior to cooking or consumption they can be dangerous. There have been cases where some foods were found to contain trace levels of heavy metals such as Pb, Hg, As and Cd and these were shown to be accidentally contaminated. There have been instances where problems arose with metal containers and ceramic cups and plates particularly those manufactured in Third World countries. Processing equipment, cheap canned packages used to store food and long-term storage can also be blamed for metal contamination. Most food safety bodies have recommended that the presence of trace metals, e.g. Al, Fe, Cr, Cu and Fe from canned foods does not constitute a health hazard and as a result have agreed that statutory limits for the cumulative poisons of Pb and As now apply to all foods, as well as higher levels of other metals, i.e. Cu, Sn and Zn.

Knowledge and understanding of heavy metals in human toxicity is important because the body processes only inactive mechanisms for their excretion, and chronic low levels can eventually accumulate to toxic proportions. In high concentrations the metals listed in Table 7.12 may contribute to many ill effects.
A deficiency in foods or supplements of the metals listed in Table 7.13 may also contribute to ill effects.

7.14.1 General Methods of Metal Analysis of Foods

The first step in the analysis is that loose foods must be representative and reduced down by multiple 'squaring' to a suitably representative sample size for analysis. The wide array of foodstuffs would require different methods to achieve homogenisation. One must be applied that is practical and fits with the available resources and does not contaminate

Table 7.12 *List of effects of consumed high levels of metals. (Note: some of these effects are based on statistical evidence and must only be treated as such)*

Metal	Effect
Al	Alzheimer's disease, nervous disorder, colic
As	Hair loss, allergies
Cd	High blood pressure, emphysema, kidney damage, prostrate cancer
Cu	Hyperactivity, liver disease, migraine, depression
Pb	Learning disability, anaemia, headache, epilepsy
Mn	Nervousness, disorder
Hg	Infertility, asthma, depression, nervousness
Ni	Allergies, dermatitis, lung cancer, vertigo, dizziness

Table 7.13 *List of effects of deficient levels of metals in foods*

Metal	Effects
Ca	Muscle cramp, fatigue, lack of growth, insomnia
Cr	Diabetes, muscle weakness, high cholesterol levels, weight problems
Cu	Heart disease, infertility, fatigue
Fe	Fatigue, anaemia, anorexia, poor oxygen supply
Mn	Pancreas, enzyme deficiencies, low metabolism
Mg	Diabetes, fatigue, emotional stress, heart disease
Mo	Obesity, gout
Se	Cancer, heart disease, cataracts
Zn	Deficient nail and hair growth, poor immune system

the sample. All-in-one food can simply be mixed thoroughly with a suitable spatula prior to sampling for analysis. Metals in foods are generally not problematic with volatility in preparation for analysis and the results obtained after careful sample preparation and analysis are generally good. The level and type of metal determine the method of sample preparation and is carefully selected because of the ease of use and the accuracy and reproducibility obtained with the results. Sample preparation techniques applied to foods for metal content are generally carried out using simple direct measurement or after dilution or extraction for most liquid samples, e.g. beverages, vinegar, beers or other liquid foods. Some samples must be prepared by destructive methods, i.e. dry ashing with a retaining compound (PTSA), microwave acid digestion, Kjeldahl digestion and oxygen bomb combustion in order to analyse the sample for metal content.

The ICP-OES fitted with dual radial and axial viewing plasma is an excellent modern instrument allowing detection of very low levels of metals in most food samples. The main advantage of ICP-OES is the increased sensitivity when compared with AAS and the ability to read several metals at the same time using a simultaneous instrument or using sequential instruments against multi-elemental standards. Matrix effects caused by sample preparation of some foods can be readily corrected using standard addition or internal standard(s). Multiple samples can also be routinely analysed for metal content using the flow injection method.

Table 7.14 *Brief list of common metals and metals in foods, method of sample preparation and measurement method used. The list and sources of effects are by no means exhaustive*

Metal	Sample preparation	Techniques	Major source
As	MAD, BC	ICP-OES/Axial + Hydide	Seafoods
Pb	MAD, BC	ICP-OES/Axial + Hydride	Most foods
Cd	MAD or Chelex 100	ICP-OES/Axial	Fertilisers and fish
Hg	MAD, BC	ICP-OES/Axial + CVT	Hg fungicides
Sn	MAD, BC	ICP-OES/Axial + Hydride	Canned foods
Zn	Dry ash	ICP-OES/Axial	Fish + cereal
Cu	Dry ash	ICP-OES/Axial	Beverages, ketchups
Se	MAD, BC	ICP-OES/Axial + Hydrode	Most foods

MAD, microwave acid digestion; BC, bomb combustion; Chelex 100, chelator extraction by column or separating funnel.

Detection of extremely low levels of metals may be possible by the use of hyphenated techniques such as hydride generation, ICP-OES/graphite furnace, ultrasonic nebuliser and cold vapour trap for Hg, and by utilising the axial viewing mode of the ICP-OES could achieve results close to ICPMS levels. Table 7.14 shows a brief list of metals and methods that are commonly considered for food analysis.

7.14.2 Conclusion to Food Analysis

During the last decade, food containing potentially dangerous toxic metals has become a major topic of public interest. Responsibility for ensuring that the foods produced for sale must comply with legal requirements in retailing, safety, quality and composition lies with the local authorities. These authorities must carry out programmed inspection and sampling throughout the food chain from growers, farmers, through to the manufacturers to ensure that safety, quality, labelling, advertising and presentation of foods meet the legal requirements. Most foods are analysed for heavy metals as part of the overall quality.

Metals that enter our bodies via food, drink and air at high concentrations are dangerous because they tend to bio-accumulate by increasing the concentration in the biological organism over time. Most countries are obliged to randomly sample the local produce and analyse it for toxic metals as well as other contaminants and report the results for public awareness. The list of literature and information available is voluminous including specialist texts concerning a wide range of contaminants in food. Food poisoning from heavy metals is rare and if it occurs it is usually accidental or due to environmental pollution.

References

[1] Nagy, G., Feher, Z. and Pungor, E. (1970) Application of flow injection analysis, *Analytica Chimica Acta*, **52**, p47.

[2] Rizicka, J. and Hansen, E.H. (1975) Flow Injection Analysis, Part 1, A new concept of fast continuous flow analysis, *Analytica Chimica*, **21**, p377.

[3] Tyson, J.F. (1984) *Analytical Proceedings*, **21**, p377.
[4] Boorn, A.W. and Browner, R.F. (1987) *Inductively Coupled Plasma Emission Spectrometry, Part II, Applications and Fundamentals*, P.W.J.M. Boumans (Ed.), New York: Wiley-Interscience, pp151–216.
[5] Brennan, M.C. (1992) *Novel Electroanalytical and Atomic Spectrometric Techniques in the Characterisation of Anaerobic Adhesives*, PhD Thesis, Cork: University College Cork.
[6] Kahn, H. (1977) 'Effect of Interfering Elements using ICP-OES', XX Crime Scene Investigation and 7th International Council Advanced Studies, Prague.
[7] Shizhong, C. Lu Dengb, Hu Zhixong and Wang Zhan (2005) The use of electrothermal vaporisation with ICP-OES for the determination of trace elements in human hair using the slurry sample technique and PTF as a modifier, *International Journal of Environmental Analytical Chemistry*, **85**(7), pp493–501.
[8] Bohme, H. and Lampe, H. (1951) Uber das Verhalten von Alkaloidsalzlosungen an Aluminiumoxydsaulen. I. Mitteilung: Chinin-hydrochlorid, *Arch. Pharm.*, **284**(5-6), p227.
[9] Minamikawa, T. and Matsumura, K. (1976) Elucidation of heavy metal including (mercury) contamination caused by Cisso Chemical Corporation in Minimata Bay in the 1950s, *Yakugaku Zasshi*, **96**, p440.
[10] Hassan, S.S.M. (1984) *Organic Analysis Using Atomic Absorption Spectrometry*, London: Ellis Horwood, pp318–322.

Index

Note: Figures and Tables are indicated by *italic page numbers*, footnotes by suffix 'n'